RAIN GOD

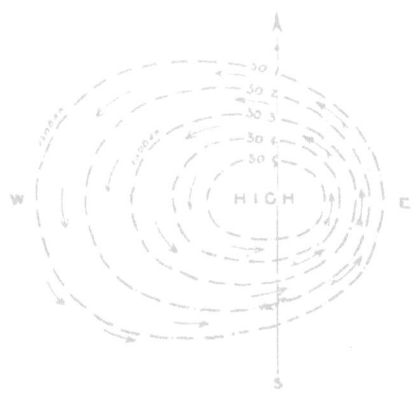

RAIN GOD

The Highs and Lows of Clement Wragge,
meteorologist with a mission

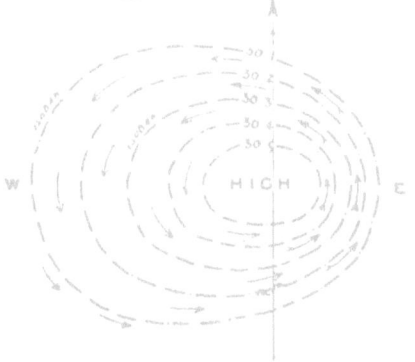

IAN JAMES FRAZER

Published by Silverbird Publishing.

First published in Australia 2022
This edition published 2022
Copyright © Ian James Frazer 2022

Cover design, typesetting: WorkingType (www.workingtype.com.au)
Proudly printed in Brisbane, Australia by Clark & Mackay

The right of Ian James Frazer to be identified as the Author of the Work has been asserted in accordance with the Copyright, Designs and Patents Act 1988.

All rights reserved. No part of this publication may be reproduced, stored in a retrieval system, or transmitted, in any form or by any means without the prior written permission of the publisher, nor be otherwise circulated in any form of binding or cover other than that in which it is published and without a similar condition being imposed on the subsequent purchaser.

Frazer, Ian James
Rain God
*The Highs and Lows of Clement Wragge,
meteorologist with a mission*

ISBN 978-0-6456941-0-9 (pbk)
978-0-6456941-1-6 (ebook)

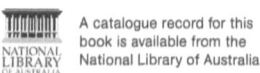

A catalogue record for this book is available from the National Library of Australia

For Diane, my super sleuth

CONTENTS

Foreword		1
Introduction		3
Chapter 1	Auckland, 1920–22	9
Chapter 2	North Staffordshire, 1852–62	25
Chapter 3	Uttoxeter to Utah, 1862–75	41
Chapter 4	In the Court of King Aeolus, 1874–77	57
Chapter 5	North Staffordshire, 1878–81	81
Chapter 6	The Hero of Ben Nevis, 1881–83	105
Chapter 7	South Australia, 1883–85	133
Chapter 8	Queensland calling, 1885–86	147
Chapter 9	The Big Picture, 1889–92	165
Chapter 10	Sunspots, 1889–90	179
Chapter 11	New Caledonia, Munich and Brisbane, 1890–91	191
Chapter 12	The Mystical Potter, 1891–92	205
Chapter 13	The Buninyong Hurricane, Cyclone Beta and beyond, 1893–97	217
Chapter 14	Britain, 1900	241
Chapter 15	Federation Fiasco, 1901	257
Chapter 16	Farewells and a Retreat, Brisbane and Mount Kosciuszko, 1902	283

Chapter 17	Charleville Debacle, Egeson Revisited, the Federation Drought Ends, 1902–03	295
Chapter 18	The Whole Kosciuszko Story, 1903	319
Chapter 19	Good Riddance, Australia, 1903–04	341
Chapter 20	Theosophy, Radium and the Eternal Dynamo, 1906–08	355
Chapter 21	New Zealand, 1908–10	367
Chapter 22	Australia, 1911–14	387
Chapter 23	New Zealand, 1916–22	401
Epilogue, 1922–2022		413
Acknowledgements		443
Bibliography		448
Index		455

FOREWORD

I first met Ian Frazer more than ten years ago, when he visited Staffordshire to see for himself where Clement Wragge spent his orphaned early years with his beloved grandmother. In his granny's home, the young Wragge lived in opulent style, rather spoilt by the servants at Oakamoor Lodge. He spent some happy years discovering the surrounding beauty, set amid the industrial presence that was then Oakamoor in the Churnet Valley, North Staffordshire.

This was also where he began collecting curiosities later exhibited in the Wragge Museum, Stafford. He also learnt the rudiments of meteorology, astrology and the natural world, and formed opinions about the accepted Old Testament teachings of the Church of England, leading him to Theosophy. In this same area, he made his own weather observations, which preceded his famous efforts on Ben Nevis in Scotland. He would return to Oakamoor often in the decades after those epic observations on Ben Nevis. Wragge's affection for the place and its people of all classes was only matched by his love of New Zealand in his later years. His was in some ways a tragic, unfulfilled life, but time and time again he proved his strength of character and willingness to be controversial whatever the cost.

I discovered the enigmatic Clement Lindley Wragge in 1995, during a local history study of Cheadle, Staffordshire, with Keele University, and read in the *Cheadle Herald* of his early world travels. Behind the rather formal, rather academic writings was an exciting personal story waiting to be found. Sadly, his lifelong diaries were lost in a fire after his death, but cryptic notes for his unwritten memoirs enabled me to piece together his formative years in this beautiful corner of England. I shared this with Ian, who has painstakingly pieced together Wragge's amazing life story, followed his international travels, achievements and the frustrations of this extraordinary, complex character. It is a story which deserves to be told.

John Williams,
Oakamoor, Staffordshire

INTRODUCTION

In 1902, the pall of Australia's Federation Drought tipped Clement Wragge from orthodox meteorology to speculative sunspot-watching and rainmaking rocketry. By the end of the year, he was certain the sun held Earth's weather in an absolute electromagnetic thrall. He boasted having foreseen the seven-year drought's waning in a trough of solar activity.

In Wragge's lifetime, 1852 to 1922, the likelihood of human activity causing climate change was seldom studied. The popular press barely noticed speculation by the Nobel Prize-winning Swedish scientist Svante Arrhenius on the global impact of industrial growth on climate. Wragge was unconcerned about fossil fuel-powered industry, but as Queensland's first government meteorologist, 1887–1902, he deplored massive land-clearing and ringbarking. His warnings that ringbarking inhibited rainfall echoed those of US conservationist George Perkins Marsh, who began campaigning in the 1840s against the reckless felling of forests and swamp-draining. Wragge argued in 1902 that wasteful interference with nature demanded remedies "in accordance with science, which is wedded to Nature" — such as revegetation. Hence, wise farmers must sustain nature's equilibrium or face "punishment to fit the crime". But his counterparts in New South

Wales and Victoria disagreed, asserting there had been negligible impact on rainfall in ringbarked districts.

From his earliest days as an intrepid weatherman in Scotland in 1881, Clement Wragge was nicknamed "Inclement". His forecasts and meteorological homilies rippled with melodrama that endeared him to fans and infuriated his peers. Rivals scorned his speculative science, farmers banked on his scent for rain, sea captains trusted his storm warnings, and all, to an extent, indulged his eccentricities. In 1894, in Queensland, he began naming tropical cyclones, a bookkeeping innovation now general practice around the world. After exhausting letters of the Greek alphabet, he named certain low-pressure systems after political friends and foes. In 1902, one of the latter, an MP scornful of barometric blarney, ridiculed Wragge in federal parliament as a "Hottentot rain god".

Australia's first celebrity weatherman, Wragge began forecasting for the whole continent soon after opening his Brisbane office. The colonial meteorologists and astronomers of New South Wales, Victoria and South Australia detested his intrusion, but grudgingly exchanged barometric data for his reckonings. Widespread public trust in his forecasts, which in the 1890s were published in each of the Australasian colonies, triggered clamouring after Federation for a true national agency. Wragge's ambition to take charge was thwarted, but he left a blueprint for Australia's Bureau of Meteorology in 1908. Unfortunately, Queensland's weather records were mangled in the six years between the Brisbane office's demise in 1902 and the opening of the federal bureau.

Data from Wragge's ambitious Kosciuszko weather station, collected between 1898 and 1902, vanished as well. Nevertheless, he and his colonial peers left a network of observation posts and a reserve of graziers, postmasters and railway stationmasters who were well-schooled in reading rain gauges, barometers and thermometers. Unfortunately, colonial-era weather data is regarded as too patchy for use as a benchmark in present Australian climate-change studies. The earliest observations used by the national bureau are from 1910.

Wragge's Brisbane observatory staff remembered him as a perfectionist. He demanded faultless observation of temperature, barometric pressure, relative humidity, ozone levels and rainfall in each of the colony's 18 first-order weather stations. His largely self-taught meteorology was grounded in the certainties of physics. He scolded clergy for naively recommending prayers and acts of repentance to break droughts. His God was not susceptible to pleading. He believed humanity — particularly the British variety — had been endowed with intelligence and moral consciousness to achieve a perfect world through science. Perhaps having lost both his parents before he was five years old made him wary of any rumoured miracles from a merciful Creator.

Born in the English Midlands in 1852, Clement Wragge grew up in a coal-fired, copper-spangled world. Coincidentally, his father's family prospered from making copper wire. He spent his boyhood in the smoky shadow of their factory in the English Midlands, which, in 1857 under new ownership, rolled out 60 tons of wire for the first Transatlantic Cable. In 1873, inheriting a small fortune enabled him to quit law studies, train as a midshipman and sail around the world as a self-styled "scientific traveller". He learnt principles of weather science and astronomy at sea that underpinned the career he made for himself as a meteorologist.

As Queensland's Government Meteorologist, he relied absolutely on copper wire for his telegraphed observations. He was retrenched well before the wireless age, but in 1915 Wragge predicted a future replete with personal wireless telephones: "We will be able to communicate with our friends as easily as we now talk face to face."

Between 1903 and 1922, he journeyed around Australia, New Zealand, Britain and India as an itinerant lecturer with lantern slides of the firmament. Throughout all his travels, he remained at heart a curious boy with a restless need to prove himself. His life was balanced between the tangibles of having been born British, privileged and wealthy, against the intangibles of his true identity and purpose. A founding member of the Queensland Theosophical Society in 1891, his God was engaged in refining galaxies of myriad other sentient beings into an inscrutable perfection. He and other Theosophists saw the carnage of World War I, followed by the Spanish flu pandemic, as collateral damage in a preordained winnowing of humanity. But on a personal level, he attended séances to try to reach his eldest son killed at Gallipoli in 1915, trusting his boy's spirit lived on in the Earth's electronic mantle.

Wragge rejected the idea of a personal God, yet cast the difficulties of his first marriage, followed by a less-fraught second, as a kind of divine reward for exemplary stoicism. There was no *Paradise Lost* in Theosophy's *Secret Doctrine*. The lost continents of Lemuria and Atlantis were reckoned stepping stones towards a distant Nirvana. Wragge, however, praised New Zealand as "God's Own Country" and imagined pre-colonial Australia as a paradise since disfigured and desiccated by farmers and pastoralists. His recommended first step to atonement was massive tree-planting and cessation of wasteful land-clearing. But, in common with other colonists, he was blind to his violence to the First People he studied and catalogued on his rounds of outback Queensland. More than 1000 of the spears, shields, string bags and other possessions he collected from them are held in Australian museums.

This book is Wragge's first in-depth biography. Keith Dunstan's *Ratbags* (1979) got me going on what has been a 10-year project. Dunstan introduced him as a failed rainmaker — a perfectionist with New Age beliefs, flaming red hair and temper to match. Paul Wilson's sketch for *The Australian Dictionary of Biography* (1990) was similarly colourful. Both drew on *The Life and Work of Clement Lindley Wragge,* reminiscences by Inigo Jones, his long-range forecasting protégé, published in 1951. More recently, two other historians have reassessed Wragge's impact on Australian meteorology: David Day in *The Weather Watchers* (2007); Denis Cryle in *Behind the Legend: The Many Worlds of Charles Todd* (2017) — a biography of Clement Wragge's South Australian contemporary, Sir Charles Todd.

My version of Wragge's life story is based on Australian, New Zealand and British archive records, interviews with descendants, and many thousands of newspaper articles written by and about him in his heyday. I've looked for the idealistic man behind the "Rain God". This is the first published account of his sometimes-fraught family life, his practice of Spiritualism, the full story of his mountain-top observatories on Ben Nevis and Mount Kosciuszko, and of his final 14 years in New Zealand as a travelling science lecturer. It's also a history of weather watching — scientific and speculative — from the days of Old Moore to today's global-warming demonstrators and deniers.

Ian James Frazer, Townsville, August 2022

CHAPTER 1

AUCKLAND, 1920–22

He was not a brilliant speaker: his voice betrayed a peculiar stutter at times, his speeches were delivered jerkily and hurriedly, but his lustrous eyes showed the listeners that here was an eccentric who was a genius. — Tribute to Clement Wragge, *Timaru Herald*, December 11, 1922.

Clement Wragge looked frail and wistful to his fellow Spiritualist Arthur Conan Doyle the day they met in Auckland in December 1920. They chatted for an hour in the tiny ridge-top kingdom in Birkenhead where Wragge and his de facto wife, Edris, ran their pleasure garden and museum.[1] Doyle remembered him as tall and skinny, clad in black, "with a face like a sadder and thinner Bernard Shaw". His eyes, peering from beneath a blue turban were heavily pouched, dim and distant. Here in his home patch was Auckland's most remarkable personality — dreamer, mystic and expert on all matters of ocean and air.

Doyle's memoir of his eventful 1920–21 tour of Australia and New Zealand suggests their meeting began awkwardly. Wragge

---

1   Wragge's de facto wife Louisa Horne was known as Edris Wragge throughout their 24-year partnership. He called her "Ed" or "Edris", a Welsh variation on the Persian "Idris".

was miffed Doyle seemingly valued his knowledge of Pacific Ocean currents and breezes above his many other accomplishments, such as pioneering meteorology on Britain's highest mountain, Ben Nevis. But Doyle made amends by asking how, given prevailing headwinds, Aotearoa's Maori people could have rowed 2000 miles from Hawaii and how they could ever have known their destination. Wragge loved these questions. "The dim eyes lit up with the joy of the problem and the nervous fingers unrolled a chart of the Pacific," Doyle wrote. Their banter quickly veered to lost continents. Wragge believed New Zealand's islands were vestiges of Lemuria, a great southern land settled by enlightened refugees from doomed Atlantis one million years before any Maori migrations. Doyle concurred. Later, he complained his excitable host had barely given him time to expound his theories on the destruction of Atlantis and his fear World War I had been a warning bell for the Aryan race, which both agreed had supplanted the Atlanteans on the cutting edge of human progress.[2]

Their shared belief in civilisation as a fastidious God's-work-in-progress was akin to the teachings of the Russian-born Spiritualist Madame Helena Blavatsky, co-founder in New York in 1875 of the Theosophical Society. German philosopher and Theosophist Rudolf Steiner later enlarged on Blavatsky's doctrine of evolving root races.[3] Wragge was a long-time follower of Theosophy which,

---

2   Doyle, Arthur Conan, *The Wanderings of a Spiritualist: On the Warpath in Australia, 1920–1921*, Chapter VIII, originally published in London, 1921, A Project Gutenberg of Australia eBook, http://gutenberg.net.au/ebooks13/1307001h.html, accessed September 28, 2022

3   Blavatsky, Helena Petrovna, *The Secret Doctrine*, Volume Two, Anthropogenesis, London, 1888; Steiner, Rudolf, *The Submerged Continents of Atlantis and Lemuria, Their History and Civilization, being chapters from the Akashic Records [authorised translation from the German]*, London, 1911

writing in 1919, he defined as "the wisdom of the infinite". He asserted every thinking man was at heart a Spiritualist yearning for Theosophy. In his memoir, Doyle joked Theosophists were really officers of the Spiritualist movement who had deserted their army, formed an officers' corps and removed the money, brains and leadership from struggling everyday people.

Doyle made no mention of their war griefs. Both had reported contact through séances with the spirits of their dead soldiering sons. Doyle's eldest son, Kingsley, died of pneumonia in October 1918, aged 24, while convalescing from wounds sustained on the Somme two years earlier. Doyle later wrote of having had six conversations, through mediums, with Kingsley's spirit. In 1916, Wragge told a full house in the South Island port city of Timaru, he could give many instances of men killed in the present war having revisited the earth in spirit. His eldest son, Egerton, was killed in action at Gallipoli in May 1915, aged 34.[4] In 1919, a correspondent to the *Greymouth Evening Star* recalled Wragge telling during one of his war-time lectures that he had direct communication with his dead son. The anonymous writer recalled having met Wragge at a séance in Greymouth nine years earlier, where he had become excited over a message from a deceased personal assistant from his time in India.[5] In 1917, Wragge matter-of-factly told an audience in the South Island city of Nelson, he had communicated, through a medium, with Leo Tolstoy and Lord Kitchener.[6] And in Melbourne in 1922, a Ceylonese psychic arrested for deception by use of "subtle craft" claimed in court that Wragge was among

---

4   *Timaru Herald,* April 13, 1916, p11
5   *Greymouth Evening Star,* September 9, 1919, p3
6   *Nelson Evening Mail,* August 31, 1917, p4

many well-known people who had used his services — in his case in connection with the weather. The defendant, Romeal Perera, used this alleged endorsement in newspaper advertising. He was fined £25.[7]

Doyle, his wife and three children, toured Australia and New Zealand from November 1920 to February 1921, invited by Australian supporters to promote Spiritualism as a healing force for a bereaved world. He understood his task as the most important conceivable — confronting doubts over religious promises of immortality with, "The actual experience of those who have made the change from the natural to the spiritual bodies." His lectures in New Zealand were popular, but he disappointed the press by seeming bored with Sherlock Holmes. "It is just as if one confronted a man with a bit of his past life which he would willingly repudiate," one journalist complained. Doyle reportedly said he was unlikely to write any more novels unless he could help people understand Spiritualism more clearly. In fact, he wrote 12 more Holmes stories between 1921 and 1927.[8]

In December 1920 and January 1921, Wragge wrote to the *Auckland Star* countering New Zealand church leaders' attacks on Doyle as a satanic agent who must be restrained from his dangerous promotion of Spiritualism:

> With regard to Sir Conan Doyle and his critics ... we watch history repeating itself. Christ was said to have a "devil", and we recollect Newton, Galileo, William Harvey, and others, and the abuse that was their lot.

---

7 *Express and Telegraph*, Adelaide, September 19, 1922, p4
8 *Taranaki Daily News*, December 22, 1920, p10

Writing as president of the Auckland Spiritual Scientists' Church, and a Theosophist of many years' standing, Wragge denied Spiritualism and Theosophy spread evil teachings. Rather, Spiritualists and Theosophists alike tried to foster a loftier and nobler conception of religion than those of their critics who gave an alleged devil more power than God. He advised readers to, "Stick to the Christ within, and hold onto God as a limpet clings to a rock."[9] Doyle sent the Wragges a signed copy of *The Wanderings of a Spiritualist* in thanks for their hospitality. In the book, he recalled his last sight of them wreathed in a riot of flowers in their hillside garden — "a blue-turbaned, eager man, half Western science, half Eastern mystic" — and his dark-eyed wife. He praised Mrs Wragge as one of the most gracious personalities he had met during his Antipodean wanderings and assumed she was a holy woman from India's Brahmin caste. In their farewells, Edris encouraged Doyle to extend his journeying to the Indian lecture circuit.[10]

* * *

The New Zealand Observer, in Auckland, summed up Wragge in January 1920, as one of New Zealand's favourite institutions. He had then been in the news for weeks, earnestly denying he had predicted the end of the world. To the contrary, he maintained that in nominating December 17, 1919, as a day to beware of

---

9 *Auckland Star,* December 18, 1920, p12
10 Doyle's sole visits to the Indian subcontinent were in transit to and from Australia and New Zealand in 1920 and '21; his final novel, *The Maracot Deep*, published in 1929, is about the discovery of a sunken Atlantean city.

earthquakes and tidal waves, he had not suggested either threat would be necessarily hazardous to Kiwis and certainly not Earth-destroying. "I simply gave the mathematical figures, and suggested what might or might not occur here," he wrote to the *New Zealand Herald*.

Despite his denials, storekeepers in the Fijian capital, Suva, boarded up their shops, and residents rushed to take out £40,000 in property insurance. Beyond the South Pacific, dread of December 17 fed on similar predictions by a Californian astronomer, Professor Alberto Francisco Porta who, like Wragge, believed cyclical surges of solar energy triggered earthquakes and volcanic eruptions. The sun was then in a state of maximum electromagnetic activity, and he and Wragge both calculated a rare planetary conjunction on December 17 would unleash an electromagnetic chain reaction with dire consequences. In the United States churches were said to have been crowded. Lead and zinc miners in Oklahoma skipped work for fear of being buried alive. In New Zealand the honorary government seismologist George Hogben discounted any link between sunspots and earthquakes. He had first disagreed publicly with Wragge's sunspot theories in 1909. A former maths and science teacher, then head of the dominion's Department of Education, he rejected Wragge's much-publicised view that solar storms had caused a disastrous earthquake in Sicily in late December 1908. The truth was much more mundane, he told Dunedin's *Otago Daily Times*.[11]

---

11 Clarke, Kim, *Professor Porta's Predictions*, M Heritage Project, University of Michigan, 2021, https://heritage.umich.edu/stories/professor-portas-predictions, accessed April 29, 2021; *New Zealand Herald,* November 24, 1919, p6 and December 16, 1919, p11; *Auckland Star,* December 19, 1919, p8

A *Melbourne Punch* story on the apocalyptic panic prompted *The New Zealand Observer*'s aside on Wragge's near-national treasure status. For the benefit of younger readers, the Australian weekly valorised the "one-time official astronomer and Lord High Weather Prophet of Queensland" once known as Inclement or Wet Wragge. Recalling his talent for naming storms, *Punch* concluded: "We were all sorry to lose him — he was the only low-comedy astronomer on earth."[12] It was then 17 years since cyclonic systems with names like Deakin and Forrest had vanished from Australia with their creator. In 1909, after settling in New Zealand, Wragge briefly revived his whimsical cataloguing in a syndicated weather column. He resumed the practice in 1918 when hired to supply Auckland's *New Zealand Herald* with daily forecasts — using a hybrid system of atmospheric and sunspot observations and giving Antarctic lows enigmatic titles such as Evoid, Zagol, Sethad and Fethad.

* * *

In February 1919, Wragge boasted an affinity with the famous Italian physicist Guglielmo Marconi. He wrote in the *Auckland Star* that Marconi's widely reported description of radio waves as eternal coincided with his own views.[13] Marconi reportedly told a London daily that wireless messages he sent into space ten years earlier had yet to reach the nearest stars and might last

---

12 *The New Zealand Observer,* Auckland, January 17, 1920, 'Pars About People', p10, citing an undated issue of *Melbourne Punch*

13 *Auckland Star,* February 3, 1919, 'Wireless Constitution of the Universe', by Clement L Wragge, p9

forever. He also said he had received strange signals seemingly from outer space and hoped for contact some day with wise beings on planets much older than Earth. "Science in these days demands a flexible mind," he told journalist Harold Begbie of the *Daily Chronicle*.[14] Wragge concurred in a letter to the *Auckland Star*, later republished in provincial New Zealand and Australian papers. Of course, he wrote, Earth occupied only a skerrick of the fathomless abysses of the endless universe — the thousandth part of a drop of water compared with all the oceans of the globe. Somewhere in the cosmos, there were bound to be blue, orange and green suns with planets tenanted by divine entities to whom war was unknown. Planets where love reigned and peace ruled. He believed as a Theosophist and Spiritualist in God's all-pervasive spark "from the influenza germ and flowers in our gardens to the giant sun Canopus". He imagined the universe awash with myriad electromagnetic waves, each of them drawing on the Bank of the Absolute Dynamo. From this point of view, why dread the trifling demise of one's mortal shell? "You can never be destroyed and no sane man can be an atheist," he concluded. "All is life, all is spirit, all is electricity." In 1915, during a lecture in Auckland, he predicted one day Earth's ether would buzz with myriad calls from mobile phones:

> I am certain that the time will come when with a wireless apparatus, each can carry 'round with him, we will be able

---

[14] Harold Begbie's Interview with Marconi was republished by the *Irish Independent*, January 21, 1919, p2

to communicate with our friends as easily as we now talk face to face.[15]

* * *

On June 15, 1920, the Australian soprano Dame Nellie Melba sang *Home Sweet Home* to an audience stretching from North America to the Middle East. Her recital transmitted from the Marconi wireless factory in England launched radio as an entertainment medium and inspired many experimental public broadcasts. In November 1921, Robert Jack, a University of Otago physics professor, transmitted New Zealand's first radio program of live and recorded music and a year later established the dominion's first station, Radio Dunedin. By then, Wragge had plans for a wireless station too. A visitor to Auckland in June 1922 recalled the silver coin admission to one of his Sunday evening lectures had been advertised in aid of a proposed station at his Birkenhead observatory. Charles Cleveland Nutting, a University of Iowa zoology professor, remembered Wragge as a rambling speaker on meteorology, geography and gardening. "The proceedings were opened with prayer and a hymn which gave them the necessary religious tone to meet the requirements of the law, as I understand it," Nutting wrote in 1923. He praised Wragge as active and alert for his age and well-respected as a weather forecaster far beyond New Zealand.

* * *

---

15   *The Colonist*, Nelson, New Zealand, July 15, 1915, p4

But in New Zealand, fellow meteorologists still rejected Wragge's long-range forecasting as unscientific. In 1921, Daniel Cross Bates reaffirmed his opposition to his sunspot-cycle theories and William Gaw, a visiting British meteorologist, described his seasonal forecasts as inaccurate and his methods faulty. Nevertheless, Gaw — an ex-naval weather man — conceded in a letter to the *New Zealand Herald* that Wragge's standing was respected throughout the dominion. Wragge replied that he, too, cordially agreed to differ and sincerely hoped for enlightenment from further debate. Surely his 42 years' standing as a Fellow of the Royal Meteorological Society entitled his views to some regard. While mistakes were inevitable in the fledgling science of seasonal forecasting, in his opinion the basis of what he termed "progressive meteorology" was as certain as the mathematics of the universe. He was happy to count among his peers Norman Lockyer, Camille Flammarion, Eliot of India, Meldrum of Mauritius and, latterly, Sir Oliver Lodge.[16]

* * *

Wragge gave more than 2000 scientific lantern-slide lectures around the British Empire between the demise of his Queensland weather bureau in 1903 and his death in New Zealand in December 1922. In October 1922 — a month after his 70th birthday and having recently recovered from bronchitis — he launched his final lecture tour, in the Waikato region of New Zealand's North Island. Basically, his rationale was constant through those 18 years. He spelt it out in a testimonial for Doan's backache

---

16   *New Zealand Herald,* Auckland, August 15, 1921, p3, and August 16, 1921, p8

pills, published in Australian and New Zealand papers in 1921 and '22. The pills' manufacturer paid him twenty guineas for a brief memoir, *Some Reminiscences of an Eventful Life*. Before endorsing their ubiquitous remedy, Wragge summarised his mission since 1903:

> All that time, I have lectured on the marvels of the universe, in England, also all over India, Ceylon, Australia, Fiji, Tahiti, Tonga, and New Zealand, doing all I could to lift mankind to a nobler conception of God and what the eternal universe really means.[17]

I don't know his subject the night he collapsed in an end-of-the-track town near Palmerston North in November 1922, but guess it included an affirmation of the spiritual consolations of stargazing. His talk on "The Grandeur of the Universe" — the centrepiece of the Waikato district lectures — featured new pictures from California's Mount Wilson Observatory of the sun, moon and planets. He was sure the awesomeness of these images eclipsed any other astrophotography seen in New Zealand. In late October, visiting the King County township of Te Kuiti, he said he wanted to lift noble minds to the loftiest-possible awareness of the universe. The *King County Courier* reported other familiar themes: his scorn of atheism, belief in God as the Infinite Dynamo at the Core of all Things, and his understanding of "religion" — from the Latin *religio* — as humanity's way of connecting with God's Absolute Energy. The Te Kuiti lecture also included a less-familiar issue, his warning that the sun's actinic vitality — that is

---

17 *Evening Star*, Dunedin, April 20, 1921, p4, from *Some Reminiscences of an Eventful Life*, an endorsement for Doan's Pills

radiant energy in visible and ultraviolet bands of the spectrum — was at a low ebb and that, in view of a spate of deaths that spring, the elderly should beware of chills and over-exertion. Of course, he added, he did not fear death because to him it meant liberation, not obliteration; the flight of one's immortal spark — or "ego" — to mesh with God's Absolute Energy.

He died at home in Auckland at sunrise on December 10, 1922, aged 70, a fortnight after a stroke in the township of Sanson, near Palmerston North. A post-mortem found he also had heart disease and cerebral sclerosis.[18] An old friend at Wragge's bedside with Edris and their son Kismet recalled his last words had been, "I must rest now", and that he had emphatically spelled out "r-e-s-t". Matthew Walker (1867–1935), a Spiritualist church pastor and health educator, imagined his friend's soul flying into the starry sky. "Nature seemed to make a mighty effort to appropriately welcome her loving son to a fuller life," he wrote, remembering a breeze through the bedroom windows. He continued:

> Over his active mind swept all the storms and brooded all the calms of human nature. His capacity for work was prodigious, and to those who do not understand the power of mind it was a mystery how that frail frame accomplished half the work it did ... Mr Wragge ... believed love, not hatred, was the great controlling power of the universe.[19]

Two days later, Walker officiated at his funeral and burial in

---

18 'Clement Lindley Wragge 1922', certified copy of New Zealand Death Certificate for Clement Lindley Wragge, 10 December, 1922. Registration No 1923002440, Auckland

19 *Macleay Chronicle*, Kempsey, New South Wales, January 3, 1923, p4

Birkenhead, and on December 17 spoke at a memorial service organised by Auckland's Higher Thought Centre, attended by 200 people.[20] In his tribute, he said they had known each other for 28 years, indicating they had met in Australia in the 1890s. He and Wragge shared a mutual acquaintance in barrister George Houston Reid, a former New South Wales Premier and Australia's fourth Prime Minister. In 1902, Reid acted for Walker who had pleaded not guilty to having made a false statement under oath during a district court case in Orange earlier that year. Reid failed to convince the jury otherwise and Walker served five months with hard labour in Bathurst jail.[21] In 1905, Walker moved to Auckland with his wife and family. By 1910, he was well established as a masseur, psychotherapist, marriage celebrant, minister of the Church of Our Father, and office-bearer of the Auckland branch of the International Arbitration and Peace Association. In January that year, he introduced Wragge, then recently settled in Auckland, to his circle as an Apostle of Advanced Thought,[22] and

---

20   The constitution of the Auckland Higher Thought Centre listed its objects as providing a Universal Church without a fixed creed, dogma or ritual: "Where people may worship God and may study principles and fundamentals of Higher Thought in all its phases ... and the practice of sane spiritual living as taught and demonstrated by Jesus Christ."

21   The widely reported case stemmed from Walker having defaulted on paying £200 damages to a plaintiff who, a year earlier, had successfully sued him for slander. In February 1902, when the litigant took him to court for his £200, Walker stonewalled by denying under oath having received an £85 property damage insurance payment after a recent house fire. This precipitated the perjury case in Orange District Court in July 1902.

22   *Daily Telegraph*, Sydney, July 2, 1902, p8; *Barrier Miner*, July 3, 1902, p3; *Macleay Argus*, Kempsey, December 24, 1904, p12; *The Macleay Chronicle*, Kempsey, February 9, p3; *Marlborough Express*, New Zealand, November 24, 1905, p2; *Auckland Star*, May 1, 1906, p3 and July 7, 1907, p6; *New Zealand Herald*, Auckland, January 7, 1910, p10

in 1915 helped him establish the ambitious Wragge Institute and Waiata Tropical Gardens at Birkenhead. Walker was prominent at the opening ceremony in January 1916, and a regularly advertised speaker at the Wragge Institute that year on topics such as "The Power of Silence" and "Thought Power in Daily Life". His Rangatira Health Institute specialised in massage, hydrotherapy, remedial exercise and psychotherapy with a mantra of: "Change the habits that cause the disease." He seems not to have been involved with the Auckland Spiritual Scientists' Church, which Wragge led in 1920, nor was Wragge a follower of freemasonry, one of Walker's many passions.[23]

* * *

Three weeks after his death, the *Daily News* in Perth rated Wragge's pioneering Australasian Weather Bureau as his greatest accomplishment. Opened in Brisbane in 1887, it predated the Federal Bureau of Meteorology by 30 years. "His weather forecasts were for many years a distinctive feature of Australasian meteorology," the leading West Australian newspaper argued. "When he predicted stormy weather along the coast every skipper got ready for it." His audacity in forecasting for all of the Australian colonies and New Zealand was also acknowledged by papers in Brisbane, Sydney and Adelaide, the radical *Australian Worker*, conservative agricultural journals *The Land* and *Sydney*

---

23   *Auckland Star*, November 15, 1935, p16. Matthew Walker's obituary asserts he attended Sydney University Medical School for three years in training to become a Presbyterian medical missionary and that in Auckland he became president of the British Union of the Anti-Vivisection Society.

Stock and Station Journal, and establishment British dailies, *The Times* of London and *The Scotsman* in Edinburgh.

But the *Daily News* conceded mixed views on his other attainments: "Even during the latter part of [his] lifetime, legend had commenced weaving itself around his picturesque personality. Small wonder is it that after his death this has been magnified a thousand-fold."[24] A number of other writers commemorated his failures and foibles. The author of the Brisbane *Daily Standard*'s tribute declared his notorious attempt to break the Federation Drought with cannon fire would always be remembered for having caused more laughter than rain. On the other hand, his wet-season forecasts had been the work of a literary acrobat, a magician in the tradition of the Fairy Paribanou in *Arabian Nights*. His uncanny detection of imminent cyclones had been as deft and mysterious as Paribanou's fabled tricks with her magic parachute — sometimes concealed in the hollow of her hand but, when needed, a tent big enough to cover an entire army.[25]

Wragge's legend survived for another decade or so in north Queensland. In 1927, a *Cairns Post* correspondent remembered him as a full-blooded, practical scientist with no time for "high-falutin airs". Old hands would surely always mark his name with the credit it was due. "He at times looked on the wine when it was red, but the greatest crime came when he became the shuttlecock for some of those ambitious mortals that are generally known as politicians," the writer concluded.[26]

---

24 *Daily News*, Perth, December 30, 1922, p8
25 *Daily Standard*, Brisbane, December 15, 1922, p6
26 *Cairns Post*, March 4, 1927, p10, letter to the editor from "Coyyan"

Rain God

CHAPTER 2

NORTH STAFFORDSHIRE, 1852–62

I hope, before I pass over, to write my life in all its detail ... and when my life is published (I am now overhauling my most voluminous journals) you, I intend, shall have a copy, this letter is the merest outline. — Draft of a letter from Clement Lindley Wragge to an unknown friend, February 5, 1921.[27]

In 1920, in notes for a never-written autobiography, Wragge identified the two most sacred places in his wandering life as his parents' graves in the English Midlands. His mother died when he was four months old and he could barely remember his father. Anna Maria and Clement Ingleby Wragge's graves are 60 miles apart, hers in the town of Stourbridge and his in the village of Oakamoor. She died of tuberculosis in early 1853, aged 30. Clement senior fell on his head from a horse in 1855, lingered for two years in a nursing home and died in 1857, aged 42. Knowing so little of his parents left Clement junior with a lifelong need to convince them of his worth. Conversely, his orphaned infancy indulged by sympathetic adults left room for imagining possession of a unique genius sprung straight from the cosmos.

---

27 Auckland Memorial Museum, Wragge Family Papers, MS1213, Box 2, autobiographical notes as basis for an article, *Reminiscences of an Eventful Life*

His jottings bundled in the archives of the Auckland War Memorial Museum research library include a draft of his reminiscences for the patent medicine maker Foster-McClellan Co. He never managed the full-scale autobiography he promised a friend the same year. His recollections, at least 60 years later, are the sole record of his childhood in the wilderness of the Churnet Valley.

After Anna Maria's death, Clement senior left his baby son in the care of his elderly, widowed mother Emma Wragge. Clement always acknowledged her as his dearest childhood friend and teacher of basic astronomy, geography and meteorology. "But for the loving care of one of the dearest of grandmothers, I, too, should have quickly passed to the astral plane," he recalled in 1911.[28] Emma was 68 years old when she added him to her household in Oakamoor. She lived in a mansion up the valley from the copper-wire factory that she and her brothers had very recently sold, after the death of her husband, George Wragge, the works' long-time manager.[29]

Clement Wragge junior was born in Stourbridge, near Birmingham, on September 18, 1852. The Midlands were unusually bleak that autumn. In November, the flooded River Stour swamped the town's high street, swept away bridges and drowned flocks of sheep. On Boxing Day, in Worcester, 20 miles away, hurricane-force winds felled elm trees, smashed window shutters and razed roofs. The *Worcester Journal* concluded on December 30, 1852: "The year about to close will long be

---

28 *The Sun*, Sydney, September 18, 1911, p1
29 Wilson, Peter L, *Oakamoor Remembered*, p15, Thomas Bolton and Sons, of Birmingham, bought the works in March, 1852, for £7750

remembered as having afforded more remarkable atmospherical and other mutations than have occurred in this country for a long period." It rained for nearly seven weeks through November to New Year's Day, and an earthquake on November 14 rattled Dublin, Liverpool and south to Gloucester.[30] Weather prophet Richard Morrison boasted of having predicted in 1834 the exact date of this shake, based on a conjunction of planets and imminent solar eclipse.[31]

Anna Maria was diagnosed with tuberculosis in November 1852, and died in Stourbridge on January 21, 1853, aged 30. Her funeral and burial a week later were at St Mary's parish church, Oldswinford. Fatal chest complaints were prevalent in Worcestershire's western districts in the final quarter of 1852.[32] Birth records show her child's first names were initially registered as William Lindley, but by his baptism on October 30 had been changed to Clement Lindley, probably with a nod to his great-uncle, Birmingham lawyer, Clement Ingleby. In the long run, Clement proved a good name for a meteorologist. Derived from *clemens* — Latin for "mercy" and the persona chosen by 14 Popes — Clement was manna for his followers and made "Inclement" inevitable.

* * *

---

30 *Liverpool Mercury*, 'Course of the Seasons', May 22, 1860, p6; Aris's *Birmingham Gazette*, November 15, 1852, p1; *Worcestershire Chronicle*, November 17, 1852, p8; *Worcestershire Journal*, December 30, 1852, p2

31 *Morning Chronicle*, London, advertisement for *Zadkiel's Almanack*

32 *Worcester Chronicle*, February 9, 1853, p6, quarterly return births, deaths and marriages to December 31, 1852

Clement senior had married Anna Maria Downing in October 1851. They came from gentrified Midlands families. His father's lineage included lawyers and captains of industry, and her late father and maternal grandfather had both been physicians. In 1849 the death of Clement senior's father, George Wragge, after a long illness, had been long awaited. It precipitated the sale two years later of the family business, the Cheadle Copper and Brass Company, and gave his three bachelor sons the means to marry. In March 1852, Thomas Bolton and sons, of Birmingham, bought the works for £7750, a bargain, considering the world's fever for telegraph wire. In those days, Oakamoor's certainties were grounded in a hunger for copper wire and brass rods. The craggy Churnet Valley — chiselled in an Ice Age 18,000 years past — was rich in iron ore, copper, limestone and coal, and had ample timber for furnace charcoal. For centuries, Oakamoor folk had used the River Churnet's North Sea-bound surge to spin waterwheels, driving furnace bellows and hammer mills. Tourists on the Churnet Valley Railway, opened in 1849, raved about the scenery. But 30 years earlier, when the mill's proprietors acquired a coal-fired steam engine, the locals nicknamed their hollow "Smokamoor". George and Emma Wragge raised their five children in this village of 600 souls with neither a church nor school until 1832.

Clement senior, born in 1814, trained as an articled law clerk at Lincoln's Inn and Great Carey Street, London, before joining his lawyering brothers, George junior and William at Wragge and Co, in Birmingham. In 1837, the *Birmingham Journal* noted Clement senior's attendance, "in a Grecian costume" among hundreds of other revellers at a fancy-dress ball celebrating the accession of Queen Victoria. Having a well-connected brother-in-law in the industrial town of Stourbridge possibly helped secure his partnership there in 1840 with a long-established lawyer, George Grazebrook. Clement senior's sister, Frances, was married to one of their distant cousins, solicitor Charles John Wragge, who was in the process of reinventing himself as a banker.

Clement senior hardly exists in local newspapers of the 1840s, beyond the advertisements for his law practice with Grazebrook. His life seems to have been free and easy. Between 1847 and 1851, the *Worcester Journal* and *Worcestershire Chronicle* noted his presence at balls in Stourbridge and Dudley. In 1848, a report of the Stourbridge horse races lists him as the owner of the second-placegetter in the Stourbridge Stakes — a three-year-old named Clairvoyance. By then Charles Wragge's bank — known as Ruffords and Wragge's — was in deep trouble. On June 26, 1851, *The Times* of London reported the bank's London agents had refused their bank bills. In Stourbridge, the bank's many customers were alarmed when the local branch failed to open. The firm's combined debts totalled £452,000, half owed to creditors of its Stourbridge branch. Charles Wragge's personal debts amounted to £5308.[33] That year, Clement senior dissolved his partnership with George Grazebrook and in 1852 advertised himself simply as Clement Ingleby Wragge, solicitor, New Road, Stourbridge. After Anna Maria's death, he began to practise law in Cheadle, four miles south of Oakamoor. In early 1854, he gave Cheadle as his address when renewing his game certificate — unlicensed hunters risked a £20 fine in those days.

* * *

---

33   *Worcester Journal*, July 19, 1851, p3; *Worcester Chronicle*, July 16, 1851, p8

In a pair of photos inherited by Wragge's New Zealand descendants, Anna Maria sits wide-eyed in a fussy frock, hair in ringlets. She's youthful and knowing. Clement senior has mutton-chop sideburns like Prince Albert's and seems contented. Six generations on, few clues remain to either life. Her careful signature, "Anna Maria Wragge" is large and legible in the smudgy legalese of the will she signed in Stourbridge on August 17, 1852. She named Clement senior as sole executor and, in leaving him her personal estate, specified he was to receive her diamond ring, watch and some domestic furnishings. It's the strongest of her traces, the others being some Public Records Office entries and a few advertisements in Midlands newspapers, such as the *Birmingham Guardian*'s notice of the death of her mother, Mary Downing, in March 1826:

> DEATH: on the 2nd inst, 1826, after a long and distressing illness, aged 32, Mary, wife of Mr Isaac Downing, surgeon of Stourbridge. Thus terminated the life of this amiable and highly respected young woman whose premature end will be ... truly severely felt by an afflicted husband and children.[34]

Anna Maria and her sister Elizabeth were then just four and seven years old, respectively,[35] and cared for by the family's domestic servant Ann Pearsall of Clent, then aged about 25. Isaac Downing's will, drawn up just before his death in November 1839, acknowledged Ann's entitlement to £200 in unpaid wages and bequeathed her his kitchen clock, a chest of mahogany drawers

---

34 *Birmingham Gazette*, March 13, 1826

35 Elizabeth Causer Downing baptised Old Swinford November 12, 1819; Anna Maria Downing baptised Old Swinford March 30, 1822.

and a mourning suit for his funeral at St Mary's, Oldswinford. Then aged about 40, she was surely the "old servant maid" with whom the *Staffordshire Gazette* reported in November 1839, Downing had entrusted a strange request:

> A few hours before the late Isaac Downing, Esq, surgeon, of Stourbridge, departed this life, he particularly requested his old servant maid, who attended him during a long illness, to visit his grave twice a day for the first fortnight after his body was interred, which request she has faithfully fulfilled.[36]

* * *

The youngest child of Bromsgrove innkeepers, Wragge's paternal grandfather Isaac Downing was born in 1787 and enrolled about 1808 as a medical student at St Bartholomew's Hospital, London.[37] By 1812, he was working in Stourbridge in partnership with surgeon John Causer Junior, whose sister, Mary, he married in 1817. Early in his career, according to Black Country folklore, Isaac was flattened by the corpse of a murderer named William Howe while trying to remove it from a gibbet at Dunsley Heath near Stourbridge. He was said to have been collecting Howe's body for anatomical study when it fell on him. Later, he carried the corpse to his surgery

---

36 *Staffordshire Gazette and County Standard,* November 11, 1839, p4
37 Journal of Sir Ludford Harvey, c. 1804–26, lists of surgical students at St Bartholomew's Hospital, 1807–26, St Bartholomew's Hospital Archives, Ref No SBHX54/1

and reduced it to an instructional tie-wired skeleton.[38] Isaac's familiarity with death left him with an unsentimental and thrifty attitude to funerals. He specified in his will that Peter Bird, the coffin-maker, make his box from unplaned wood "of the same strength of those of any of poor caste". Furthermore, he asked that his coffin be borne to the church by eight poor men, to be paid 10 shillings each for their trouble, and that each should wear black gloves which he considered sufficient mourning wear. He continued, "It is my express wish and desire that no other expense be incurred in my funeral, excepting a hatband and scarf and gloves to the clergyman who buries me and a hatband and gloves to the coffin maker."

* * *

Clement senior established his new law practice in Stourbridge in 1852, while his disgraced brother-in-law Charles Wragge and family moved to London. That English winter was abnormally mild, the snow-deprived prelude to a rainless spring and waterlogged autumn. In late November, a correspondent to *The Times* of London mulled over 1852's thunderstorms, floods and strangely oppressive atmosphere. Could it all have been caused by the freak of two full moons in July? Seers and self-styled astro-meteorologists, such as retired British naval lieutenant Richard

---

38 See 'The Ghost of Gibbet Wood', by Josephine Jasper, *Black Country Bugle*, http://www.blackcountrybugle.co.uk/ghost-gibbett-wood/story-20155455-detail/story.html, accessed April 22, 2013; Isaac Downing's rumoured theft of Howe's body was mentioned in a story titled Gibbet Lane (near Stourbridge), on the website of Astral Search, UK, http://astralsearch.webs.com/locationhistory.htm, accessed May 30, 2013

James Morrison, offered speculative answers. His annual sixpenny compendium, *Zadkiel's Almanack*, sold 32,000 copies in 1851.³⁹

* * *

Emma Wragge's history is as sketchy as her luckless son and daughter-in-law's. Born in Cheadle in 1784, she was the third of eight children, eldest daughter in a family of high-achievers in industry, law and medicine. Her grandson Clement loved her. She taught him the northern constellations and explained the agency of atmospheric pressure, visible in the weather glass as well as the storms that swept over the Churnet Valley. His first astronomy primer was *Dick's Solar System*, published by Britain's Religious Tract Society. The author, Thomas Dick — a graduate in divinity from Edinburgh University — believed astronomy pointed to the existence of "an eternal and incomprehensible Divinity".⁴⁰

In 1831, George and Emma Wragge were prominent in building Oakamoor's first Anglican church and a school for working-class children. The Cheadle Copper and Brass Company gave £100, land and stone to build the Holy Trinity Chapel of Ease, which was

---

39 Anderson, Catherine, *Predicting Weather: Victorians and the Science of Meteorology*, Chicago, 2005, p67; *Nottingham Review and General Advertiser*, November 11, 1852, p3, Sutton, bookseller, Nottingham advertises 32 just-published almanacs for 1853, average price sixpence.

40 Astore, William Joseph, *Observing God: Thomas Dick (1774–1857), Evangelicalism and Popular Science in Victorian Britain and Antebellum America*, MA thesis submitted to the Faculty of Modern History, University of Oxford, for Degree of Doctor of Philosophy, 1995, pp 193–97. The first and second parts of *Dick's Solar System* sold, respectively, 30,510 and 26,890 copies by 1850. Dick, Thomas, *The Solar System with Moral and Religious Reflections in Reference to The Wonders therein Displayed*, Vol X, Philadelphia, 1854, preface, p12, accessed online from State Library of Queensland.

opened in August 1832. The crypt was set aside for poor children to receive elementary education in accord with Church of England teachings. The population of Oakamoor parish stood at 671 in 1831, boosted by relocation of copper-wire production from Cheadle to the village.[41] George Wragge's name appeared occasionally in 1830s provincial newspapers in support of Church of England causes. But in 1841, he publicly announced his allegiance to the Catholic Church and declared a return to the Catholic truth was the only thing that could save England. A year later, he had a stroke, was permanently paralysed and died in 1849, aged 64. His tombstone in the mossy grounds of Oakamoor's Anglican church is inscribed with the bare facts, advising that: "Underneath are interred the remains of George Wragge, of Oakamoor Lodge." The stone revered by Wragge, above his father's grave, has a similarly plain acknowledgement of lineage and mortality.

* * *

In middle age, Wragge remembered his freewheeling childhood as schooling in the glories of divine nature. The Churnet valley was his Eden, his sanctuary from the stuffy world of Oakamoor Lodge. He conceded some of his elders thought him a spoilt brat. One of his favourite pranks was flight-testing Emma's old hens from a craggy peak, hundreds of feet above the valley floor. In 1861, census records show he was the sole child — eight years old — in

---

[41] *Staffordshire Advertiser*, October 1, 1831, p4; ibid, August 22, 1832, p4; Hodgkinson, Lilian, *Holy Trinity Oakamoor, A Brief History*, revised edition August 2012, p1; *A Topographical Dictionary of England*, by Samuel Lewis, 7th edition, 1848, put the population of the ecclesiastical district of Oakamoor at 700 in 1848.

an ageing household serviced by five servant maids and a 15-year-old groom. According to Oakamoor folklore, Emma assigned the groom, Henry Jackson, to steer Clement's wanderings. Henry, the son of the village's railway gatekeeper, was one of ten children.

In 1873, when estranged from his father's family, Wragge celebrated his 21st birthday with friends from his early days in the village. As a lad, well before beginning formal school classes in 1862, he often camped out in the valley while exploring and collecting rocks, plants, shells and eggs for his first natural history museum.[42] The stony ridges around the village were rich with fossils. Years later, he recalled having imagined himself the storybook castaway Robinson Crusoe. The risk of stirring the "buggarts" — ghosts in the woods feared by village folk — didn't scare him.[43] He had a rosy version of boyhood that he shared with reporters all his life. The scraggly wilderness of the Churnet Valley mingled in the home country of his dreams with mysterious Alton Towers, an aristocrat's folly on the escarpment above Oakamoor. Fortified, razed and rebuilt by generations of strongmen since the Iron Age, this retreat with extravagant, unkempt grounds was, in Wragge's day, the occasional home of the 18th Earl of Shrewsbury. By 1860, when the Earl took possession of his contested inheritance, only one enchanted turret remained — a 19th century, neo-Gothic tribute. The garden rambled downhill replete with artificial lakes, fountains, waterfalls, conservatories, statuary, shrubs, flowers and 13,000 trees planted in the era of the Napoleonic Wars. Its entanglement with the Churnet's fir-tree and sycamore woods

---

42 *The Sun*, Sydney, September 18, 1911, p1
43 A "buggart" or "bogart" was a ghost or poltergeist in 19th century dialects of northern England.

inspired Wragge's Brisbane and Auckland landscaping.[44] In the 1960s, investors began redeveloping Alton Towers as a theme park, which was ranked among Europe's top 10 amusements of its type in 2022, with the world's first 14-loop rollercoaster.[45]

* * *

The dazzling lights of the night sky gave Wragge loads of amusement. When nine years old, studying the cosmos through Emma's telescope left him grasping for answers. Decades later, in Australia, Wragge recalled, "As I looked up at the Great Bear night after night, my infant thoughts — impelled by some memory from a previous plane of existence — wandered to the mighty Universe beyond." His questions seemed beyond the ken of the Church of England. Beyond that, their God was frightening. In 1920, recalling his childhood, he jotted "O'Moor church and fainting fits". How could the Power Supreme of the Universe also be the Old Testament's petty, vengeful Jehovah? He hated the King James Bible's story of the execution of Agag, leader of the heathen Amalekites, for warring with the chosen people of Israel. In 1 Samuel, chapter 15, verses 32–33, the ageing prophet Samuel obeys God by dismembering Agag. Was this vengeful God really the Master of the Stupendous Whole? "Some voice inside said 'No'. I realised there must be, and is, a Power Supreme who is not

---

[44] 'History of Alton Towers', Wikipedia, https://en.wikipedia.org/wiki/History_of_Alton_Towers, accessed 23 June, 2022; *Derby Mercury,* April 18, 1860, 'The Earl of Shrewsbury's entry to Alton Towers'

[45] *Times Travel,* 'Europe's best theme parks', https://www.thetimes.co.uk/travel/experiences/adventure/europes-best-theme-parks, accessed June 16, 2022

locked up in churches," he wrote. In contrast, Emma's religion seemed to him simply practising the Golden Rule. Writing in 1894, he remembered her devotion to the elderly and poor:

> She spent her life in doing good and many an old villager yet alive blesses her name for tons of coal, basins of soup, tid-bits of chicken and bottles of old port which she used to distribute among the poor and needy with a free hand. Some of my happiest memories are of days when I helped to wheel her along in a bath chair on these errands of charity.

He traced his phobia for butchers' shops to Emma's alarm at any cruelty to animals. He recalled once seeing her scold a man for thrashing his donkey, snatch his stick and throw it into the river.[46]

* * *

Wragge traced the beginning of his religious angst to 1861 — the year the Scottish physicist James Clerk Maxwell calculated the speed of an electromagnetic wave at 310,740,000 metres per second, and also when *The Times* first published weather forecasts from Captain Robert FitzRoy. "Prophecies and predictions they are not," FitzRoy wrote in 1863, as chief of the British Board of Trade's meteorological office. "The term forecast is strictly applicable to such an opinion as is the result of a scientific combination and calculation." FitzRoy had founded the weather office in 1854 after several shipping disasters in storms off the British coast. In following years, he worked on making 48-hour forecasts. Between 1855 and 1860, a total of 7402 ships sank,

---

46   *The Sun*, Sydney, September 18, 1911, p1; Auckland Memorial Museum, Wragge Papers, MS1213, Box 2, 6a, 'My Life'

with the loss of 7200 lives.[47] On October 25, 1859, the iron steamer *Royal Charter* foundered in a gale on rocks near Cardiff after a voyage from Melbourne carrying 498 passengers and crew, and freight of wool, skins and gold. There were only 39 survivors. FitzRoy responded in 1860 by beginning to telegraph storm warnings to at-risk ports. These wires evolved in 1861 into his terse forecasts for *The Times*, which earlier that year had championed FitzRoy's infant meteorology as a practical and beneficial science. "Meteorology now rests upon evidence as palpable as that which confirms our theories of astronomy," the newspaper asserted in a leading article on February 13, 1861. "We cannot yet forecast the general character of the season, but it seems that we can really foretell a gale three days before it comes ..."[48]

* * *

In 1858's summer, celebrated by the author of *Course of the Seasons* as the most beautiful ever known in England,[49] the wire works' vats gleamed with molten ore for Oakamoor's part of the fledgling Transatlantic Telegraph Cable. Wragge, incidentally, in his reminiscences of the Old Country, never fretted over its industrial grime. His North Staffordshire was ever the land of

---

47 Robert FitzRoy, *The Weather Book*, London, 1863, p171, cited in Anderson, K, 'The Weather Prophets, Science and Reputation in Victorian Meteorology', *History of Science*, Vol 37, pp 179–80; *The Nautical Magazine and Naval Chronicle for 1865*, London, 1864, Wreck Register

48 *Staffordshire Advertiser*, August 1, 1857, p4. Bolton, "With his accustomed liberality, regaled about 150 children belonging to the village with a plentiful supply of tea and plum cake."

49 *Liverpool Mercury*, May 22, 1860, p6, 'Course of the Seasons'; *The Australian Star*, Sydney, May 26, 1894, p7

daisy-speckled meadows and the aromas of May blossom and fermity porridge.⁵⁰ Between January and June 1858, the Oakamoor wire drawers extruded 60 tons of copper wire to repair the famous cable, established, then broken, in 1857. In August 1858, Queen Victoria celebrated the cable's temporary reconnection with a 98-word telegram to US President James Buchanan that took 16 hours to send. Then the cable between Ireland and Newfoundland failed again. Bolton and Sons, one of several contractors for the first cable, made 113 tons of copper wire in their Oakamoor works for the second and successful link, completed in 1866.⁵¹ The Boltons prospered from this spadework for the global village. Soon, the market for submarine cable stretched east from Dover to the Persian Gulf and beyond, and west to the Americas, and east to Singapore and Australia.

---

50   *Swan Express*, Midland Junction, Western Australia, March 24, 1906, p3; Interview with Mr Clement L Wragge; *Wragge* magazine, November 27, 1902, p158, homage to fermity porridge, "an old North Staffordshire dish"

51   Blake-Coleman, Barrie Charles, *Copper Wire and Electrical Conductors: The Shaping of a Technology*, Berkshire, 1992, pp 176–78; Burns, Bill, 'Thomas Bolton and Sons', in *History of the Atlantic Cable and Undersea Communications, From the First Submarine Cable of 1850 to the Worldwide Fibre Optic Cable*, https://atlantic-cable.com/CableCos/Bolton/ accessed September 21, 2022

CHAPTER 3

UTTOXETER TO UTAH, 1862-75

Travel is fatal to prejudice, bigotry, and narrow-mindedness, and many of our people need it sorely on these accounts. Broad, wholesome, charitable views of men and things cannot be acquired by vegetating in one little corner of the earth all one's lifetime. — Mark Twain, *The Innocents Abroad*, 1870.

Wragge boasted in his 1911 newspaper memoir of having possessed from the age of seven the shapes of the world's great continents imprinted in his mind. "I yearned with an intense longing to travel and see it all," he wrote. His journey began modestly in 1862, when aged 10, he began school at Alleyne's Grammar School in the market town of Uttoxeter, 13 miles by steam train from Oakamoor.[52] Founded in 1658 by mathematician Thomas Alleyne, the school readied boys from eight to fourteen years old to enter business or a trade, or advance to higher education. Subjects included classical literature, history, geography, mathematics and drawing. The 79 boys enrolled in 1862 occupied an almost-new building erected by Alleyne's trustees — the master, fellows and scholars of Trinity College, Cambridge. Emma readied her bright boy for enrolment

---
52   *The Australian Star*, Sydney, May 26, 1894

by hiring him a Latin tutor. By then, Wragge was at odds with his father's eldest brother, George Paulson Wragge, executor of the fortune he would inherit at 21. A senior partner of the family's Birmingham law firm Wragge and Co, Uncle George saw enrolment at Alleyne's as grounding for a secure career in the law.

Through his famous years in Australia and New Zealand, Wragge curated himself a mystical childhood. In 1894, he told *The Australian Star* he had been born with a spirit of independence — "a desire to explore the wondrous little world around him". He conceded his wanderings had caused his nursemaids endless anxiety. The Sydney daily reported the romantic scenery of the Churnet Valley, "its pine-clad crests, shady dells and moss uplands", had stirred his life-long love of the beauties of nature.[53] In New Zealand, two decades later, he remembered himself as a bookish boy too. By then, the promoters of his travelling scientific shows called him "Professor Wragge". In November 1922, the *New Zealand Truth*, having heard of his serious illness, published a contradictory tribute emphasising his love of learning from an early age, fostered by his grandmother: "In the 'onomies' and 'ographies' he wandered early, finding more beauties in his text books than in the lovely Churnet Valley."[54]

Wragge took the train to and from Alleyne's school until early 1864 when Emma had a stroke. It was an unhappy year, beginning with the sudden death of his headmaster, the Reverend William Harvey, aged 34. The Wragge family photo album has a picture of Harvey's tombstone, recording his good works and Christian faith. There is also a portrait captioned "Azoul about 12" of a fair-haired

---

53   *The Australian Star*, May 26, 1894, p7
54   *New Zealand Truth*, November 18, 1922, p1

boy leaning awkwardly on a studio pedestal, posed in a straw boater, big black boots and a blazer with pocket handkerchief. Edris called him "Azaul" or "Azoul" — likely from the Spanish *azul* — for his blue eyes. The boy in the photo looks lost, as if drained by too many deaths.

Clement's jottings reveal he ran away from school in 1864 to the scorn of his 24-year-old cousin, Ellen Wragge, who had moved from London to care for Emma. She was unmoved by his homesickness and dislike of boarding in Uttoxeter with the family of a railwayman he nicknamed "Cuckoo Joseph". The town's population was 3600 in the 1862 census. London, 140 miles south-east, was the biggest city in the world with nearly three million people.

Emma's second, and fatal, stroke in February 1865, ended Clement's Oakamoor days and propelled his helter-skelter adolescence in London. Sixty years later, he scrawled "my great LOSS" and "What shall I do Granny, dear?" remembering her body in the dining room at Oakamoor Lodge and rushing away to the solitude of Toothill Rock, high above the Churnet Valley. He recalled her death as his greatest sorrow in early life, but also the prelude to discovering the metropolis which, he conceded in another reminiscence, both appalled and charmed him.

* * *

Wragge's Aunt Frances inherited him after Emma's death. In his 1911 memoir, he condensed the next eight years into a sentence: "An old aunt took charge of me and I went to London, living in and around that great city till I was a young man." He had more to say in 1894 when interviewed in Sydney and in the early 1920s when planning his life story. His Latin grind continued from 1865 to 1868 at Belvedere House, a private boarding school for sons of gentlemen, and in 1869 at Launceston in Cornwall, where he received private tuition to ready him for probation as an articled law clerk. Belvedere House School adjoined the famous relocated Crystal Palace on Sydenham Hill, Upper Norwood. The school's founder, Stephen Cousins, was among dozens of wily teachers who opened private academies in the neighbourhood. The palace, sheeted with 900,000 square feet of Birmingham plate glass, was replete with samples of industrial ingenuity and wonders of the natural world. Its core treasures came from the Great Exhibition of the Works of Industry of All Nations, assembled in Hyde Park in 1851 under the patronage of Prince Albert. In 1852, speculators bought the palace and began shifting it seven miles to Sydenham. Led by principals of the Brighton Railway, they formed the Crystal Palace Company and rebuilt the massive pavilion on a 300-acre site on the summit of Sydenham Hill. They calculated its arched roof would be seen from 40 miles away — the first sign of London for travellers arriving from Germany or France — and built a branch line into Britain's expanding railway labyrinth to maximise patronage. The project cost £1.3 million, a huge, quixotic investment.[55]

Wragge hated the school. Many years later, while not naming Cousins, he recalled "the master" had often beaten him with a ruler, hit him on the head for mistakes in Virgil, and made him endlessly rewrite active and passive Latin verbs. He recalled in 1894, "At Norwood the classics were jambed [sic] down my throat to the

---

55   Coulter, John, *Norwood: From 19th century Common Land to City Commuters*, from 'Ideal Homes: A History of South-East London Suburbs', http://www.ideal-homes.org.uk/case-studies/norwood, accessed July 1, 2017; *The Times*, London, June 23, 1852, p4

exclusion of more useful subjects and geography, with its kindred branches, was frequently studied as a pastime, so also was astronomy." Cousins ruled from 1852 to his death in 1885, aged 65. Two years later, his Oxford-educated son, Stephen Junior, took charge.[56]

The exotic pavilions and vast grounds of the palace compensated Wragge a bit for his exile from Oakamoor. "I rejoiced in Nature, and the surroundings of the Crystal Palace fostered my interests," he wrote in 1921. A hothouse full of birds, monkeys and rare plants simulated tropical clamminess. There were concerts, fireworks nights and also life-size models of prehistoric animals. Noel Coward's grandfather, James Coward, gave recitals on a massive pipe organ in the concert hall. Visitors from Norwood's numerous private boys' and girls' schools flirted around the food and grog booths. In a sketch of his life at Belvedere House, he wrote, "At 15, I chased the girls — Fatima and other pretty girls 'round the Crystal Palace." The palace opened seven days a week from 1860 when the Crystal Palace Company won a campaign to trade on Sundays, strongly opposed by the Lord's Day Observance Committee.[57]

In mid-winter 1865, Maria Isabella FitzRoy persuaded her exhausted 59-year-old husband, Robert FitzRoy, to move from the inner-East End suburb of Brompton to the cool and calm of Upper Norwood. FitzRoy, promoted to Admiral in 1863, was burnt out after 10 years running Britain's first meteorological office. He confided to an old friend, Lieutenant Bartholomew Sullivan, he

---

56   *Morning Post,* London, August 19, 1887, p1; *Blackheath Gazette,* January 1, 1892, p1

57   *Our Crystal Palace Story,* published 1962, The Norwood Society, https://www.norwoodsociety.co.uk/articles/our-crystal-palace-story, accessed September 21, 2022; *The Australian Star,* Sydney, May 26, 1894, p7; *The Times,* London, June 23, 1852, p4

was worried forecasting would cease if he quit. Sullivan, who sailed with FitzRoy on the *Beagle* expedition, 1831–36, had since been appointed to the Board of Trade as a naval representative. In April 1865, FitzRoy wrote to him that he had survived a very serious illness by God's grace, through the care of his wife and a skilful doctor. A day later he killed himself — a sudden impulse of insanity, Sullivan reasoned later in a letter to Charles Darwin. FitzRoy was survived by his wife, their seven-year-old daughter and three adult children from his first marriage.

Board of Trade president Lord Stanley believed he had worked himself to death, the victim of zealous public service as much as any soldier killed on active service. "His zeal led him to labour beyond his strength," Stanley wrote.[58] In May 1866, the weather office suspended sending forecasts to *The Times* on the advice of the Royal Society, which had always considered FitzRoy's methods unscientific. The society took control of the weather office and in January 1867 appointed a new director, Irish engineer Robert Henry Scott.[59] FitzRoy's colleagues had a footstone made for his grave, with a verse from the opening of Ecclesiastes: "The wind goeth toward the south, and turneth to the north, it whirleth about continually, and the wind returneth again according to his circuits."[60]

* * *

---

58 *The Times*, London, September 3, 1866, 'Lord Stanley, Speech to Liverpool Chamber of Commerce', cited in Anderson, Katharine, *The Weather Prophets, Science and Reputation in Victorian Meteorology*, York University, Ontario, p179

59 Wheeler, Malcolm, *History of the Meteorological Office*, Cambridge, 2012, pp 71–72

60 Old Testament — Ecclesiastes 1:6

Wragge was never as mellow as the author of Ecclesiastes. Any possible wisdom in the Old Testament sage's nothing-new-under-the-sun equanimity eluded him. In later years, he boasted of being at heart still an excitable boy, albeit a sobered one, as recommended by his friend Sir Robert Christison as an antidote to old age. Sir Robert, a long-serving professor of medicine at Edinburgh University, climbed mountains until months before his death at 84, in 1882.[61] Wragge shared his temperament. Besides, the greatest city on earth simmered with distractions. A murky Babylon to the novelist Henry James and an enormous Babel to historian Thomas Carlyle, London was said in 1870 to have 20,000 public houses visited by 500,000 customers, and to shelter 100,000 homeless people each winter.[62] In 1866, its population reached three million, including a legion of perhaps 300,000 workers mainlined daily into the heart of the city on steam trains from respectable new suburbs like Upper Norwood and Teddington. Frances Wragge moved from Hampton to Teddington after the death in May 1861 of her bankrupted husband, Charles John Wragge. In 1865, the household that Clement joined, when on holiday from boarding at Belvedere House, comprised his aunt, five cousins aged from 18 to 29, and a couple of servants. Another cousin, second-oldest son Edmund, 28, lived nearby with his wife, Lucy, whom he had married in 1861 while working in South Africa as a railway engineer.

Between 1865 and 1868, Wragge graduated from treble to

---

[61] *The Sun*, Sydney, September 18, 1911; Bettany, George Thomas, 'Robert Christison', *Dictionary of National Biography, 1885-1900*, Volume 10, https://en.wikisource.org/wiki/Christison,_Robert_(DNB00), accessed on September 21, 2022

[62] Ackroyd, Peter, *London The Biography*, London, 2000, pp 576-79

bass in the Teddington Church of England choir, taught Sunday School with misgivings about Adam and Eve, and was confirmed in the Anglican faith. He also discovered rowing on the Thames, day and night. Regular church attendance was then the norm for well-off Anglicans, but for many poor people "C of E" meant Christmas and Easter, with luck. A census of religious worship in 1851 revealed only half the Anglicans of England and Wales attended regularly. In 1881, *The Times* estimated the one-quarter of the Christian population were regular attenders and guessed they were mainly members of the gentry, storekeepers, tradesmen and artisans keeping up appearances.[63] Labourers were absent because they judged the practice useless in the struggle of life. "They have no appearances to keep up, no fortune or favour at stake, they are fed, indeed, with high-sounding promises, but have no earnest or present proof," *The Times* argued in a leading article.

For his part, Wragge wearied of his aunt's insistence on sending him to church four times every Sunday. "Who can blame me?" he wrote in one of his newspaper memoirs.[64] His notes make no mention of solace in the creeds of the church, such as the hope of eternal life. He lost his last direct link with his mother in June 1867, when her sister, Elizabeth Downing, died at the health resort of Teignmouth, Devon, aged 43. Elizabeth's death certificate describes her as a spinster gentlewoman and gives the cause of death as a spasm of the heart caused by nervous exhaustion.

---

63   *The Australian Star*, Sydney, May 26, 1894, p7; *Census of Great Britain, 1851*, 'Religious Worship in England and Wales', abridged from the official report made by Horace Mann, Esq, London, 1854, pp 86–89; *The Times*, London, December 26, 1881, leading article p7

64   From autobiographical notes for an article, Auckland War Memorial Museum, Clement Lindley Wragge, 1852–1922, MS-1213, Box 2, 6a

She left personal effects valued at under £1500, including silver heirlooms later inherited by her nephew.

* * *

In 1868, Wragge graduated from Belvedere House. Now 16 years old, he was, however, still captive to Uncle George, who sent him to Cornwall for Latin coaching from a scholarly curate. "One of the jolly parsons, you know, as parsons go," Wragge recalled in the 1894 interview. "After reading a few odes of Horace, [he] used to take me on his rounds, and afterwards allowed plenty of holidays ... treating me, in fact, as a very son." The Reverend Wickham Montgomery Birch, MA, curate of Launceston, was then about 40 years old and new to the district. He made time to take his energetic pupil tramping along North Cornwall's coastline, 10 miles or so west of Launceston. At low tide, the weather-beaten beach at Bude stretched another six miles west to the edge of the receding Atlantic. Clement collected shells from a sandy littoral that was also raided by builders for construction jobs around England.

Wragge was reluctant to return to Teddington. A picture in a family photo album shows him standing tall in Cornwall, arms folded, broad-browed, lean-shouldered and stern. His jottings suggest he made many friends during his 18-month stay. He moved back to London in 1870 with adequate Latin to become a probationary law clerk at Clifford's Inn, but his notes reveal he missed Launceston badly, especially the daughter of a lawyer in the church circle. He wrote: "The pain of leaving Launceston — trying to get back and pawning for means ..."

During his four years as an articled clerk, the urge to shoot through constantly drove Wragge out of chambers and into the real world. Midway through 1871, within months of his indenture to Lake Brothers, of Lincoln's Inn, he walked the ruins of the Paris Commune. Tens of thousands of workers had occupied Paris from mid-March to late May 1871, backed by the National Guard militia. At least 7000 died as the army retook the city in a week-long civil war from May 20 to 28 dubbed "The Bloody Week" (*La Semaine Sanglante*). Army casualties were put at 877 killed and 6454 wounded. When interviewed in 1894, Wragge confessed the horrors of the commune still haunted him. He wrote he had needed the solitude of the Ardennes wilderness, in south-east Belgium, to regain his calm. His disjointed CV between April 1871 and his 21st birthday, September 18, 1873, shows he managed holidays in Holland, Germany, Austria, Italy, Spain and Portugal in that time.

When in London, he enjoyed the brotherhood of junior law clerks in Fleet Street pubs and beyond. His curiosity carried him from The Cock tavern to the Cheshire Cheese and Rainbow through inner-city slums, down Wych Street and on through Seven Dials, Regent Circus and Belgravia. "London for a time had great attractions," he recalled. "I shall never regret the regime at Lincoln's Inn; it was good business training in a way." Along the way, he moved in with another George Wragge, the eldest of the London cousins, after his aunt kicked him out of the family home in Teddington: "My excellent relatives with whom I boarded ... regarded me as a species of black sheep and kept a tight rein," he recalled in 1894. This George worked as a clerk in the Admiralty and was almost a generation Clement's senior, at 35 years old.

Chapter 3 Uttoxeter to Utah, 1862–75

* * *

On October 20, 1873, a month after coming of age, Wragge left London for Paris on a Cook's Tour of Egypt and Palestine. He was still indentured to Lake Brothers, but took eight months of sick leave on medical grounds, probably to recover from a nervous breakdown on the Continent earlier that year. In another of his memoir drafts, he wrote, "Natural constitution very hardy and wiry but young people sometimes outgrow their strength and doctor ordered 8 months' leave from office." Uncle George was furious. He cancelled the birthday dinner he had organised at Oakamoor Lodge, leaving Wragge to celebrate with old friends in the village. His inheritance included at least £2000 from his father and perhaps £1500 and silverware from his aunt Elizabeth Downing.

* * *

Thomas Cook led his first tour of Ottoman Empire Palestine in 1869, three decades after inventing charter travel in England. He began excursions to the Continent in 1855 and his son, John Mason Cook, led the firm's first tour of North America in 1866. The opening of the Suez Canal in 1869 brought new Mediterranean steamer services, enabling the Cooks' venture into Holy Land tourism — a year after his rival, former London bootmaker Henry Gaze, who in February 1868, offered a 10-week excursion of Egypt and the Holy Land.

Turkish sultans then ruled the heartland of the New Testament, Torah and Koran under the watch of Britain and France. The sultans governed a confederacy of 300,000 souls from Northern Syria to Egypt. Holy Land tourism, archaeology and geology flourished through the 1860s despite volatile political and sectarian undercurrents. In 1862, Queen Victoria's eldest son, 20-year-old Albert (Bertie), Prince of Wales, spent four months in Palestine, two years after the massacre of 20,000 Christians in Damascus in the Mount Lebanon Civil War. Lebanon became an autonomous province with a Christian governor in the peace settlement.

In 1867, Samuel Clements, aka Mark Twain, toured the Holy Land and Egypt and wrote a bestseller, *The Innocents Abroad, or The New Pilgrim's Progress*, published in 1870. Readers around the world bought 85,000 copies in the first 18 months of publication. The book was popular in England, with public readings in provincial towns in 1871, followed by the author's first lecture tour of Britain.

Wragge's Holy Land reconnaissance coincided with surveys by Britain's Palestine Exploration Fund and the American Palestine Exploration Party, approved by the Ottoman leader Abdul Aziz, who in 1867 became the first sultan to tour western Europe. In 1865, *The Times* saw the latest archaeological digs as part of a perennial and perhaps futile quest to verify Biblical stories, or at least "provide a firm basis for Sacred History and Geography". The newspaper argued recent essays in Biblical criticism tacitly demanded an elaborate commentary on the whole Bible, based on up-to-date scholarship and science. Incidentally, by 1872, *The Origin of Species* had been reprinted six times and sold 18,000 copies. *The Times* observed that the Holy Land had for centuries drawn curious strangers — from ancient naturalists and historians to scientific travellers and pleasure-seekers. In

Britain, children were taught the map of Palestine together with the map of England, so that the cities and rivers of Judea were household names.⁶⁵ Thomas Cook wrote to *The Times* in March 1874, that his company had taken 100 visitors through the Holy Land in 1873, double the number in any previous year.⁶⁶ At least one pilgrim in Wragge's party used Samuel Clements' bestseller as a guide book.

Wragge laced his own Holy Land travelogue — published five years later — with scientific observations. In the Jordan Valley, he identified shrubs of the *Aquifoliaceae* or *ilex* variety and *Spina Christi*, (Christ's Thorn tree). He swam in the Galilee and collected from its shore shells of *Neritina jordani*, a freshwater mollusc with related species in the West Indies, Mauritius and India. Further south, he gathered white limestone from the base of Mount Tabor and snail shells of the *Helix spiriplana* and *Bulimus laborsus* species. He bought souvenirs too: from Cana, a twist of camel's hair from an Arab boy's cap; from Nazareth, a pair of ankle rings with silver bells attached.

In his stories, he and his party were ever the *Howadji* — wealthy travellers to be fleeced — and the Arab vendors invariably embroiled in *bucksheesh* bargaining. Nearly 30 years later, he remembered Damascus best as a kaleidoscope of humanity and medicine for melancholia. Writing in his *Wragge* magazine in 1902, he praised the city's cosmopolitan mix of Africans, Arabs, Turks, Syrians, Druses, Bedouins, Kurds, Circassians, Armenians,

---

65   *The Times*, editorial, April 22, 1865, p9; Browne, Janet, *Darwin's Origin of Species: A Biography*, New York, 2006, p105, for details of editions and print runs

66   *The Times*, April 3, 1874, p10

Persians, "and various others in costumes most varied and wondrously striking". He published a photo of himself disguised as a Bedouin Sheik smoking a hubble-bubble pipe. "We like the Arabs and made friends among them," he wrote. The picture showed him balanced on bare crossed legs, showing off long, skinny athletic calves, muscular feet and angular fingers. Bearded, wide-eyed and blissful, like George Harrison in the Beatles' Maharishi days.

Wragge enjoyed some transcendental moments in Syria and Palestine, such as seeing Venus and Jupiter glinting above the Lebanon Range one daybreak. One dusk in early November, he followed the Jordan River to its source in the mountains near Caesarea Philippi, where Jesus was said to have been transfigured by seeing God and returned to his disciples, "His face shining as the sun."[67] In the tradition of orthodox pilgrims, Wragge and the others washed in springwater that ran from the limestone cliffs. He recalled absorbing that landscape "with peculiar feelings akin to awe".[68] He had more joy the next day when he split from the meandering caravan and galloped across the grassy Jordan plain, heedless of rumoured bandits in the hills. Cook's Tours hired 40 or so wretched horses for the ride to Jerusalem. Wragge deplored their saddle sores. Later, he spent some wild weeks in Egypt after parting from the group in Jerusalem. He carried a camera and a few years later, back in Staffordshire, assembled an exhibition of 150 photographs of Holy Land and Egyptian sites.

* * *

---

67   New Testament, Matthew, chapter 17, verse 2
68   *The Cheadle Herald*, November 11, 1878, XXIII, p1

About Christmas 1873, Wragge decided to extend his trip around the world, despite having used most of his special leave. His sole comment, "I risk Lincoln Inn and go on by *John Tennant* by Red Sea," suggests he had not, by then, completely given up on law. The 1470-ton steamer *John Tennant* carried passengers and freight regularly between London and Bombay, via the Suez Canal and Red Sea. He joined the ship in Joppa, probably in late February 1874, intending to visit his friend, the Reverend Robert Alfred Squires, an Anglican missionary at Nasik, about 100 miles north-east of Bombay. Suez to Bombay took about a fortnight. Wragge's notes suggest he toured northern India by train before reaching Nasik, travelling perhaps as far as Allahabad, nearly 900 miles north-east of Bombay. Robert Squires was an Oxford-trained clergyman, in charge of the Church Missionary Society's Sharanpur Mission House, near Nasik, a city known as a centre of higher education and industries. They were friends from Wragge's London days.

In 1872, Thomas Cook, then aged 64, circled the world in 222 days by rail and sea and launched around-the-world excursions for a basic fare of £200. Jules Verne's novel *Around the World in 80 Days* followed in January 1873. Globetrotting had been made possible for wealthy tourists by the opening of the Suez Canal, completion of the first railway across the United States in 1869, and a railway across the Indian subcontinent in 1870. Wragge, still with the means to ramble, returned to Bombay from his holiday in Nasik in time to board the P&O Line mail steamer *Pera* on April 4, 1874, on his first trip to the United States, via Australia.

CHAPTER 4

IN THE COURT OF KING AEOLUS, 1874–77

The flag that lights the sailor on his way, the flag that fills all our foes with dismay, the flag that always has carried the day, the Union Jack of Old England. — Lyrics from the song *The Union Jack of Old England*, written and first performed by Charles Williams, 1872.[69]

Wragge reached San Francisco on August 16, 1874, on the iron clipper *Melpomene* with wind in his sails and a stuffed albatross in a potato sack. He was one of three paying passengers on the square-rigged coal ship's 12,000 miles Pacific run. She was about a week overdue with 1800 tons of coal because of storm damage south of the Pitcairn Islands, 20 days out from Newcastle, New South Wales. *The Daily Alta* newspaper, in San Francisco, noted *Melpomene's* battering on June 18, at Latitude 29S, Longitude 129W: "... lasting but a few minutes [it] carried away fore topgallant mast with fore topmast head, also main topgallant mast and sprung main topmast head, and blew away the great part of sails on fore and main."

Crossing the Pacific in the superstitious brotherhood of the

---

69   Richards, Jeffrey, *Imperialism and Music: Britain 1876–1953*, Manchester, 2001, p328

Melpomene amounted to a weather-watching baptism. In 1893, Wragge traced his love of meteorology to this voyage, "having a good tutor in the person of King Aeolus himself".[70] Getting from India to Australia by Royal Mail steamer was, in contrast, fairly tame. He reached Melbourne on May 4, 1874, after a month on the *Pera*, then travelled by steamer to Sydney before joining the *Melpomene* in Newcastle, 100 miles further north, on May 28. His curios — the albatross, a stuffed Remora, whale teeth, models of canoes and a Fijian club — came from the Pacific, not New South Wales. He continued collecting in the United States, for example, acquiring samples of sugar pine and other timber during a whistlestop in the Californian capital, Sacramento.[71] New South Wales collieries had an annual peak production capacity of 1.5 million tons for domestic and foreign markets in this era. In 1875, Newcastle exported 723,844 tons of coal to foreign and inter-colonial ports and coaling stations.[72]

Captain H O Christiansen, who was loading coal from the Wallsend mine near Newcastle, had advertised, in mid-May 1874, accommodation on his splendid ship for a few cabin passengers. Consignments of New South Wales coal to California by windjammer were common in the 1860s and '70s. For large

---

70   *Evening Journal,* Adelaide, April 20, 1895, quoting from 'How I make My Weather Forecasts', an article by Wragge in the *Review of Reviews*

71   *The Cheadle Herald*, July 12, 1879, p1, 'Afloat and Ashore: San Francisco to Sacramento', records items included South Seas clubs, models of canoes and stuffed albatrosses; 'Afloat and Ashore: Sacramento to Salt Lake City', July 19, 1879 p1, footnote records his acquisition of timber specimens for his proposed museum.

72   *The Great Circle, Journal of The Australian Association for Maritime History Inc*, vol 28, No 2 (2006), 'Bound out for Callao! The Pacific Coal Trade 1876 to 1896: Selling Coal or Selling Lives?', Part I, Michael Clark, p31, https://www.jstor.org/stable/41563218, accessed September 21, 2022

iron clippers such as *Melpomene*, the trip across the Pacific was generally part of a three- or four-stage haul around the world, picking up and delivering freight. Within a month of reaching San Francisco, she had been unloaded, repaired and sent on her way to England via Cape Horn, packed with wheat. Sail still ruled in the 1870s, despite a surge in steam-powered freighters since the opening of the Suez Canal in 1869. Lloyd's shipping insurance company put the world's mercantile sea-going sailing fleet at 57,258 in 1875, and the merchant steamer fleet at 5519. Shipping mishaps cost 1829 lives that year.[73]

Wragge was 21 years old when he made his first long voyage under sail. You can see its impact in articles on meteorology he wrote decades later, imagining Captain Christiansen on the bridge of *Melpomene* in a hypothetical storm on the Tasman Sea. Never mind that the 1869 Glasgow-built *Melpomene* was, by then, long gone, having sunk off the Andaman Islands in 1880, or that Captain H O Christiansen [sometimes Christianson] had, by then, retired to recreational sailing on Sydney Harbour. Wragge's stories in Queensland's *The Worker* newspaper conveyed to "bushies" his love of sailing:

> Some think that a voyage in a sailing ship is monotonous. To us it is of the fullest interest — from the "mollyhawks" of high latitudes to the Bos'n birds of the tropics — always something to do — observations — every "eight bells", and oh, luxury of all, coffee before sunrise.

In a newspaper article written in 1898, Wragge placed himself

---

[73] *Journal of the Statistical Society of London*, Vol 39, No 33, September 1876, pp 577–79, quoting from *Lloyd's Statistics for Marine Losses for the Year 1875*

aboard the imaginary *Melpomene*, battling from Melbourne to some unknown New Zealand port. He reported the captain's vigilance of barometer and canvas, his whistling for the wind and the cook's efforts at shark catching, to "breed a breeze of wind". He recorded the approach of the storm, marked by a sudden fall in atmospheric pressure and retreat of the anticyclonic calm into cyclonic bluster. His hypothetical storm probably differed a lot from the squall he experienced on the *Melpomene* in 1874, most significantly in having been forecast and named. But in both cases, the clipper was reduced to "bare poles" — by the wind in 1874, and in 1898 by order of the captain, whom he imagined damning weather prophet Wragge for the humbug of the gale, while rueing not having heeded his forecast. Wragge's writing was informed by his first Pacific adventure, in details like the battered *Melpomene*'s vortex of seabirds and the skipper's warning, as the storm centre passed, "First rise after low indicates a stronger blow," heralding hours of scourging southwesterlies, despite a rising weather glass. His homage to Virgil's King Aeolus, ruler of the elements, was a bit overblown:

> Oh, how well we know the scene, oh, how our spirit is there and how we revel in a storm in the Roaring Forties! The Infinite is with us and we feel ourselves face to face with Him. Oh, the poetry and the majesty of the thing! Grand though the English language is, it fails to describe it ...

* * *

By September 1874, he had been travelling east for nearly 11 months. He enjoyed the cachet of being British and wealthy in

Chapter 4 In the Court of King Aeolus, 1874–77

Cairo, the Holy Land, India, New South Wales and California. At times his scientific trekking veered into a spiritual odyssey. For example, he detoured to Salt Lake City while traveling the new transcontinental railway to New York to test a rumour that the ageing Church of the Latter Day Saints' leader Brigham Young was willing to greet any visitor to the Utah Territory capital. Young, then aged 73, obliged with a brief audience and impressed him as "a fine, hearty, John Bull farmer" with a firm handshake:

> The light brown hair, beard and whiskers were largely tinged with grey and his light grey eyes lit up a sallow complexion, the more striking in contrast to his suit of black. He requested my autograph ... After telling him about my wanderings and general small talk, he again took my hand and departed.

The massive, recently completed dome of the Mormon Tabernacle symbolised Salt Lake City's religious pretensions, but Wragge was more interested in stories of Young's harem of 60 wives: "18 of his own and 42 others — widows of deceased members of the Mormon flock, held ... under the sealing-over process." He met Young in Brigham's Block, a complex said to contain the notorious Beehive House, loaded with women and children. In 1911, writing for *The Sun* in Sydney, Wragge recalled having seen numerous nurse-girls and babies in Salt Lake City and joked he had considered becoming a Mormon. In 1921, he wrote that Young had nearly convinced him to convert, but "the green fields and hawthorn of England" had prevailed.[74] Young was said to have been survived by 23 wives on his death in 1877, aged 76. An obituary in *The Times*, London, acknowledged his

---

74   Auckland War Memorial Library, Wragge papers, MS1213, Box 2

genius in developing Utah's material resources, but concluded death had saved him from punishment for unspecified crimes that had advanced his "most marvellous career". In 1873, writing to the *New York Herald*, Young estimated he had brought 100,000 people to Utah and asserted he had founded more than 200 cities, towns and villages inhabited by Latter Day Saints, from Idaho in the north, Wyoming in the east, Nevada in the west and Arizona in the south.[75]

If Wragge's nostalgia for England had helped him resist becoming a Mormon, he nevertheless loved Utah's mysterious landscape. Writing in 1879, he recalled dawn on the east-bound Atlantic Express the day he had chosen to test Brigham Young's word — September 3, 1874. He saw the sun rise over the Great Salt Lake, blazing over fields of sunflowers and the far-off Wahsatch Mountains. On the platform at Ogden, where he changed trains for Salt Lake City, the early morning air was pure and invigorating. By then, he was 900 miles east of the Pacific coast. "The scenery and surroundings filled me with a fascination almost indescribable," he recalled. A day or so later, he returned to Ogden for the next train east, with Salt Lake City embedded in his heart as "the very picture of peace, repose and quiet beauty". In 1902, he argued Queensland's drought-blighted settlers should use their God-given intelligence like Utah's Mormons to make the desert bloom. If only the people of Cunnamulla had a leader like Brigham Young to capitalise on sweet-scented native grasses.[76] Wragge found on his 1874 train ride, a new and, at times, surprisingly strange

---

75 Brigham Young, Salt Lake City, Utah, April 10, 1873, letter to the *New York Herald*, republished by *The Times*, London, April 30, 1873, p12
76 *Wragge*, October, 1902, p91

world in the reintegrating United States, a decade after their civil war. In 1879, recalling this trip, he juxtaposed his admiration for Mormon enterprise with the realisation that even aboard the New York-bound Atlantic Express, he had been in the "Wild West". He wrote of having been warned by a soldier on the train that the revolver was the only law in America's West and reflected:

> I find that one cannot be too careful in being scrupulously polite and civil to all strangers in the West, card sharps especially, in those states where there is no law to supress them: one may always expect courtesy in return.

He broke his trip at Buffalo and ventured north to Toronto via Niagara Falls to visit his cousin, railway engineer Edmund Wragge, before continuing to New York and returning to England by steamer.

* * *

Interviewed in Sydney in 1894, Wragge said he had grown up wanting to see the whole wide world as a perfect scientific traveller "following, however humbly, in the footsteps of the illustrious Humboldt". The Prussian naturalist and author Alexander von Humboldt (1769–1859) explored Latin America between 1799 and 1804, and recorded his experiences in travel diaries later published in England under the title *Personal Narrative of Travels to the Equinoctial Regions of America, During the Years 1799–1804*. Charles Darwin considered him the world's greatest scientific traveller. He took *Personal Narrative of Travels* on his famous expedition on the *HMS Beagle* (1832–36) and late in life revered Humboldt as, "The parent of a grand progeny of scientific travellers who taken together have done much for science."[77]

Wragge came of age in a coal-fired world. Steam engines and the electric telegraph — the nervous system of late 19th century commerce — stimulated a mass of new wants and needs, conjuring millions of new jobs.[78] In 1874, San Francisco's *Daily Alta* newspaper celebrated the engineers of North America's transcontinental railway as modern revolutionaries engaged in recasting world geography and the currents of commerce. New York was now just a week away by rail, compared with seven weeks by steamer via Cape Horn, and at half the cost.[79]

In the 1860s and '70s, Scottish physicist James Clerk Maxwell (1831–79) built the theoretical foundation for wireless communication by recasting space as a tapestry of vibrating electric fields — the most obvious yet unrecognised of which was light. Maxwell refined his classical theory of electromagnetic radiation in the 1860s while working with other researchers on a British Association for Advancement of Science-sponsored project to improve conductivity in the fraught transatlantic submarine

---

[77] Darwin Correspondence Project, University of Cambridge, quoting from an 1881 letter from Darwin to his friend Joseph Hooker, https://www.darwinproject.ac.uk/alexander-von-humboldt, accessed September 21, 2022

[78] Government telegraphs: Argument of William Orton, president of the Western Union Telegraph Company, on the bill to establish postal telegraph lines, delivered before a select committee of the United States House of Representatives, p24, cited by Wolff, Joshua D, *Western Union and the Creation of the American Corporate Order*, 1845–1893, Cambridge, 2013, p200

[79] *Daily Alta, California*, March 13, 1874, p4. The newspaper estimated that since 1869, when the central route was completed, 75,000 East Coast tourists had visited California by train and that 54,000 Californians had visited the eastern states.

Chapter 4 In the Court of King Aeolus, 1874–77

telegraph cable.[80] In 1871, he warned of growing confusion over scientific doctrines as the gadgets of science multiplied:

> Such, indeed, is the respect paid to science that the most absurd opinions may become current, provided they are expressed in language, the sound of which recalls some well-known scientific phrase.[81]

In later years, Wragge avoided electromagnetic-physics jargon when writing and lecturing on solar radiation, climate and weather. Instead, he settled on recurring declarations of divine revelation in everything from radium to rainbows, almost as if trying to convince himself as well as his audience. In Belfast, in August 1874, Irish physicist John Tyndall advocated, in a keynote address to the British Association for Advancement of Science, a first-principles interrogation of the natural world, free from religious dogma and guided by scientific discipline. Decades later, Wragge praised Tyndall, Thomas Huxley and Charles Darwin as founders of the unity of nature movement, which he also followed.[82]

* * *

---

80 Mackenzie, Dana, *The Universe in Zero Words – The Story of Mathematics*, 2012, Maxwell, Theory, pp 142–47; Schaeffer, Simon, *The Laird Of Physics*, in *Nature*, March 17, 2011, Vol 471, p291

81 *Introductory Lecture on Experimental Physics*, James Clerk Maxwell, 1871, cited in Niven, W D, (ed), *The Scientific Papers of James Clerk Maxwell*, Vol II, 1890, p242

82 *West Coast Times*, New Zealand, September 8, 1909, p2. Report of Wragge's lecture on the Majesty of Creation, including lantern slides of Queen Victoria, Huxley and Tyndall as promoters of the unity and beauty of all creation.

Wragge returned to England in late 1874 and quit law at once. This meant visiting Birmingham, where Uncle George, then aged 63, lived in semi-retirement. In 1871, his uncle had guaranteed his five-year indenture to Lincoln's Inn attorney Thomas Beaumont. Wragge wrote later of their meeting: "To cut a long story short — chitty old contracts and musty old deeds were not for me ... I meant to see the world and none should deny me."[83] While in the Midlands, he made his public-speaking debut with a series of talks on Jerusalem and the Holy Land, in aid of Oakamoor's public reading room. *The Staffordshire Advertiser* noted his use of lantern slides and Arab costumes in the second talk, held at the Oakamoor Mills Schoolroom on December 21. He drew a crowd despite severe winter weather.[84]

On March 1, 1875, Wragge applied to become a Fellow of the Royal Geographical Society, endorsed by three prominent members: honorary secretary Clements Robert Markham, ex-president Sir Henry Bartle Edward Frere, and Indian Raj civil servant Robert Michell. Being a gentleman of independent means probably clinched his election three weeks later. To secure election, candidates needed to show prospects of advancing geographical science, and to have been proposed and seconded by two existing Fellows. Wragge wrote later Markham and Frere had godfathered him. Markham, a renowned Arctic and South American explorer, had helped foster rapid growth of the society since becoming secretary in 1863. His declaration of having personal knowledge of Wragge was probably based on a single interview. He judged the 22-year-old disenchanted ex-law clerk had the makings of a useful

---

83 *The Sun*, Sydney, September 5, 1911, p9
84 *Staffordshire Advertiser*, January 2, 1875, p1

Fellow. Ex-governor of Bombay Sir Bartle Frere and Russian-born Robert Michell then signed off on the secretary's endorsement.

In May 1875, Markham reported membership had grown to 2960 Fellows during 1874, excluding honorary individuals. They were literally all fellows, as women were not eligible for election until 1913. The society recognised Australian explorers Peter Egerton Warburton and John Forrest in 1874 and 1876, respectively — Egerton Warburton for his trek through the desert lands west of Alice Springs, and Forrest for surveying a route from the Murchison River to the Overland Telegraph Line in Central Australia. Sir Bartle Frere's successor as president, ex-British East India Company army officer Sir Henry Rawlinson, stressed in a speech in 1875 that the society's fostering of science had rebounded "to the honour of England".[85]

By late March, when formally elected a Fellow, Wragge had moved to Stafford Lodge in the London suburb of New Hampton. Boarding in Teddington with his Aunt Frances and cousins had always been difficult and became worse after his travels, culminating one daybreak in an argument after a night out. In 1911, he still remembered being stuck outside without a key and the shouting match that led to renting a place with room for his curios: "[I] cut the painter one morning at 6 o'clock as my old aunt and cousins would not allow me a latchkey," he wrote. In 1921, he titled this incident "My Rebellion" and recalled having carried his inherited silver plates as well as trunks of souvenirs when moving into the new lodgings.

During 1875, he began remaking himself as the type of Britisher

---

85   *Illustrated London News*, June 27, 1874, p11 and May 27, 1876, p10; *Yorkshire Post and Leeds Intelligencer*, May 25, 1875, p8

admired by his new-found geographical society peers. Twice a week, he joined medical student friends in viewing surgical operations at St Bartholomew's Hospital. He also enrolled in navigation classes at Mrs Janet Taylor's Nautical Academy, inspired by his 80 days at sea on the *Melpomene*. Located near the Tower Bridge and the St Katherine Docks, the school operated in the heart of maritime London in the Minories, so-called after the devotees of a monastery that once stood on this site. Mrs Taylor, a mathematician, author, inventor and astronomer, opened the school in 1835 with her business-minded shipwright husband, George. She carried on after George's death in 1853 until selling up in 1866 because of poor health. The new owner, chronometer maker John Bryer, sustained the school's reputation for grounding navy and merchant-service masters and mates in navigation, astronomy and mathematics. Janet died of bronchitis in 1870, aged 65, survived by six of her eight children and remembered as a first-class mathematician and a remarkable woman.[86]

Wragge's published memories from 1875–76 are full of bluster; finished with Lincoln's Inn and his Teddington kin, he was his own master, "Full of energy, life and go, with a nervous system vibrating like Marconi's wireless." But his jottings hint at stir-craziness and unexpected scraping, with references to "Arab nights" at his lodgings, hocking the silver, trips back to Stourbridge and Oakamoor, and his being chipped by a clergyman over a dalliance with a girl named Florence. In September 1875, he upset a reviewer in the *Surrey Comet* newspaper by singing *The Union Jack of*

---

[86] *Athenaeum*, February 5, 1870, obituary cited by Croucher, John S and Croucher, Rosalind F in *Mistress of Science*, Stroud, UK, 2016, p256, and also p190 and p236, respectively, for details of navigation training and sale to John Bryer.

Chapter 4 In the Court of King Aeolus, 1874–77

Old England, unannounced and badly, at a New Hampton charity concert. "Mr Wragge had mistaken the audience's laughter for encore calls and had sung another couple of verses of the Crimean War song without accompaniment or the slightest expression," the reviewer sniffed.[87]

* * *

Nearly a year later, in early August 1876, Wragge left London as a rookie crewman on the 1000-ton wooden clipper ship *Wimmera*, carrying general freight for Melbourne. He signed on as a midshipman after finishing his navigation studies. Captains regularly advertised in London newspapers for midshipmen and third officers for East Indies and Australian trips. In the merchant navy, a midshipman needed competency in celestial navigation, map-reading, reefing sails, and splicing ropes. Wragge never explained exactly why he began his life of science and travel by sailing to Australia as a junior officer on a windjammer. The *Australian Star* reporter who interviewed him in 1894 concluded, "His wish was to become a skilful navigator and intrepid bushman." The prospect of 12,000 miles under sail via Tristan de Cunha and the Roaring Forties was definitely more exciting than catching a steamer to Australia in 1874. Besides, he had a family connection in South Australia, which gave him hope of exploring the interior of that colony in some official capacity. He found storage for his treasures at Millwall, bought a midshipman's top hat, frock coat and britches, and met Captain Cooper of the *Wimmera* at London Dock. There were no passengers for Melbourne. Freight included

---
87   *Surrey Comet*, September 18, 1875, p4

dressed timber, marble slabs, ironmongery, saddlery, books, paper, brandy and beer. They left on August 5 and reached Melbourne on November 5.

The trip was unexpectedly tedious. "Dirty weather for 6500 miles from Tristan da Cunha to Western Australia had been dreadful," he wrote later to his uncle William Wragge. The non-stop Southerly Busters meant sailing with constantly close-reefed topsails and a main deck awash with salt junk. The flight of shadowing sea birds had been some consolation, though. Molly hawks, albatrosses, speckled Cape petrels and whale birds tracked them to the Great Australian Bight, with the Cape petrels persisting until well past Cape Otway, Victoria. He recalled the crew's final sea shanty, *We're All Bound to Go*, as he was paid off in Melbourne. The stars had shone with a brilliancy unique to the southern latitudes. His souvenirs from the *Wimmera* included another stuffed albatross and samples of crimpled kelp weed that old hands used to predict changing weather.[88] In a letter published a few years later, he described Melbourne as the Metropolis of the Southern World, the product of British enterprise and a restless, San Francisco-like Yankeeism. Melbourne-born girls seemed to him wonderfully precocious. He continued, recounting Saturday night in central Bourke Street:

> [They] appear to be exceptionally well educated and I am bound to add that I think them very pretty — some are really [more] handsome and finer than any I have seen, barring perhaps young Arab women. Bourke Street on a Saturday evening is verily a parade of female beauty.[89]

---

88 *Wragge*, August 21, 1874, p44, *A Blow off Tristan da Cunha*
89 *The Cheadle Herald*, September 13, 1879, p1

Chapter 4   In the Court of King Aeolus, 1874–77

The *Wimmera* docked during a parched Melbourne Cup week, when the city's aristocracy, country visitors, mug punters, showmen and shysters all hit Flemington racecourse. Wragge joined a 15,000-strong Derby Day crowd on the Prince of Wales' birthday. He was astonished at punch-ups outside the drinking tents that continued without intervention by police or bar staff. After the last race, dusty northerly winds showered punters queueing for cabs and carriages with soil from the backblocks. He found the city's public baths to wash himself clean. A fortnight later, he boarded the coastal steamer *Penola* for Adelaide, the South Australian capital. A type of drought that scientists now classify as an El Niño event gripped central and southern Australia from 1876 to mid-1877. In India and southern China, the failure of the expected monsoon in 1877 caused a famine that killed an estimated 17 million people.[90]

There's a picture in a family album of Wragge, circa 1876, apparently captioned by Edris as "Azoul, the sailor period". He's wispy bearded, smoking a pipe, with a gaze from hooded eyes as bland as the stories he used to tell reporters about his adventures in South Australia. It is difficult to believe these suave versions of events, or his own glosses published in later years. His unpublished notes and little-known travel essays suggest these 15 months in South Australia were bewildering and bizarre. He arrived in Adelaide in 1876 as a Walter Mitty law clerk, then left in 1878 as a would-be meteorologist with a pregnant wife. Undoubtedly, meeting Nora Thornton, sister-in-law of his cousin Rupert

---

90   Nicholls, N, Bureau of Meteorology Research Centre, Melbourne, 'El Niño and the Southern Oscillation Index', in *Encyclopaedia of Atmospheric Sciences*, Vol 2, pp 713–19

Ingleby, QC, curtailed his bachelor days. But I doubt his return to Oakamoor in May 1878 with barometers, thermometers and an Australian wife was connected with the Scottish Meteorological Society's campaign — first reported in May 1877 — for daily observations atop Ben Nevis, Britain's highest mountain. In 1894, recalling his marriage and subsequent trip back to England, Wragge joked his partnership with Nora had put an end to two of his pet schemes: visiting Russia's Kamchatka Peninsula and exploring Tierra del Fuego.

* * *

His distant South Australian cousins, brothers Rupert and John Ingleby, were old hands in the colony. The former, a lawyer, had emigrated from England in 1847 with his first wife and a cachet of respectability. The latter, an auctioneer and member of the South Australian House of Assembly, followed in 1849 and settled in Mount Gambier, 200 miles south-east of Adelaide. Wragge left a muddled record of his time in Adelaide and its droughty hinterland. Within a week of arrival, he joined a boatload of muttonbird "eggers" on the Neptune Islands, at the mouth of Spencer Gulf, 150 miles north-west of the capital. Writing to William Wragge in November 1876, he justified having joined in the slaughter of numerous docile short-tailed shearwater birds — better known as muttonbirds — explaining he had wanted a brace of six specimens for his museum.

Rupert Ingleby welcomed him into his household in Adelaide. He and his second wife Margaret Thornton, their four young children and servants lived in a large house on Hurtle Square.

Chapter 4 In the Court of King Aeolus, 1874–77

Wragge's only note on their hospitality was a terse: "Adelaide — Rupert Ingleby and Halton House". He liked Adelaide, though, which he described to his Uncle William in December 1876, as better laid-out than Sydney and Melbourne. Established in 1836, Adelaide had a population of 80,000 in 1876. His notes indicate he met Margaret Ingleby's sister Nora early in his stay, perhaps at a Christmas party. He wrote under 1875–76, "Nora Thornton!!! The middy's dress and party!" Her name does not appear again until their wedding on September 13, 1877.

In January 1877, he introduced himself to the *South Australian Register* as a visiting Fellow of the Royal Geographical Society. He gave details of the British organisation's joint venture with the Royal Geographical Society of France to charter a steamer as a floating classroom for wealthy young adventurers. There was room for 50 first-class passengers on the first of the round-the-world voyages, leaving Havre in May and expected to call at Adelaide before the end of the year. The professors hired for the journey would teach natural philosophy, meteorology and other practical subjects. What's known of his South Australian adventures suggests Wragge saw himself engaged in a do-it-yourself version of this endeavour: "For young men who have finished their education [to gain] a practical and superior instruction while seeing the wide world."

His status as a visiting fellow probably helped him secure membership of the Adelaide Philosophical Society in July 1877. South Australia's Postmaster-General and Government Astronomer Charles Todd and retired explorer Colonel Peter Egerton Warburton were both long-time members of this club, established in 1853 to further the natural sciences. So was Rupert

Ingleby, a founder and president of the Adelaide Gardeners' Mutual Improvement Society.

Todd was the famed architect of Australia's Overland Telegraph Line, completed in 1872, and founder of South Australia's 80-strong network of weather stations. The son of a London tea merchant, he joined the Royal Observatory, Greenwich, as a statistical assistant in 1841, practised and studied astronomy at Cambridge University, became South Australia's electric telegraph superintendent and government astronomer in 1855, and Postmaster-General in 1870. He was a regular at philosophical society meetings, including the night of Wragge's election, when he gave the latest anemometer statistics from Adelaide Observatory.

Colonel Peter Egerton Warburton sounds like Wragge's type of old boy. Aged 64 in 1877, the former South Australian police commissioner was half blind from having nearly starved while exploring the Great Stony Desert in 1872. Philip Egerton Warburton, who in 1882 joined Wragge in a well-publicised risky ascent of Ben Nevis, is likely to have been his nephew. They might have met through Thornton family's network. Another of Nora's sisters, Mary Ann, was married to ex-mounted policeman John Vereker Lloyd (1842–1927), a contemporary of the colonel's and a Top End explorer. They lived on the outskirts of Adelaide, where Vereker Lloyd had taken up farming after retiring from the police force. Wragge and Nora had their eldest son baptised Clement Lionel Egerton in 1880, but always called him Egerton, "Eggie" for short, presumably in honour of the colonel.

Early in 1877, Wragge joined the South Australian Surveyor-General's Department as a field assistant. He worked the Flinders Ranges, Murray Scrub and the north-west bend of the Murray

River, 100 miles north-east of Adelaide, where the township of Morgan was being established. Led since 1861 by the energetic George Woodroffe Goyder, the department's chainmen pegged out the country of First People, who fascinated Wragge much more than any survey map or plan. He wrote to his uncle that while on the North-West Bend survey, he had stayed five days at Blanchetown and visited a nearby Aboriginal camp, which he found "wild, grotesque and entertaining". He recalled finding under towering, ancient gum trees a campfire communion of "wurlies, growling dogs, black naked figures of the natives with waddies, tin cans, skins and pieces of rag pipes and remains of last feed scattered around". He judged the people friendly and superstitious. There was a beneficent creator in their ken and an evil spirit that lurked after sunset in their campfire shadows.

Later, he told journalist George E Loyau, author of *Notable South Australians 1885*, that police corporal John Ewens had shared what he knew of Aboriginal customs of the Morgan district and had let him collect weapons and other artefacts. Earlier, he made a private visit to Point McLeay Mission on Lake Alexandrina, near the mouth of the Murray River, and collected weapons with permission of the settlement's founder, the Reverend George Taplin. Walking the beach at the Coorong, where the river meets the sea, Wragge picked up some perfectly spherical rocks that he identified as cooled fireballs, some larger than a cricket ball.[91] Loyau wrote that another of Wragge's new-found friends, *South Australian Register* editor John Howard Clark, had obtained for him specimens of nearly every shell found on the

---

91 *Nature, a Weekly Illustrated Journal of Science*, Vol XXVIII, May to October 1883, p31

colony's coasts.[92] Loyau portrayed him as a gentleman scientist whose feats as a meteorologist on Ben Nevis had been preceded by hard work in South Australia. He praised his willingness "to impart to others the outcome of his experiences, gained often under the most untoward circumstances, without fee or reward". *The Illustrated Sydney News* ran a similar story in 1891, noting that while working in the Flinders Ranges and Murray Scrub, Wragge had pursued his ethnographical, zoological and geological observations so well that he had been able to add considerably to his already large museum. His goal, according to another report, had been amassing sufficient specimens to illustrate the geology and zoology of South Australia.[93]

The dates of his appointment and departure are missing from the Surveyor-General's 1877 records, which only show he applied for five weeks of leave between April and September. Neither is there confirmation of his having been second-in-command on any of the surveys, as sometimes reported. To the contrary, in 1896, in a story kindled by his growing fame as a meteorologist, an engineer from Broken Hill, New South Wales, recalled him as his eccentric cadet, easily distracted from the survey chain by butterflies and beetles. The unnamed engineer remembered one day finding Wragge had noted in the field book the foremost of his doings as "saw a black, stumpy-tailed grasshopper".[94] In 1911, writing for *The Sun* in Sydney, Wragge abbreviated his service to "went on the survey in South Australia". But in 1921,

---

92   Loyau, George, *Notable South Australians 1885*, Adelaide, 1885, p95

93   *South Australian Register*, January 10, 1880, p5, citing story in *Nature*; *Illustrated Sydney News*, March 28, 1891, p14

94   *Barrier Miner*, Broken Hill, November 28, 1896, p2

Chapter 4 In the Court of King Aeolus, 1874–77

apparently past caring, he made a surprising disclosure in jottings for his Doan's Pills testimonial: he had become lost in the Wild Dog Ranges while distracted by a love letter. His notes include this statement: "Joined the South Australian Survey, got lost in the bush over a love letter and had to suck stones and eat saltbush." Writing to a friend in 1921, he elaborated on the saltbush, crediting it for having saved his life, but said nothing of the disturbing letter.

* * *

I've seen only three photographs of his first wife, Nora Thornton: young and strong with her two eldest sons in 1884, inscrutable in middle-age, and gaunt and grim approaching her 80th year, circa 1930. In the last-mentioned, she sits on her front step in Brisbane, craggy and broad-chinned, beside a similarly square-chinned son. Edward and Anna Thornton registered their second-youngest daughter simply as Nora after her birth in Adelaide on March 30, 1852. But by the time of her marriage, she had changed her given names to Leonore Eulaliecia Edith Florence d'Eresby and dropped her age by a year.[95] These inventions were probably meant to impress her husband, who was persuaded by her trustees

---

[95] Between 1839 and 1852, Nora's father, Edward Thornton, advanced from training as a law clerk to admission to the South Australian Bar, aged 35. He worked as a teacher in England before emigrating in late 1838. Her mother was the daughter of ship owner Captain Adam Mackay of Warren Point, Northern Ireland, and had three given names: Annabelle Macaulay Rae. It's possible the Mackays had a family link with the Willoughbys, landed gentry in possession since the 14th century of Eresby Manor, Lincolnshire. Nora and Wragge included Willoughby among the given names of their fourth son Reginald.

to enter a £2000 marriage settlement deed in provision for their hoped-for children.

Nora probably wrote the letter that giddied him in the Wild Dog Ranges. Reputedly, a wealthy and eligible bachelor, Wragge must have been a much-anticipated guest at Rupert Ingleby's home. Rupert's wife, Margaret, had three younger unmarried sisters — Blanche, aged 29; Nora, 25; and Frances, 24. Forty-five years later, Wragge tacitly acknowledged in the draft of the letter canvassing his chances of finally getting a divorce from Nora, that Rupert and his brother-in-law Edward Thornton junior had seen him coming. The letter to his lawyer in Auckland in September 1921, gives no insight into their engagement, only a blunt: "Married at Adelaide Leonore E. E. F dL. Thornton, sister-in-law of my relative the late Rupert Ingleby, QC, of Adelaide."

Rupert Ingleby's friendship with his second father-in-law, Edward Thornton senior, began in Adelaide in the early 1850s, when both were associated with George Milner Stephen, a controversial barrister and sometime faith healer — Edward as chief clerk and Rupert as a partner in his law practice. Their families were close. In 1854, when Edward died suddenly of epilepsy aged 36, Rupert was listed on his death certificate as the informant. Edward senior's wife, Annie, was left with seven children — Margaret the eldest having just turned 12. Rupert and his first wife, also named Margaret (nee Dutton), had no children and readily helped Annie with hers. This included taking her eldest daughter into their household in an unknown capacity. In a sensational court case in 1857, the then 15-year-old Margaret Thornton was described as "A young lady living at Rupert Ingleby's residence." She gave evidence against the

Chapter 4 In the Court of King Aeolus, 1874–77

Ingleby's neighbour, Frederick Ranger Rutland, who was charged with having assaulted and libelled Rupert Ingleby. This case and another in 1858 stemmed from an alleged affair between Rutland and the first Mrs Ingleby.

The first Margaret Ingleby died of apoplexy in December 1866, aged 37, and a year later Rupert married Margaret Thornton — they were aged 47 and 25, respectively. By mid-1877, they had four children, all under 10 years old, and another on the way. In October 1877, a few weeks after Nora married Wragge, Margaret advertised in the *Evening Journal* for a housemaid for general duties and a young person to become their nursery housemaid. By then, Rupert had mellowed into a well-known City Coroner, occasional Special Magistrate and Vice-President of the Philosophical Society.

Clement and Nora wed on September 13, 1877, at St Paul's Church of England Cathedral, Adelaide, with Blanche Thornton and Herbert Foster Stedman as witnesses. Stedman, a dispenser at Adelaide Hospital, was possibly a fellow member of the Philosophical Society. The officiating minister, Alexander Russell, Dean of Adelaide, also belonged to this scientific brotherhood. Wragge scrawled his name with large, extravagant loops. Nora wrote hers carefully, her multiple initials slender and upright. Four decades later, reflecting on his wedding, Wragge remembered the date as Friday 13 and joked, "Luck? What is it?!" Their marriage had been unhappy from the honeymoon onwards, he wrote to his lawyer in 1921, when hoping to finally divorce Nora. He accused her of jealous rages, the first on their honeymoon when she challenged him with having flirted with a servant girl. He complained:

In a word, my wife turned out a replica of Dr Jekyll and Mr Hyde, but not in an immoral sense. Short periods were as sweet and as bright as the new moon and all the rest was like the moon in eclipse.

The honeymoon began at Victor Harbor, a seaside resort 50 miles south of Adelaide, and ended at a hotel in Mount Barker in the Adelaide Hills. Wragge found time to write to the Royal Geographical Society in London, enclosing eight pages of his copy of instructions, issued in July 1877, to surveyor Henry Vere Barclay for his exploration of country east of Alice Springs. He also sent secretary Clements Markham a pamphlet on Aborigines from the district where he became lost when separated from his survey party.[96] In January 1878, he and Nora left for London via Cape Town aboard the sailing ship *Hesperus*. She was a couple of months pregnant. Her first child was expected in the Northern Hemisphere summer. There is no evidence Wragge intended becoming a meteorologist. Interviewed in September 1880 by *The Staffordshire Sentinel*, he still aspired to become "the recognised head of an exploring expedition in some unknown region".

---

96 Royal Geographical Society-IBG Collections, ar RGS/CB6, Wragge, C, Copy of instructions to leader of surveying expedition on Herbert River

CHAPTER 5

NORTH STAFFORDSHIRE, 1878–81

It is in the high regions of the air that meteors are formed, rain, snow and hail. There the thunder rolls and the lightning traces its furrows. — Jean-Baptiste Biot (1774–1862), pioneering hot air balloonist.[97]

One late-summer's day in 1878, a trainload of citizen scientists arrived at Oakamoor railway station on an excursion to the village of Wootton. These 30 men and women — amateur geologists, botanists, meteorologists and archaeologists — were led by William Spooner Brough, a silk maker from Leek and an authority on the French philosopher Jean-Jacques Rousseau. Brough had steeled his friends for an eight-mile hike to mark the centenary of Rousseau's death near Paris in 1778. He devised a rambling out-and-back route to Wootton Hall, where Rousseau and his mistress, Thérèse Le Vasseur, spent five months in 1766, as guests of a sympathetic nobleman. Rousseau's critique of the French ruling class in *The Social Contract*, published in 1762, led to death threats that forced him into exile in England in 1766. He began writing his famous *Confessions* at Wootton Hall and met

---

97   *Nature*, January 16, 1879, p237, quoting Jean-Baptiste Biot

Midlands naturalist Dr Erasmus Darwin as each of them explored the neighbourhood's forest glades.[98]

Brough's report of the North Staffordshire Naturalists' Field Club's expedition on August 21, 1878, coincidentally sketches Wragge's Churnet Valley heartland. Three months earlier, Wragge and Nora had found lodgings in Oakamoor in readiness for their first child, May, born on July 11.[99] Then they rented a farm cottage about a mile from the valley, on the moorlands at Farley, where he set up his first weather station in June 1878. He felt at home in the country between Farley and the Weaver Hills, the limestone range separating Staffordshire from Derbyshire. The way to the hills was flanked by rough stone walls, wildflowers, moss and lichen, then ascended through conifer plantations and sub-alpine moorlands, then up the range for the vista of the Dove and Churnet River valleys. The village of Wootton rested in shadows of these limestone hills.

Brough was a vice-president of the club founded in 1865 for the practical study of all branches of natural history and "a fuller knowledge of the antiquities in the neighbourhood". In 1878, its 341 members included 67 women and 17 ministers of religion. Wragge joined in 1879. Brough introduced his party to a lesser-known, rustic Rousseau whom, he asserted, had zealously studied moorlands' flora and sometimes — in random acclimatisation experiments — scattered seeds and embedded cuttings. Quoting from a revisionary tribute in the *London Daily News* in July 1878, Brough praised the revolutionary philosopher's awe of nature: "He

---

98 *North Staffordshire Naturalists' Field Club 1879 Annual Report*, 'Excursion to the Weaver Hills and Wootton', p30

99 Baptised Leonora May Emerline Ingleby Wragge, but known as May

was strangely and deeply moved by the beauty and unconscious life of inanimate nature ... as no man before him had been ..."[100] According to folklore, locals who saw him at large wearing an Armenian hat and caftan had believed him to be a fugitive king. Children had fled from him in fright.[101]

Wragge cut a similarly mysterious figure across the moorlands between 1878 and 1881. The *Staffordshire Sentinel* confided in September 1880, villagers from Farley were mystified by his labours. He could be seen almost any hour, investigating some springy recrudescence, mid-summer's flush or wintry hiatus for London's learned societies. Beyond that, he hiked daily to and from his new weather station on Beacon Stoop, 1150 feet above sea level in the Weaver Hills, and also took details at least twice a day from meteorological instruments at Farley. During the most recent autumn equinox, he had sent Farley's readings to the Meteorological Office every hour, day and night, for nearly 60 hours. No wonder he was looking for an understudy.

The newspaper reported domestic staff were wary of Mr Wragge because of his interest in astronomy. A young woman hired through a Leek labour registry was said to have fled when told by villagers he made his money looking at the stars, which she interpreted as the dark art of astrology. In July 1880, in the neighbouring county of Derbyshire, a collier and aspiring

---

100 *North Staffordshire Naturalists' Field Club 1879 Annual Report*, 'Excursion to the Weaver Hills and Wooton', p30. Brough's views on Rousseau appear to have been taken from the *London Daily News*, leading article on the centenary of Rousseau's death, July 5, 1878, p4.
101 *Rousseau in Wotton*, a translation by Malcolm Crook and Stephen Leach of pp 35–72; *Le Sejour de Jean-Jacques Rousseau en Angleterre* (1910), by L J Courtois, p22 https://philosophyk.files.wordpress.com/2008/10/rousseauinwootton_courtois.pdf, accessed September 26, 2022

fortune-teller named Levi Cooke was jailed for three months for having conned a fellow worker out of two shillings with bogus advice. The prosecution case included Cooke's possession of notebooks on astrology and the *Weather Almanac 1880*.[102]

The Stoke-on-Trent-based Sentinel published an inside story on Wragge in early September 1880, having a fortnight earlier praised his leadership of a 10-hour-long North Staffordshire Naturalists' Field Club expedition, in the wake of William Spooner Brough.[103] He took the party of 60 through his childhood patch — the Churnet Valley, the moorlands and Weaver Hills. Their excursion was grounded in meteorology with trimmings of botany, geology and history. He spent three hours showing off the battery of instruments in his garden at Farley, then marched the party to Beacon Stoop, then finally back to Oakamoor railway station, where he had monitored the weather since August 1879. The *Sentinel* rated his leadership energetic, practical and painstaking, and his scientific knowledge catholic. Beyond these attributes, he was a local gentleman of great perseverance and discernment who, having prevailed in Australia, now longed to lead expeditions into truly unknown regions. More obeisance followed. But none of the newspaper stories that flowed during his six years in Britain, 1878–83, nor his jottings on this period explained why, six years after fleeing from a career in law, he had taken up meteorology. His memoir notes mark the start of his new career simply with "to Oakamoor — my observatory".

The unpredictability of English weather was topical when he

---

102 *Derbyshire Advertiser and Journal,* July 16, 1880, p7
103 *Staffordshire Sentinel,* August 21 and September 4, 1880

and Nora arrived in London by sea in May 1878. Eight weeks earlier, off the Isle of Wight, a freakish snow storm capsized the shallow-hulled Royal Navy training frigate *Eurydice*, drowning 319 crewmen. A coronial inquest was still under way and appeals to aid the victims' kin continued, with £17,000 in kitty. Gerard Manley Hopkins expressed the nub of the tragedy in his poem *The Loss of the Eurydice*. Storm-warning flags flown in English ports were most effective when hoisted early on advice from the meteorological office of approaching low-pressure systems from the Atlantic. In the case of the *Eurydice*, Hopkins wrote:

> No Atlantic squall overwrought her/ Or rearing billow of the Biscay water/ Home was hard at hand/ And the blow bore from the land.

Giving evidence to the *Eurydice* court martial, held in Portsmouth in August 1878, a Royal Navy captain from Salisbury, 40 miles from the catastrophe, recalled the squall had burst over his house from the north-west like a clap of thunder, heralding five minutes of fury. While his barometer had been falling for 48 hours beforehand, the weather that day had been unusually fine.[104] The court martial found the *Eurydice*'s captain and crew blameless. *The Times* concluded: "However seamanship may be cultivated and however able and devoted may be our seamen, there are rough blows of natural forces which no skill can parry and against which no foresight can provide." In 1884, the Royal Meteorological Society published an analysis suggesting the deadly trough had been scimitar-shaped, more than 400 miles

---

104   *The Guardian*, London, August 29, 1878, p6

long, less than two miles wide and travelling at 50 miles per hour, in short "unusually complex and of exceptional intensity".[105]

* * *

Interviewed in 1906, Wragge recalled having become fascinated with meteorology while at sea, beginning with observing the stars.[106] His return to England in 1878 enabled him to further that interest. Compared with colonial life, Britain offered autodidactic gents of his ilk a smorgasbord of scientific and cultural fellowship. England and Scotland both boasted well-established meteorological societies and a world-class weather service offering beginners guidance and discipline. And London's scientific instrument makers were famed for their state-of-the-art thermometers and barometers. Where better than England, famed for having lots of weather and no discernible climate, to learn the secrets of meteorology? Stoke surgeon and amateur botanist Robert Garner, a founder of the North Staffordshire Naturalists' Field Club, summed up his 70 years of experience of Midlands' gloom in the group's 1879 annual report: "England, I love thee well, but not thy clime/ 'Tis only good for ducks and frogs ... and rhyme."

Wragge tried to explain his yen for meteorology writing in 1885 for *The Brisbane Courier*, a year before becoming Queensland's first government meteorologist. He introduced himself as a

---

[105] *The Times*, London, September 3, 1878, p7; *Quarterly Journal of the Royal Meteorological Society*, Vol. X, No 51, July 1884, 'The Origin and Course of the Squall that capsized the *HMS Eurydice*, March 24, 1878', by Ralph Abercomby, FRMS, p180

[106] *Swan Express*, Midland Junction, Perth, Western Australia, March 24, 1906, p3

scientific traveller interested in all branches of physical geography and especially solving "the abstruse problems of ever-changing weather". He conceded knowledge of Earth's atmosphere was as yet minimal, as tantalising to meteorologists as the human mind must have seemed to psychologists in their inchoate discipline. The meteorologist laboured at immense disadvantage, "Anchored as we are to the bottom of this deep atmospheric ocean":

> Weather laws, like those of the mind, are hard to determine on account of their complexity ... I have been so impressed by my own observations of cosmical and biological phenomena, with the idea of the sublime harmony of Nature and fixed laws of the universe, that I can hold no other ... Of what immense value to the civilised world would be sure forecasts of coming changes of weather.[107]

In 1894, Wragge told a Sydney newspaper he had specialised in meteorology because of his fascination with the workings of weather and climate. His other scientific interests were mere recreations.[108] In 1902, he wrote that as a practical meteorologist he worked on the broadest lines and with the widest sympathies. His weather-watching dovetailed so often into kindred sciences that it was almost impossible to determine boundaries. He joked that in medieval times when tempests were blamed on devils of the air, he would have been flayed alive or cast into burning oil for his prescience. While complaining the tag "weather prophet"

---

107 *The Brisbane Courier*, August 11, 1885, p3, 'Notes of a Scientist, No. 1 Gossip on the Weather'. Psychology emerged as a scientific discipline in the 1870s. German doctor of medicine and anthropologist Wilhelm Wundt pioneered scientific psychology in Leipzig in the 1870s.
108 *The Australian Star*, Sydney, May 26, 1894, p7

undervalued his scientific skill; in later years, as an unorthodox, sunspot-savvy, long-range forecaster, he played up to his persona of "Old Wragge, the oracle".[109]

In early June 1878, Wragge had arrived in Oakamoor loaded with exotic curiosities and a brace of weather-measuring devices made by the London scientific instrument manufacturer Negretti and Zambra. The long-established partnership of Henry Negretti and Joseph Zambra catered to professional and amateur weather-watchers alike. In 1880, they advertised a £22–£25 package deal on sets of equipment for "second-order" observatories, to monitor barometric pressure, temperature, humidity, rain, wind and electrical phenomena. A standard marine barometer in a bronze frame cost four guineas — equivalent to three weeks' pay for a school teacher. The firm put the cost of equipping a "first-order" observatory at £330 to £450.[110] Membership of the Royal Meteorological Society, a rough guide to upper middle-class interest in weather and climate science, almost doubled between 1873 and 1882, from 308 to 571. Wragge joined in November 1879. By 1880, Britain's Meteorological Office had charge of seven first-order observatories, 63 climate stations reporting once or twice a day, and 30 telegraphic reporting stations. In 1880, England teemed with 2100 rainfall observers, an increase of 600 since 1870. The rainfall network was run by George James Symons, a meteorological office and meteorological society stalwart.

---

109 *Wragge*, December 25, 1902, 'Meteorographica, The Art of Meteorology', pp 285–86; *Wragge*, November 20, 1902, p149, medieval weather prophecies

110 Turner, Gerard l'Estrange, *Nineteenth Century Scientific Instruments*, Berkeley, 1983, quoting from Negretti and Zambra's 1880 catalogue, pp 235–36, https://ia902607.us.archive.org/0/items/bub_gb_FaAYfJYVNXQC/bub_gb_FaAYfJYVNXQC.pdf, accessed September 23, 2022

Chapter 5 North Staffordshire, 1878–81

On September 1 in 1878, Wragge began taking twice-daily readings at his Farley observatory and sending regular returns to the meteorological office. By then, he was also negotiating with Stafford Town Council to rescue his hoard of travel treasures from damp storage in Oakamoor and London's docklands. He offered the municipality a permanent loan of most of the natural history and ethnographic artefacts. The council subsequently found space in the Stafford Borough Hall, on the recommendation of a Mr R Flamank — said by the *Staffordshire Advertiser* to be a great authority on such matters — who predicted Wragge's collection of Aboriginal weapons could be worth a mint if the First People of Australia died out as expected.[111] A reporter who viewed the Aboriginal artefacts a few years later noted numerous photographs of men, women and children and a sample of leaves from children's exercise books, "Which for neatness and writing and correctness in spelling and arithmetic would not discredit English children of the same age."[112] Good attendances at a public viewing, in February 1879, persuaded the council to accept the permanent loan and to include the museum in a new public library and art gallery, which was formally opened in November 1881, at a cost of £2000.

Wragge advertised his return to the district as a scientific voyager with a series of adventure stories in the *Cheadle Herald*, which continued on and off for five years. In November 1878, he showed 150 photographs from his Cook's Tour of Palestine in a fine art show at the Stafford Borough Hall, and was invited to join civic leaders in a march to the official opening behind the county's

---

111 *Staffordshire Advertiser*, August 10, 1878, p7
112 *Staffordshire Advertiser*, September 15, 1883, p3

militia band. The *Cheadle Herald* also carried reports from him on meteorology and astronomy, including his first weather forecast, published on September 7, 1878 — a prediction of an imminent south-west change based on rising barometric pressure after a week of gentle fluctuations. In August 1878, he wrote to London's *The Times* on crop damage in the Churnet Valley and Staffordshire moorland, caused by a downpour of two inches over two days.

* * *

Nora's life in Farley is poorly recorded. I've found no trace of letters to her mother or sisters in South Australia and virtually nothing from Wragge on the early years of their marriage. In his jottings, he stressed Nora's griefs over the hired help with three jabbed exclamation marks — "the nurse from Stoke and the result!!!" In 1921, he wrote to his solicitor that marital miseries marred their six years in England and Scotland, and continued on their return to Adelaide. His writings between 1878 and 1883 make minimal mention of Nora. A footnote in his meteorological papers acknowledges she sometimes read the rain gauge at Farley. The North Staffordshire naturalists' account of their Churnet Valley excursion with him records a vote of thanks to Mrs Wragge. Nora's first English winter was long, bleak and dreary. In July 1879, the medical journal *Lancet* reported Britain was under siege from the worst spell of cold and damp since 1838–39. Spring and summer crops had failed for want of sunshine and the weekly death rate among the frail and sickly was well above average for the preceding 10 years. The *Lancet* asserted any westward spread of the plague then threatening the Continent might at least be slowed by the frequently freezing temperatures, westerly winds and, "An atmosphere incessantly washed by rain." The journal took heart in recalling Britain's recovery from two equally cheerless years in 1838–39, when awash with record rainfall.[113]

In a letter to *The Times* in February 1879, Birmingham weather observer Thomas L Plant claimed the previous December had been the coldest on record, followed by the coldest January since 1838.[114] But *The Times* also observed that a year earlier, the mild winter of 1877–78 had stirred nostalgia for King Winters of old, said by some to have since been emasculated by the modern evils of forest-clearing and coal-burning. Old-fashioned seasons would be ever remembered as more testing of moral and physical strength than the present. "We find nothing to suggest the average winter 80 or 100 years ago was severer than that which we now experience," *The Times* concluded. None of the climate-change theories stood up. There had been far less land-clearing in the UK and Europe than the United States, and the coal burnt in England's industrial towns was likely to have had only a minuscule effect on air temperature.[115]

Wragge recorded the raw weather he and his family experienced in December 1878 in a precis of his Farley observations published later in *The Cheadle Herald*. Snow fell on nine days, including Christmas Day. In August 1879, he watched the ruinous rain from his observatory, imagining the rumbling River Churnet beneath clouds of fog and mist rising from the valley. He took readings daily at 9 am and 9 pm, checking the maritime barometer in the cottage hallway and the cage that hung in the porch with

---

113 *Derby Mercury*, July 30, 1879, p7, quoting from *Lancet*
114 *The Times*, February 4, 1879, p10
115 *The Times*, January 1, 1879, p11

ozone-sensitive paper saturated in water, starch and iodide of potassium.[116] Between October 1879 and January 1880, he tested, with inconclusive results, for the presence of wind-borne ozone from the Atlantic Ocean. This form of oxygen, with three atoms to the molecule, was then prized as a disinfectant and air purifier.[117]

In 1880, Wragge gave two talks on the previous year's peculiarities to his friends in the North Staffordshire Naturalists' Field Club. He honed his self-taught meteorology during 1879's miserable summer, parched autumn and frosty early winter. His grasp of weather theory seemed confirmed — at least to his peers' understanding — in papers subsequently published in the naturalists' 1881 annual report. And in March 1881, the Royal Meteorological Society inspected and approved his weather stations, endorsing his practical knowledge of observation procedures. The observatory at his cottage in Farley, 646 feet above sea level, was the sole source of data for the first of his papers on what he called the "most unsummer-like summer" of 1879. The *Staffordshire Sentinel*'s report of his first paper to the naturalists at Newcastle-under-Lyme on February 27, 1880, introduced him as a recognised volunteer observer connected with the meteorological society and meteorological office.

In November 1880, his second paper dealt with the abnormally dry closing months of 1879 on the Staffordshire moorlands and Churnet basin. His observations, presented to a naturalists' meeting at Hanley, near Stoke, drew on data from Farley and the

---

[116] This ozone test was devised by Christian Schonbein, Professor of Chemistry at Basel University.

[117] *Staffordshire Sentinel*, March 1, 1880, p4, from Wragge's report to the North Staffordshire Naturalists' Field Club on his observations at Farley on Summer in the Staffordshire Moorlands, 1879.

new station he set up at Oakamoor in August 1879. He explained he had taken simultaneous weather readings to test his theory that the Churnet Valley's climate was "somewhat peculiar to itself". He managed this by recruiting Oakamoor's railway stationmaster as an observer. Comparing four months of data, he concluded his old home village experienced an excessive range of temperatures and extremes of heat and cold peculiar to its cleft in the narrow valley floor. In contrast, Farley, only a mile distant, and 300 feet above sea level higher on the moorlands, had a more equitable climate.[118]

Wragge was baffled by the story of 1879 embedded in his barometric readings. In a topsy-turvy year, the wet summer had been characterised by minimal fluctuations in air pressure, compared with a procession of high atmospheric crests through autumn into winter. In October, a great anticyclone had parked over the south of England, drawing dry air from the west side of a retreating depression, resulting in 10 extremely dry days with savage frosts. He conceded meteorologists had yet to satisfactorily explain Britain's extreme monthly and periodic changes in barometric pressure. He continued:

> Many problems in meteorology, of which this is one of the chief, have yet to be solved; and it is the part and duty of the earnest worker searching for the *truth* when dealing with the science of the atmosphere to advance cautiously ... having the facts of precise observation before him, that will ultimately lead to the perfect unfolding of the laws that govern the weather.[119]

* * *

---
118 *North Staffordshire Naturalists' Field Club 1881 Annual Report*, p67
119 Ibid, p55

In December 1880, less than a month after affirming his faith in mysterious yet ultimately knowable weather laws, Wragge unveiled a chivalrous plan to advance meteorology on Ben Nevis, Britain's highest mountain. He wrote to the Scottish Meteorological Society volunteering to make observations on the often-snowbound 4406-feet-high peak of the Ben in return for a pony, accommodation and instruments to set up a base station in Fort William. Next summer looked feasible. He envisaged hiking up and back daily, a round-trip of 14 miles, for 9 am readings to be made simultaneously on the summit and in the town, subject to finding an assistant. On January 14, 1881, the council of the meteorological society considered this offer and also a proposal from friends of the late Western Highlands' steamship operator David Hutcheson for an appeal to build a permanent weather station on the Ben in Hutcheson's memory.[120] The society quickly met his requirements. On March 23, members offered him two or three rooms in Fort William rent free and the use of a pony, promised to get him a tent from the War Office, and assured him 10 am observations would be acceptable, "Considering the difficulty of the ascent of the mountain and the uncertainty as to the time that might be required."[121]

I doubt his boast a few years later that he volunteered impulsively, galvanised by an article in *Nature* magazine

---

[120] David Hutcheson, a founder of the David Hutcheson and Co shipping company, died in Glasgow on December 18, 1880, aged 79. His Scotch Royal Mail steamer service opened the west coast of Scotland to tourists from the mid-19th century onwards. *Ardrossan and Saltcoats Herald*, October 13, 1877; *Edinburgh Evening News*, December 20, 1880, p2.

[121] Abstracts from the Scottish Meteorological Society Council minutes 1878–81, relating to the setting up of the Ben Nevis Observatory and the offer of Clement Wragge to make daily ascents in 1881.

criticising the British government's neglect of high-altitude weather observations. *Nature* first raised this issue in January 1879, arguing Britain had a duty to the Continent to telegraph year-round storm warnings from Scotland's highest peaks — Ben Nevis and Ben Macdhui:

> Great Britain's duty, interest and credit as a nation concur therefore in this matter ... unless, in fact, we are prepared to see these problems presented by the phenomena of the highest air worked out in other countries.[122]

In February 1879, *Nature* welcomed the London Meteorological Council's offer of £100 yearly towards running expenses of a future station. The establishment cost for a building and instruments was then estimated at £800 and annual running expenses £300. I feel these two articles motivated Wragge's two years' study of practical meteorology in North Staffordshire. In August 1879, he extended his monitoring from Farley to Oakamoor and, in July 1880, on to Beacon Stoop, suggesting he knew of the Scots' interest in measuring vertical gradients in atmospheric pressure as well as standard horizontal gradients.

A few years earlier, Scottish lighthouse engineer and meteorologist Thomas Stevenson had begun studying means of measuring vertical barometric gradients. He theorised that taking air-pressure readings simultaneously at closely grouped, descending altitudes might yield data to improve the accuracy of storm forecasting. Scottish Meteorological Society secretary

---

122   *The Brisbane Courier*, October 1, 1885, p3: "Happening to see a paragraph in *Nature* lamenting this state of affairs, I at once wrote to the society offering to place my own standard instruments on the summit." *Nature*, January 16, 1879, 'Scottish Meteorological Mountain Observatory', p237.

Alexander Buchan strongly advocated making simultaneous high-altitude, intermediate and low-level observations on the Ben.

Wragge's daily trek to and from North Staffordshire's highest peak was both a test-run for taking synchronised observations and training for the forays above the snowline of the Ben that won him his greatest fame in Britain. In March 1881, a Royal Meteorological Society inspector acknowledged the rugged conditions on the Weaver Hills, observing that the south-westerlies that scoured Beacon Stoop sometimes blew the index off the minimum thermometer causing unreliable readings. Writing in 1882, after his first summer on the Ben, Wragge summed up the rationale of his offer:

> I had devoted myself to a life of travel and research and was convinced that in endeavouring to promote the establishment of meteorological stations at high levels, I was turning my energies in the right direction.[123]

Meanwhile, during 1880, friends of David Hutcheson had begun campaigning to build a Ben observatory in honour of his services to west coast tourism. His friends included Edinburgh barrister David Milne Home (1805–90), the Scottish Meteorological Society's co-founder and long-time chairman. Hutcheson died on December 18, 1880, aged 79. In January 1881, the meteorological society's council dealt simultaneously with Wragge's letter and the Hutcheson memorial proposal, and subsequently gave priority to Wragge's comparatively concrete

---

123 *Chambers's Journal of Popular Literature, Science and Art,* conducted by William and Robert Chambers, Fourth series, No 957. — Vol XIX, April 29, 1882, p265

plans. In February 1882, the *Glasgow Herald* reported the Hutcheson appeal had fallen through, subsumed by fund-raising for widows and children of 189 fishermen killed in a storm off Scotland's south-eastern coast five months earlier. Sixty-three of the victims were from the village of Eyemouth on the Berwick Coast. Milne Home, then also chairman of Berwickshire County Council, led the relief effort.

In March 1882, *The Scotsman* reported a meeting of Inverness gentry had backed the meteorological society's campaign for an all-weather observatory on Ben Nevis, but had shelved the idea of honouring Hutcheson in its naming. Wragge wrote to *The Times* immediately after the Eyemouth storm on October 14, 1881, with three steps for improving warnings of storms from the Atlantic, including building an all-weather observatory on the Ben linked by an indestructible underground telegraph to the world beyond.[124]

* * *

The eldest son of Admiral Sir David Milne, of Edinburgh, David Milne Home was called to the Scottish Bar in 1831 and a year later married Jean Home, eldest daughter of William Foreman Home. He acquired her family name in 1852 after her father's death and her inheritance of the Home estates in Berwickshire. Milne Home's wealth allowed him to indulge his interests in geography, geology and meteorology through the Royal Society of Scotland, the British Association for the Advancement of Science, and the

---

[124] *The Times*, October 17, 1881, p10, *Dundee Advertiser*, October 18, 1881, on the storm; *The Times*, October 25, 1881, p10, for Wragge's advocacy of signal ships and underground telegraph cables; *Glasgow Herald*, February 2, 1882, p6, *The Scotsman*, March 2, 1881, p4, on the demise of Hutcheson appeal

Scottish Meteorological Society, which he helped found in 1855. Thomas Stevenson, incidentally father of the author Robert Louis Stevenson (1850–94) was a co-founder. Their zeal in promoting observation and classification of meteorological minutia had, by 1879, yielded a network of 107 weather stations through Scotland and beyond. The Scottish Registrar-General published quarterly reports of the observers' data for use in atmospheric research. In September 1878, the athletic 73-year-old Milne Home walked to the snowy top of Ben Nevis to survey possible observatory sites. His hike from Fort William with local schoolmaster Colin Livingston, 50, was widely reported on account of their age. Milne Home's stamina was matched in 1880 by a Captain D Brotchie, of Greenock, described by the *Aberdeen People's Journal* as "in his 74th year" and then the oldest man to ascend the Ben.

In February 1883, speaking at a public meeting in Glasgow, Milne Home shared his faith in high-level weather readings. Seeking pledges of cash for a permanent observatory on Ben Nevis, he said it would be superior to any weather balloon in yielding high-level clues to weather forecasting. Furthermore, a permanent observatory would be better placed than any of Europe's recently built high-level stations in detecting approaching Atlantic cyclones. In France, high-level stations had been established on the Puy de Dome at 4809 feet, on Mount Veutox (7267 feet) and the Pic da Midi (10,788 feet). There were also another five in Austria and Italy, at altitudes from 7057 to 9652 feet. Scotland's existing highest, set up by the society in 1878 at Dalnaspidal railway station on the Inverness-to-Perth branch line, was a mere 1450 feet above sea level. Milne Home believed that readings taken simultaneously from adjacent high- and low-level stations would bring greater knowledge of storms and other weather changes. His campaign lasted 10 years, triggered in 1873 by the opening of the world's highest weather station at Pike's Peak — 14,216 feet above sea level in the Rocky Mountains, near Colorado Springs. He told the Glasgow meeting he had been ashamed of Britain's neglect of high-level weather watching.[125] In 1876, rumours proved unfounded that six US Army Signal Service soldiers and staff had perished on Pike's Peak after becoming snowbound.

In 1877, Britain's Chancellor of the Exchequer refused a request from Milne Home for £400 towards construction of a Ben observatory. At the time, the English Meteorological Society was promoting the Lake District's 3054-foot Skiddaw Peak as a suitable site for monitoring the "movements of the gaseous envelope of the Earth".[126] A year later, Milne Home made his reconnaissance of the Ben and, in October 1878, reported to the committee of the Scottish Meteorological Society that construction of a permanent observatory would be quite practicable if the money could be found. He asked Thomas Stevenson to draw up plans for the building. Finances were tight with little more than token government support for the network of voluntary observers who recorded daily weather data for the British meteorological service. Milne Home warned in his 1878 annual report the society had been near extinction before a recruiting drive the previous year yielded 75 new members, lifting total membership to 649.[127] But numbers dipped again to 596 in 1880–81. Members' annual one

---

125 *Glasgow Herald,* February 15, 1883, p3
126 *Ben Nevis Observatory, 1883–1904,* The Royal Meteorological Society, 1983, 'Ben Nevis Observatory, 1883–1904', by James Paton, Edinburgh University, pp 1–2
127 *The Scotsman* February 22, 1878, p3

guinea subscriptions covered inspections of only a handful of the society's 102 volunteer-run weather stations.

Hence Milne Home and colleagues gladly accepted Wragge's offer. They were content to postpone meeting him in person until May, when in Edinburgh on his way to the Highlands. Public awareness of the case for an observatory on Britain's highest peak had grown in 1879 and 1880, through Milne Home's campaigning and general debate over weather-related catastrophes. A correspondent to *The Scotsman*, in January 1880, argued public funding for storm-watching on the Ben might have saved the lives of 75 passengers and crew drowned in the Tay Bridge disaster on December 28, 1879. The victims had been aboard an Edinburgh-to-Dundee train that plunged into the Firth of Tay when gale-force winds blew over a section of the new two-mile steel railway bridge.[128]

The likelihood of the Ben being harnessed for science produced some wild stories, such as the *Aberdeen People's Journal*'s report in August 1880, that a local artist had volunteered to take charge of the proposed observatory through winter, subject to installation of a pipeline to the Nevis distillery. In August 1879, Milne Home's campaigning in London drew a mock rebuke from the journal *Graphic*, warning against disfiguring the mountain with an ugly tower: "Please, you restless scientists, leave Ben Nevis and Snowdown alone."[129] In 1880, the society rejected an offer from journalist and storm-warning campaigner John Sands to set up and man for 12 months a weather station on the bleak granite stack

---

128   *The Scotsman*, January 15, 1880, p6
129   *Graphic*, August 2, 1879, p2; *Aberdeen People's Journal*, August 28, 1880, p5

Chapter 5 North Staffordshire, 1878–81

of Rockall in the Atlantic Ocean, 290 miles west of Scotland.[130] The US weather service's patchwork of high-level readings was further expanded in September 1880, when builders at Lick Observatory, on Mount Hamilton (4209 feet) in California's Diablo Ranges, began informal but regular observations for the US Army Signal Service.[131]

* * *

The Wragges' second child arrived on September 2, 1880. Registered as Clement Lionel Egerton, their eldest son was nicknamed "Eggie" — pronounced "Edgy" — to save confusion. Nora gave birth in the cottage at Farley a week or so after the North Staffordshire naturalists' visit on their Churnet excursion. In Wragge's notes, "Eggie born" is in the margin adjacent to "Applied re Ben Nevis", with no mention of Nora. His assertion in 1921 that intense unhappiness drove him to excesses as a young man might in part explain the rigour of his meteorological fieldwork from mid-1880 onwards, beginning frenetically on the North Staffordshire moorlands and continuing on Ben Nevis in 1881 and '82.

I think he at first intended leaving Nora and family in Farley if the Scots bought his Ben Nevis dream. His initial offer was

---

[130] *Daily Review*, Edinburgh, July 11, 1881, p6; John Sands (journalist), From Wikipedia, https://en.wikipedia.org/wiki/John_Sands_(journalist), accessed on September 23, 2022

[131] *Monthly Weather Review*, June 1914, p340, *Meteorology at the Lick Observatory*, by William Gardner Reed, American Meteorological Society, Journals online, ftp://ftp.library.noaa.gov/docs.lib/htdocs/rescue/mwr/042/mwr-042-06-0339.pdf, accessed on September 23, 2022

conditional on supply of quarters in Fort William for himself and an assistant, with no mention of housing his wife and children. The assistant was needed to take readings at sea level, at the same time as the high-level observations. Nora was surely insulted at the idea of being left behind. Besides, she had proven at Farley quite capable of collecting weather data as his occasional assistant. On March 4, 1881, he wrote again, clarifying his housing requirements, and the society replied offering two or three rooms, rent free, during the proposed summer observations.[132]

In late May, 1881, Wragge and his family left for Scotland without fanfare. He packed his Farley barometer and thermometers, gave his Oakamoor railway station instruments to the Royal Meteorological Society, and arranged for the Beacon Stoop observations to continue for a short time.[133] The last of his occasional North Staffordshire weather reports appeared in the *Sentinel* on April 23, with an aside on the death of Benjamin Disraeli, aged 76. He speculated on the cause of the frosty easterly change that was thought to have hastened the death of the ailing ex-Prime Minister.

A few weeks earlier, a census collector compiled the following snapshot of the Wragges at Farley: Clement, 28-years-old, of independent means from dividends and shares and engaged in comprehensive geography; Nora, 25-years-old and South Australian-born; their children May, aged two, and Clement Lionel Egerton, eight months, both British-born. A 19-year-old servant

---

[132] Abstracts from 'Scottish Meteorological Society Council Minutes, 1878 to 1881', relating to the setting up of the Ben Nevis Observatory and the offer of Clement Wragge to make daily ascents in 1881.

[133] *Quarterly Journal of Royal Meteorological Society*, Vol VIII, London, 1882, 'Report of assistant secretary on inspection of stations, 1881'

maid, Sarah F Moreton, was then also part of their household. Wragge gave his true age but registered Nora's as 25, despite her then-recent 29th birthday.

CHAPTER 6

THE HERO OF BEN NEVIS, 1881–83

There is one proper way of doing everything, and the rule is as applicable in the climbing of mountains as in anything else. To make the ascent comfortably and with least output of exertion, one should wear a pair of strong tackety boots, provide himself with a stout walking stick and be clad as lightly as possible. — William T Kilgour, *Twenty Years on Ben Nevis*, 1905.

It's 290 miles north as the crow flies from the North Staffordshire moorlands to Fort William in the lee of Ben Nevis. The Jacobite rebellion-era fort that gave the town its soldiering name was decommissioned 15 years before the Wragges' arrival. Its *Gaidlig* name, *An Gearasdan,* recalls the garrison built by Oliver Cromwell on Loch Linnhe in 1654 to counter the Cameron clansmen of Lochaber. Rebuilt and renamed in 1690 by the government of William of Orange, Fort William withstood a five-week siege by 1500 Jacobites before English troops routed them at Culloden in April 1746. Today's Western Highlands' tourist town grew from the port village that emerged beside the fort in those old days. Thousands of displaced crofters were shipped from Fort William to Canada during clearances of the Highlands from the turn of the 18th century.

Ben Nevis drew adventurers to the town from the early 19th century onwards. Tourists found the Highlands in the 1820s, following the Celtic whimsy of James MacPherson, Walter Scott and William Sharp (aka Fiona MacLeod). Scott later wrote of these excursions, "Every London citizen makes Loch Lomond his washpot and throws his shoes over Ben Nevis."[134] Generations of trampers preceded Wragge up the Ben's rutty tracks. In 1812, Thomas MacKnight, a mathematician, geologist and Church of Scotland minister alerted fellow explorers to the frightful precipices he saw on his survey of the Ben's flattened crown. The poet John Keats reached the summit in the summer of 1818 during a walking tour of the Scottish Highlands and was underwhelmed. He wrote in a lesser-known sonnet, beginning: *"Read me a lesson, Muse, and speak it loud/ Upon the top of Nevis, blind in mist!"*[135]

A decade later, Fort William guides vied for footloose tourists. Recalling one such supervised ascent in 1831, a correspondent of the *Glasgow Free Press* likened the mountain's primacy among surrounding peaks to the supremacy of an Amazonian queen in the midst of her crouching subjects. In 1835, Charlotte Ann Montague Douglas Scott, the 24-year-old Duchess of Buccleuch, won praise for pluck when her young guide got lost in a mist. They were rescued by John McDonald, proprietor of the Ben Nevis Distillery, on horseback with a handbell. In 1842 and again

---

134 Fort William History, www.castlesfortsbattles.co.uk, accessed October 3, 2022; *Aberdeen Free Press*, December 30, 1880, p5 for statistics on migration from Fort William in 1801–02; Macleod, John, *Highlanders, A History the Gaels*, Edinburgh, 1996, for Walter Scott quote, p246 and pp 242-47 for 19th century Scottish tourism craze

135 John Keats, *Read me a lesson, Muse and speak it loud*, from Mapping Keats's Progress, A Critical Chronology, 1818, http://web.uvic.ca/~pwak3/keats/1818-08-02.html#/ accessed October 3, 2022

in 1871, patriotic townspeople built bonfires on the summit to celebrate, respectively, Queen Victoria's first tour of Scotland and the marriage of her daughter Princess Louise to the Marquis of Lorne. In 1864, a party of Royal Engineers confirmed the Ben's pre-eminence by finding it pipped Ben Macdhui, Scotland's next tallest peak, by 72 feet.[136]

On Saturday 28 May 1881, Wragge and family reached Fort William by steamer from Oban.[137] He was anxious to begin his Ben observations on June 1 as he had promised. Already running late after two days of talks in Edinburgh with the council of the Scottish Meteorological Society, he faced a further day's delay in setting up because labourers in the Highlands insisted on their Sabbath rest day. His own Sunday observance was limited to attending an evening service at the Episcopalian church. On Monday 30 May, aided by Colin Livingston, the town's public-school headmaster, he found a joiner and eight labourers prepared to start work that night. Livingston was an experienced trekker, a volunteer observer for the Scottish Meteorological Society and a Ben observatory project insider, having joined David Milne Home

---

[136] *The Scots Magazine*, May 1, 1812, p22, 'Narrative of a Mineralogical Excursion to the top of Ben Nevis by Dr MacKnight', from Memoirs of the Wernerian Natural History Society, p345; *The Scotsman*, December 3, 1831, p4, quoting from an undated story in the *Glasgow Free Press*; *The Observer*, London, October 8, 1838 and August 30, 1847, *The Guardian*, August 9, 1864; Kilgour, William T, *Twenty Years on Ben Nevis*, facsimile of second (1906) edition, pub Gwynedd, Wales, 1854, p52

[137] Wragge wrote about his first summer in Fort William and on Ben Nevis in *Good Words*, a popular religious journal specialising in uplifting stories: Altick, Richard D, Nineteenth Century English Periodicals, 'The Newberry Library Bulletin 9 (1952)', 255-64, https://www.newberry.org/british-periodicals-19th-century-altick-richard-d-nineteenth-century-english-periodicals-newberry, accessed October 3, 2022

in 1878 on his survey of likely sites. He was renowned for his interest in geology, meteorology and the Gaelic language.

The work party left at 10 pm to carry instrument fittings, tools, posts and stanchions to the mountain top. Wragge and Livingston followed three hours later, carrying thermometers, a rain gauge and a barometer from the meteorological society, custom-made for use in extreme weather. They followed a little-travelled route through the northern foothills, over heathy moors, following the *Allt Coire an Lochain* burn to the tarn known as *Lochan Meall an t-Suidhe* (Loch of the Hill of Rest), which they reached at 4.15 am. After crossing swampland, they began a stiff climb to the snowline, about 2400 feet above sea level, then gingerly took a rocky, diagonal short-cut with their fragile cargo. He and Livingston took turns at tucking the Fortin's-principle barometer under an arm, cistern end upright, knowing any jolt could force air into the tube and stymie observations. At 6.30 am they drank from the spring Wragge later named Buchan's Well, after Scottish Meteorological Society secretary Alexander Buchan. Then, following cairns piled by hikers past, they made the final ascent of 850 feet, reaching the summit about 7.30 am. Snow lay in patches shaded by the rocks and boulders that litter the Ben's flattened crown. Livingston led them through a thin mist to the chosen site of the weather station — an army-ordnance cairn marking the mountain's 4406-feet peak. They rested at the trig marker, wary of the sheer drop from nearby cliff tops. It was a fine, clear morning. Wragge described the vista as range after range, peak behind peak, "Sombre moors, ancient valleys, deep glens and blustering torrents, all beneath heaven's clear blue vault, all charged with Nature's majesty."[138]

---

138   *Good Words*, 1882–83, 'Watching the Weather on Ben Nevis', Part 1, pp 343–47

Livingston divided the work crew into two teams, one to build a seven-foot-tall cairn for the barometer, the other to erect a wooden thermometer screen to Thomas Stevenson's specifications. By 3 pm, Wragge declared the makeshift station ready for official readings at 9 am on Wednesday 1 June. That gave him a day's grace for a trial run on Tuesday 31 May. He hung the barometer — capable of reading as low as 23 inches of mercury (psi) — in a locked, protective box attached to the sturdy new cairn. His dry- and wet-bulb thermometers and maximum and minimum self-registering instruments from Farley were mounted and locked in the Stevenson Screen. Nearby, he set up solar-radiation and terrestrial-radiation thermometers and a capacious rain gauge. Their descent to Fort William took just two hours in abnormally sunny weather — Wragge and others tied handkerchiefs over their heads for sunstroke protection in the dry mountain air.

Details of Nora's life in Fort William are scant. Later, Wragge wrote his wife had offered to take the sea-level readings at Fort William — 11 sets of data daily, from 5 am to 9 pm. The Fort William base station was in the grounds of a furnished villa that the meteorological society found for them beside Loch Linnhe. Nine days into his soldiering up and down, he recruited 16-year-old William Mackenzie Whyte to assist him on the mountain and to back up Nora as needed.[139] In late July, society secretary Alexander Buchan inspected the station and congratulated Nora on her carefully and intelligently made observations, "Made as nearly as possible at the same instant of time that observations

---

139 *Good Words*, ibid; *The Times*, August 28, 1881, pp 6–7. The reporter from *The Times* who joined Wragge on a terrifying excursion up the Ben on August 26, 1881, noted his assistant sometimes made the home station readings.

are made on the top."[140] She had by then surely hired a nursemaid for the children. May was three years old that summer and Eggie nearly one year. Beyond her meteorological chores, Nora sometimes baked oatmeal cakes for snacks on the mountain.[141]

The meteorological society paid William Whyte's wages. The eldest son of Imperial Hotel proprietor Robert Whyte,[142] he served in 1881 and '82 as an assistant observer and in 1883 took joint charge of the temporary stations with 21-year-old student teacher Angus Rankin. Neither Whyte — later ordained as a minister of the United Free Church of Scotland — nor Rankin, Argentina's future chief weather forecaster, was willing to work on Sundays.[143]

Wragge pushed himself to take every reading on time, but reaching the summit by 9 am took some luck. For example, he noted in his logbook on June 8, 1881, a delay caused by his stable boy sleeping in.[144] His daily ritual began at Fort William with a 4.30 am breakfast and 5 am observations at the home station. At first, he set aside four hours to ascend the Ben, but later found two-and-a-half hours feasible in fine weather. Typically, he rode to the *Meall an t-Suidhe* tarn, 2000 feet above sea level, hobbled the pony there and tramped on to make his readings on the summit at 9 am, 9.30 and 10 am — from eight instruments. In all, he took seven sets of readings during each of his daily, 16-mile excursions,

---

140 *Nature,* November 3, 1881, p11
141 *Good Words,* 1882, pp 377–78, Part II, 'At Work'
142 *The Aberdeen Journal,* June 17, 1881, p6
143 *The Times,* June 18, 1883, p5, reported Wragge deplored their stand as an unnecessary disruption of observations: "He considers the work one of so much public utility and mercy that it involves no Sabbath breaking, but he has failed to convince the Highlanders of that."
144 *The Beaudesert Times,* July 8, 1955, p11, quoting an extract from Wragge's Ben Nevis and Fort William Observatories' logbooks, cited by B H Humble in an undated story published in *Scottish Field,* 1955

the others being the temperatures taken at Buchan's Well, 3500 feet above sea level, and the tarn, 2000 feet, on the way up and down. Back in Fort William by 4 pm, he sent the daily data to the British Meteorological Office and entered into his logbook the day's readings — each to three decimal points. Then he had some dinner and slept for three or four hours. Up again at 9 pm for another set of readings, he ploughed into more bookkeeping and writing notices for newspapers before turning in at midnight for a second sleep. He wore as a badge of honour his zeal for the cause. Recalling he sometimes had to delegate the Ben observations to Whyte for as many as four days a week, he wrote in 1882, "But I made a point of making up for it, and in consequence have sometimes climbed the Ben on eight days in direct succession."[145]

* * *

Nora is almost invisible in his Ben jottings. The grizzles shared with his solicitor many years later seem based on his belief neither he nor she had been capable of the empathy he imagined possessed by soulmates. His memoir notes make no mention of the death of his father's cousin Rupert Ingleby in Adelaide in December, 1881. Rupert's death, aged 63, left Nora's eldest sister Margaret widowed with five children aged from three years to twelve. A father figure to the Thornton girls, Rupert had secured for Nora a small fortune in her marriage settlement. He and Nora's brother Edward junior, were joint guardians of

---

145 *Chambers's Journal of Popular Literature, Science and Art,* conducted by William and Robert Chambers, Fourth series, No 957. — Vol XIX, April 29, 1882, p265

the £2000 trust fund. Wragge and Nora honoured him in the naming of their third child, Rupert Lindley Wragge — known as Bert — born in Fort William in August, 1882. Nora's yearning for South Australia grew with Bert's birth.[146]

* * *

Wragge's perfectionism was soon apparent to the local press, which had at first dismissed him as an enthusiastic mountaineer. In a letter published on June 8, 1881, he scolded the *Dundee Advertiser* for muddling a description of his array of meteorological instruments on the Ben, and inferring his data would be irregular and unreliable. The newspaper's Fort William correspondent apologised, conceding in a statement also published in *The Oban Times* that Mr Wragge was anything but a mere adventurer. He had devoted his whole life and fortune to the cause of science, and expected to be able to give 24 hours' warning of approaching weather using observations from the Ben. But a month later, a correspondent of Edinburgh's *Daily Review* tipped the young Englishman would soon run out of puff and abandon the project. Wragge's frustration with slapdash science erupted again in September 1881, when he blasted the *Standard* newspaper in London for frequently sloppy transcription of his

---

146 I have used the nicknames "Eggie" and "Bert", respectively, for Wragge's eldest and second-eldest sons, Clement Lionel Egerton Wragge (1880–1915) and Rupert Lindley Wragge (1882–1952). Wragge uses both names in his memoir notes. It's probable Eggie was pronounced "Edgy". A story in *Queensland Figaro*, April 19, 1906, p26 refers to him as "Regie" Wragge.

telegrams from the Ben. He rejected the newspaper's excuse that errors occurred in transmission.[147]

His nickname "Inclement" was inevitable as the Scots twigged to his mission. William T Kilgour, one of his weather-watching successors on the Ben, traced its coining to that first frenetic summer. Writing in 1905, Kilgour recalled mockers dismissing Inclement as a misguided enthusiast who was fast riding his hobby to death. But far from such self-indulgence, Kilgour wrote in his book, *Twenty Years on Ben Nevis*, Wragge had actually followed the discipline of his project in "even tenor". He revered him as a founder of the permanent observatory opened on the mountain in the winter of 1883, then decommissioned, starved of government support in October 1905.[148]

* * *

William Anderson, Edinburgh correspondent for *The Times* of London, conceded mixed public opinion over whether Wragge's daily exertions were heroic, daft or a bit of both. In late August 1881, Anderson joined Wragge for a sleety morning on the Ben, with stops for a smoke and swig of Long John whisky, and later Anderson lionised Wragge in *The Times* for his imperishable devotion to science. A storm that delayed their 9 am readings on the summit also flayed the herring fishing fleet at Fraserburgh, 150 miles north-east on the North Sea coast. Six fishermen drowned.

---

147 *Dundee Advertiser*, June 8, 1881, p3; *The Oban Times*, June 25, 1881; Wragge's complaint against *The Standard* was reported in provincial British papers including the *Abergavenny Chronicle*, September 2, 1881, p2.

148 Kilgour, Wm T, *Twenty Years on Ben Nevis*, Paisley, Scotland, 2nd edition, 1906, facsimile copy, Holyhead, Wales, 1985, pp 14–15

Wragge told Anderson it was his worst day in three months of Ben ascents. Locals could not recall a fiercer autumn blow.[149] Anderson, then aged 48 and known as a champion of Edinburgh's slum-dwellers, introduced him as a wiry young enthusiast. "The regular exercise seems to have strengthened a frame not naturally very robust," he wrote. Based on their shared ordeal, he praised his endurance, willpower and enthusiasm for science. The value of his project was obvious from the British Meteorological Office's ready acceptance of data from the Ben and Fort William. In North America, observations from high-level stations had improved storm warnings. Anderson argued the storm on August 26 proved the need for a permanent observatory. His reporting created an adventurous and more sympathetic persona than the *Staffordshire Sentinel*'s eccentric portrait a year earlier. Anderson took for granted the younger man's horsemanship on the slippery lower slopes of the Ben and introduced his big, black Newfoundland dog, Robin Renzo, as a constant companion through that wet and windy day. He praised his public-spiritedness in seeking nothing more than a furnished house, a pony and an assistant. As for Nora, Anderson acknowledged her devotion to the sea-level readings and her housekeeping skill in supplying the consolations of a hot bath, hot drink and change of clothes that awaited them on their return from the Ben.

During the next 24 months, Wragge used his contact with

---

149   *The Times*, September 1, 1881, pp 6–7; The Archive and Record Office of News UK lists William Anderson (1883–99) as *The Times*' Edinburgh correspondent between 1877 and 1891. His obituary in *The Montrose, Arbroath and Brechin Review*, December 15, 1899, p3, describes him as a native of Brechin, "For many years a journalist in Edinburgh and all along took a very warm interest in the poor of that city."

Chapter 6   The Hero of Ben Nevis, 1881–83

Anderson to promote the Ben project through *The Times* — a sounding board for worthy causes.[150] Towards the end of 1881, Wragge's labours on the summit yielded him space on the letters page and Anderson another story, stressing the difficulties of on-the-fly meteorology in ice and snow. A catastrophic storm on October 14, 1881, that killed 189 fishermen off Scotland's south-eastern coast rekindled alarm at the Meteorological Office's seeming inability to forewarn events once accepted as Acts of God. The Scottish forecast published in *The Times*, Friday 14 October, drastically underestimated the ferocity of hurricane-force winds that smashed the haddock-fishing fleet off Berwick on Tweed and further north that day. Regarded as one of 19th century Britain's deadliest storms, it's remembered as the Eyemouth Hurricane, in memory of the deaths of 129 fishermen from that tiny port.[151]

In a letter published on October 17, Wragge complained that because of a telegraph failure, his report of massive rainfall on the Ben three days before the storm had not reached London in time for use by the Meteorological Office. His letter had a footnote from Anderson that the deadly storm on October 14 had stymied readings on the summit that day. Mr Wragge had retreated home at 2000 feet, beaten by 70 mph winds and suffocating snowdrifts.

---

150   19th century literary critic Matthew Arnold described *The Times* in 1864 as "an organ of the common, satisfied, well-to-do Englishman". 'The Function of Criticism at the Present Time', *The National Review*, November, 1864, http://fortnightlyreview.co.uk/the-function-of-criticism-at-the-present-time/ accessed October 3, 2022. See also Hobbs, Andrew (2013), 'The deleterious dominance of *The Times* in nineteenth-century historiography', *Journal of Victorian Culture*, 18 (4), ISSN 1355–5502.

151   John Doull, Fishery Officer, Eyemouth, to D F Primrose, Fishery Board, Edinburgh, October 15, 1881, National Records of Scotland reference: AF23/47 pp 169–70

On November 3, *The Times* ran a longer account from Anderson — based on Wragge's notes — of those fraught hours as the storm swept east. Anderson reported the storm had in effect ended the summer and autumn observations for 1881, leaving an unbroken sequence of observations over 19 weeks and a day. Wragge had sent *The Times* a precis of rainfall and maximum and minimum temperatures for both the Ben and Fort William, extracted from data sent to the Scottish Meteorology Society and Meteorological Office. Anderson also recounted a further ascent, on October 27, by Wragge and his occasional guide Colin Cameron to retrieve precious thermometers and a rain gauge from the summit. Despite ice-encrusted signs warning tourists to keep their distance from meteorological instruments, the Ben attracted intrepid vandals year-round.[152]

* * *

In 1955, Scottish mountaineer and author B H Humble counted Wragge as a kindred spirit, a mountain man whose yen for transcendence could be seen in his daily logbooks. He cited, for example, his observation of the atmospheric flux on October 12, 1881, two days before the Eyemouth Hurricane. On the plateau that morning, he had seen the first stirrings of this storm — snow whirling from roiling cumulus clouds, shafts of autumn sunlight fitfully shimmering on Loch Linnhe far below. "Condition of atmosphere majestically grand," Wragge wrote. "Indeed, I may say that the views of this morning were the finest I have ever seen at home or abroad." He loved sharing his high-country raptness

---

152   *Dundee Evening Telegraph*, July 17, 1882, p4

Chapter 6   The Hero of Ben Nevis, 1881–83

— whether recalled from Ben Nevis, Mount Kosciuszko or the Holy Land. In 1882, writing for the religious family magazine *Good Words*, he acknowledged the God of Creation when recalling the mesh of ice, wind and fire on the Ben one Sabbath morning as his second Scottish winter approached. The autumn sky was spangled with bands of stratus clouds tipped with wisps of cirri, and yellow streaks on the horizon forewarned of an imminent storm. Anyone with a soul could hardly help fervently thanking God for the gift of life. A few days later, winter gales and cloud-fog curtailed his five months of observations for that year.

On the whole, in *Good Words* stories published in June 1882 and June 1883, Wragge left his thoughts on the sublime to his readers' imagination. Summing up the stormy curtailment of his 1881 observations, he simply affirmed contentment with his active, open-air life and restated the need for a permanent, all-weather station to ensure winter data. In passing, he confessed neither he nor his friend Colin Cameron, who battled through the storm with him on October 14, had experienced weather as furious as that day's. The blizzard froze their beards and clothing stiff and turned his black dog a ghostly white. Wragge typically wore thick lamb's-wool underclothes and a sailor's jersey under his oldest suit. He scorned gloves but shod his feet and ankles in lace-up boots and leggings. Anderson's version of that fraught, freezing October acknowledged his achievement in assembling the valuable weather data, stressing the project was "as humane in its ends as it is scientific in its methods". By then, scientific and public opinion concurred that many lives could be saved by year-round observations from a properly manned weather station connected by cable to London.

In a letter to *The Times* published on October 25, 1881, Wragge gave three ideas for improving warnings of disastrous hurricanes like the one that doomed the Eyemouth fishing fleet. At the least, he argued, the weather station on Valentia Island off the south-west coast of Ireland must be connected by cable with the Meteorological Office in London. But since cyclonic storms often advanced rapidly from the Atlantic, a self-recording weather station should also be set up on a ship or buoy 500 miles west of Ireland and connected by cable with the coast. Beyond that, priority must be given to establishing high-level observatories throughout Britain and installing underground telegraph cables between the nation's chief weather stations and the head office. He reminded readers that in the absence of a permanent observatory, he was unable to continue his reports from the Ben through winter.

* * *

The Wragges wintered in North Staffordshire, leaving Colin Livingston to make monthly checks on the thermometers still on the summit. Nora and the children are likely to have stayed in Stafford at least until the beginning of 1882. Letters between Wragge and the Stafford town council on the relocation of his museum show that he asked the council to cover the cost of four furnished rooms in November and December. Citing family reasons, he wrote asking for the free accommodation in return for his labelling, arranging and cataloguing the thousand or so exhibits. The council declined but agreed to waive for a year the £10 annual subscription he had previously offered. The memoir notes suggest he and the family moved back to Farley in the new year. He spent some of December in Edinburgh, enjoying what a Staffordshire friend later called "the pleasant society of the philosophers of that Athens". His meteorological society connections gave him access that winter and the next to physics classes at Edinburgh University with Alexander Buchan and Natural History Professor Peter Guthrie Tait (1829–1907).[153] Years later, Wragge remembered Tait as kind, sympathetic and a perfect master of physics.[154]

Wragge enjoyed many affirmations of his public-spiritedness while away from the Ben. On November 3, 1881, he was guest of honour at a dinner celebrating the reopening of his museum on the top floor of Stafford's new library and art gallery. A few days later, a correspondent to the *Glasgow Herald* praised his contribution to the British Association's debate over the intensity of the Tay Bridge Disaster Hurricane in December 1879. Mr Wragge's calculations of likely wind pressure and velocity deserved a handsome reward, the writer, a Stirling engineer, J P Walker, argued.[155] Provincial and colonial newspapers spread Wragge's renown. In December 1881, praise of his devotion to meteorology reached Australia in reports of the Eyemouth disaster. *The Leader*, in Melbourne, argued the hurricane's hammering of Mr Wragge's makeshift observatory showed the need for adequately funded high-level observatories. But *The Leader's* science writer, Oedipus, considered him foolhardy to have extended his watch through

---

153 Goss, Eva Adeline, *Fragments from the Life and Writings of William Henry Goss*, Stoke on Trent, 1907, pp 112–13; Wragge, November 6, 1902, p133; *The Australian Star*, May 26, 1894, p7

154 Wragge, November 6, 1902, p133

155 *Glasgow Herald*, November 9, 1881, p10; *Dundee Advertiser*, December 13, 1881, p6

autumn into winter and felt the demise of the flimsy hut probably saved him from perishing for his passion.[156]

In January 1882, possibly inspired by stories in *The Times*, a London-based Christian Socialist campaigner Mary Hart hailed him as "The Hero of Ben Nevis". She told a rally in Warwickshire he had adhered to the Law of Love in his mission on the Ben, unbothered by scoffers. She predicted his search for the Law of the Winds would one day save the lives of hundreds of sailors.[157] On March 22, 1882, councillors of the Scottish Meteorological Society took for granted his chivalrous intent when awarding him a gold medal for his honourable and useful work on the Ben. Congratulating him at the society's half-yearly general meeting in Edinburgh, chairman David Milne-Home tacitly attributed his zeal to a vigorous exercise of the code of *noblesse oblige*. Officially, the medal recognised his great skill in organising the project, "his fertility of resource in emergencies [and] his indefatigable energy and undaunted devotion to his work". Secretary Alexander Buchan also acknowledged Nora's "scrupulous punctuality and care" in her base station duties. Wragge replied he felt well repaid by being able to further the cause of physical research. The society's recognition compelled him to press on.

* * *

The councillors voted to immediately launch a public appeal to build a permanent observatory. Proposing the appeal, Sir William Thomson, long-serving Professor of Natural Philosophy

---

156 *The Leader,* Melbourne, 'Science Gossip', December 31, 1881
157 *Buckingham Advertiser and Free Press,* January 21, 1882, p5

at Glasgow University, praised Wragge for having helped confirm the importance of high-level stations to weather forecasting. The station on the Ben must have a secure telegraph link, Sir William added. Elevated to the peerage as Lord Kelvin in 1892, Sir William was, among his many other achievements, a pioneer of telegraphy and scientific advisor in the first Transatlantic Cable project, 1856–58.[158] In July, the society's council agreed to manage the proposed observatory with oversight from the Royal Societies of London and Edinburgh. Mrs Cameron-Campbell, owner of the western side of the summit, had offered a two-acre site, to be vested in the Council of the Edinburgh Royal Society. The society postponed officially launching the appeal to the new year, having decided to first test the largesse of wealthy Scottish friends of meteorological science.

On March 28, 1882, a few days after receiving the medal, Wragge celebrated on the Ben's snowbound summit with a South Australian friend, Philip Egerton Warburton, and the ever-ready Colin Cameron. Warburton, a nephew of explorer Peter Egerton Warburton, had recently completed medical studies at the University of Edinburgh. I feel Wragge included "Egerton" in his eldest son's string of first names in admiration of the elder Egerton Warburton's survival nous in Western Australia's Great Stony Desert. The friends' expedition up the Ben on March 28 took about nine hours, through wind, rain and fog. Wragge still found time to identify and collect a load of moss and lichen specimens between the *Lochan Meall an t-Suidhe* and the summit. He took Philip Egerton Warburton with him again nine months later, on January 2, 1883, to retrieve

---

158   *Glasgow Herald*, March 23, 1882, p6

meteorological instruments in worse wintry conditions. Cameron, who missed this second trek, was drowned in a boating mishap near Fort William in 1885. He was aged 43. Warburton later worked in Western Australia as a medical officer and butcher and was killed in a riding accident in 1892, aged 38.[159]

Between trips to London and Scotland, Wragge lobbied the Stafford council to let him set up a first-class weather station in the town. The authorities at first declined, anticipating having to pay an observer, but relented when the borough surveyor and his brother volunteered.[160] In public-speaking engagements that the gold medal yielded, Wragge argued for greater government investment in British weather forecasting — particularly storm warnings. For example, his talk in London on April 21, to the Balloon Society of Great Britain, included advocacy for a network of high-level observatories, and establishment of special meteorological schools for ship owners and skippers. His comparison of anchored and piloted weather balloons as a source of upper-atmosphere data was topical. In early December 1881, Walter Powell, a British Member of Parliament, vanished in a truant War Office balloon off the coast of Dorset, feared drowned. An experienced balloonist, Powell, 39, was last seen being swept out to sea, trapped in the basket of a calico-fabric weather balloon named *Saladin*. The pilot and a passenger both escaped. They had been attempting a coastal landing after measuring air temperature and the height and movement of

---

159 *Alnwick Mercury*, 'Ascent of Ben Nevis', April 8, 1882, p3; *Dundee Advertiser*, January 6, 1883, p5; *Glasgow Herald*, 'Ben Nevis in Mid-Winter', January 15, 1883, p4; *Western Mail*, Perth, Western Australia, 'News from the Nor' West', October 15, 1892, p18

160 *Staffordshire Advertiser*, January 14, 1882, p2 and May 6, 1882, p4

clouds on the outskirts of fog-bound London.[161]

During May 1882, Wragge continued arguing for greater government investment in meteorology. On May 2, he told the Staffordshire Field Club in Stoke that he was sure weather could be forecast with certainty if data was available readily from observatories around the North Atlantic. The existing link between the Orkney and Shetland Islands needed to be extended to relay readings from Iceland, Greenland and Newfoundland, and an anchoring ship 800 nautical miles west-south-west of Valencia.[162] On May 19, he shared with fellow members of the North Staffordshire Naturalists' Field Club, in Stafford, his dream of a chain of high- and low-level stations throughout Britain to save life and property. He was among friends in both groups. Their fellowship entwined upper middle-class bonhomie with the mutual respect of earnest amateurs, each affirming the other's dissection of nature. Wragge found a kindred spirit in fellow North Stafford club committee member William Henry Goss, a famous Stoke potter who, like him, saw God everywhere in nature.[163]

---

161 *Sheffield Daily Telegraph*, December 13, 1881, p3; *The Standard*, London, April 19, 1882, p4; *Glasgow Herald*, April 22, 1882, p5. Using balloons for upper-atmosphere data had been pioneered in the 1860s by James Glashier, revered by Wragge as the father of British meteorology. Glashier made 29 balloon ascents between 1862 and 1865, while superintendent of the magnetical and meteorological department of the Royal Observatory at Greenwich. In September 1863, he survived blacking out, starved of oxygen, at 29,000 feet over Wolverhampton, during an ascent believed to have reached a world record altitude of 37,000 feet. Wragge, April 2, 1903, p298, tribute to Glashier after his death in February 1903, aged 93.

162 *Staffordshire Advertiser*, May 6, 1882, p6 and May 20, 1882, p4

163 The *Liverpool Courier*'s obituary of Goss, January 9, 1906, described him as a "Student of the hidden wisdom of the world [with] a noble reverence for every form of life from the humblest to the highest."

* * *

Back on the Ben on June 1, 1882, Wragge resumed exploring visible and invisible forces of nature with renewed zeal. By then he had convinced his mentors in Edinburgh he had the stamina and skill to more than double the daily observations on summit and mountainside, with help from assistants Angus Rankin and William Whyte. This meant adding four new observation points to the summit track and installing a barometer and thermometers near the existing *Meall an t-Suidhe* tarn rain gauge. In theory, the seven stops between the base station and peak were spaced for readings at precisely every 30 minutes on the way up and down. He and his offsiders used pocket thermometers and a portable aneroid barometer for these observations. Their five summit readings were scheduled half-hourly between 9 am and 11 am — an hour longer than the previous summer. Each observation entailed recording data from six instruments and estimating wind speed and cloud cover. All of the daily readings on the Ben were later matched with half-hourly weather observations made at the Achintore base station. He and his assistants took turns at this chore. They also checked the weather at 5 am, 6 pm and 9 pm.

Wragge traversed the Ben at least four days a week for five months between June 1 and November 1, 1882, ritually leaving home at 5 am, riding his horse to the tarn and walking and running the final two hours to reach the top by 9 am. In summer, the fog, rain and snow that enveloped the summit sometimes dissipated into a dead calm, briefly awash with sunlight. When his fingers were too frozen and notebook too sodden for writing, he scratched the data on the door of the barometer cairn to copy the next day. Beyond the cold, he wrote later, another difficulty had been steering the horse around ruts and swamps while ascending and descending. "The latter are so very treacherous and deep that the poor animal has a trying time of it," he wrote in *Symons's Monthly Meteorological Magazine* in July 1882.[164] His kit included instruments, notebook, lunch, pipe, tobacco, keys and matches.[165] He generally began the descent by 11.20 am and reached home by 3 pm, "covered in mud and dirt and drenched to the skin", he recalled in *Good Words*. Next, he routinely bathed his feet in hot water, ate dinner, smoked his pipe and slept a few hours before transcribing the 21 sets of weather notes into log books. He slept again from midnight to 4 am.

Nora probably supervised these cosy homecomings, at least until Bert's birth on August 12, 1882. There's scant evidence of her life in England and Scotland that year, beyond Bert's birth certificate and the birth notice Clement placed in the *Pall Mall Gazette*, published in London a few days later. She and the children possibly retreated to Farley in November or December, 1882. Notifying his travel plans in *The Midland Naturalist* in April 1883, Wragge gave Farley as his forwarding address until May 25, then an address near the centre of Edinburgh until August 1, when he anticipated leaving for Australia. He later postponed departure until October, having instead returned to Stafford to finish reorganisation of the museum on what he termed a sound scientific footing. Nora probably remained in Edinburgh in this period, in their large rented home at 6 East Mayfield, about a mile from the university.

---

164 *Symons's Monthly Meteorological Magazine*, July, 1882, 'Resumption of the Ben Nevis Meteorological Observations, 1882', by Clement L Wragge, p84

165 *Good Words*, 1883, p358

* * *

The Scottish Meteorological Society officially launched the Ben observatory appeal in Edinburgh and Glasgow in February 1883. By July, an estimated 2000 people had between them given £4400, enabling the first stage of construction to begin in August — a new six-foot wide bridle path to the summit, on a route suggested by Wragge with scenic views of Glen Nevis. Donations ranged from single pennies to £200. He was proud Queen Victoria and the Prince of Wales had given, respectively, £50 and £26/10/–. The timber-lined observatory with four-feet thick granite walls was formally opened on October 17, 1883, by Mrs Cameron-Campbell, in two feet of snow. Later, at a banquet in Fort William Town Hall, Alexander Buchan proposed a toast to the absent Clement Wragge, declaring that the observatory bore testimony to his achievements as an observer on the Ben. *The Scotsman* reported Mr Wragge's toast had been heartily drunk by banquet guests, an estimated 100 ladies and gents, but he had unfortunately been unable to attend, having announced he was leaving for Australia the next day.

Wragge reinvigorated and sustained the case for a permanent observatory through the spectacle of his zealous monitoring. The British press reported with alarm and disbelief the Spartan routine he had set himself and his assistants. In July 1882, the *Glasgow Herald* called their daily rigours scandalous and argued the government must at once find the paltry £5000 needed to build a permanent observatory, at least to keep up with France, which had three high-level stations situated 5000 to 7800 feet above sea level.[166] William Anderson took a similar stance in *The*

---

166 *Glasgow Herald*, July 18, 1882, p4

Chapter 6   The Hero of Ben Nevis, 1881–83

Times in November 1882, arguing friends of science owed it to Mr Wragge to lighten his "extraordinary and self-imposed labours" by repairing the pony track to the summit and building a decent observatory.[167] But a satirist writing in the *Sheffield Daily Telegraph* was less sympathetic, describing Wragge as an accomplice of Scottish scientists determined to extract £5000 from the public purse for a Taj Mahal on the Ben. Why not build an alpine-style tourist railway to the summit if he insisted on stationing himself like the Herald of Mercury on that heaven-kissing hill?[168]

* * *

Wragge stayed in Scotland through the winter of 1882–83 with conscientious intentions. In early November 1882, he told the man from *The Times* he planned weekly ascents of the Ben to check the frozen rain gauges and self-recording thermometers. He had already begun a rigorous daily schedule of observations at the base station, taking readings every three hours, including 3 am, of sea temperature and ozone as well as temperature, air pressure and rainfall. He also found time to indulge his interests in astronomy, physics, meteorological theory and travel writing. For example, during November 1882, seeing brilliant and continuous displays of aurora prompted him to study the link between solar storms and terrestrial magnetism. Writing later on the meteorology of the English Midlands, interspersed with observations from Fort William, he speculated on the impact of sunspots on weather patterns and geophysical activity — a matter that decades later strongly influenced

---

167   *Dundee Advertiser*, July 31, 1882, p3
168   *Sheffield Daily Telegraph*, July 22, 1882, p2

his meteorological practice. The same solar storms crashed telegraph services in Britain and the United States. On December 6, 1882, he observed the transit of Venus from Fort William and reported in *Nature* having seen and sketched a peculiar dark band or ligament stretching from the planet's disc to the edge of the sun — connected with sunspot activity. He stayed in Fort William until at least mid-January 1883, when he sent the *Glasgow Herald* a shipping warning of an imminent storm from the south-west. In February, he attended physics lectures at Edinburgh University, looking in particular at correlations between electromagnetic disturbances on the sun and fluctuations in barometric pressure.

Wragge never explained precisely the trigger for his surprising announcement, in April 1883, that he intended leaving for Australia in August. In April, the Birmingham scientific journal *Midland Naturalist* reported his British projects were finished and that, during his coming voyage to Australia, he intended following up the ocean meteorological work of the *Challenger* expedition, under the auspices of the Scottish Meteorological Society.

Wragge reached Adelaide on December 6 loaded with souvenirs — marine shells from Ismalia, pumice from the Krakatoa volcano, and his Ben medal — and giving mixed reasons for his return. *The South Australian Register* reported he was back on family business and the *Adelaide Advertiser* referred vaguely to his owning property in the colony and his wife's family ties. In 1894, well established as Queensland Government Meteorologist, he elaborated in an interview with *The Australian Star*, in Sydney, that his wife's delicate health and wish to revisit her native country had obliged him to leave. He told almost the same story during a visit to Scotland in 1896, blaming Nora's illness on the

Chapter 6 The Hero of Ben Nevis, 1881–83

cold weather and conceding her need to return to "her old sunny land". By 1911, when Sydney's *Sun* newspaper interviewed him at length as one of Australia's great personalities, he had bundled his Ben years into a kind of music hall refrain, culminating in his departure for Australia, finer details no longer of concern:

> Up the old Ben I went, and down again in all weathers, day by day, and in meteorology I became a specialist. Oh! The music of the winds — the keen delight of a tearing gale. Thin, but with a constitution like wire nails, my lungs were, and are, of the very best. Back to Australia ... and started the meteorological station on Mount Lofty.[169]

If delicate family business forced his retreat from the Ben, it's likely to have been connected with Bert's birth in August 1882 and Nora's next pregnancy a year later. Twenty years later, she confided to Queensland Premier Robert Philp that she had never properly recovered from the chill she suffered while helping her husband on Ben Nevis.[170] Wragge's 1921 jottings link Bert's arrival cryptically with a "toothless nurse" and the statement, "I resign Ben Nevis!! And why?" By early autumn 1883, when Nora's new pregnancy was apparent, he had already announced his travel plans. There was also one other family concern hastening his return to Adelaide — an apparent theft from the marriage settlement trust fund controlled solely by Nora's brother Edward Thornton, since Rupert Ingleby's death in 1881.

---

169 *South Australian Register*, December 7, 1883, p5; *The Australian Star*, May 26, 1894, p7; *Sun*, Sydney, September 2, 1911, p1

170 Leonore Wragge to Queensland Premier, August 5, 1903, Queensland State Archives, PRE/A154, SR66402/1/147, covr05900, 11/8/1903

Based on Wragge's newspaper interviews and articles in 1883, he's unlikely to have wanted to run the new station. In her 1983 centenary history of the Ben station, Marjory Roy found he had been one of nineteen applicants for the position of superintendent awarded in September 1883 to Edinburgh physicist Robert Trail Ormond (1858–1914). She assumed he had been disappointed, but doubted this spurred his exit. Speculation that he left in a huff ignores widespread reporting from April onwards of his impending voyage to South Australia.

In early May, back in Fort William with summer approaching, Wragge announced he would soon resume his world travels, including revisiting Australia and that his assistant William Whyte would take his place as Ben observer. *The Times* reported in June that because of having to leave soon for Australia, Mr Wragge was resigned to doing little more than re-establishing the project. For example, he and Whyte were experimenting with carrier pigeons to send data from the summit, aiming to streamline telegraphing of warnings to Edinburgh and London. The three pigeons were still in training in July to "home" to Fort William and later won a reprieve when a telegraph cable was laid to the new observatory.

There's also evidence of his chronic restlessness with domestic routine. In September 1882, just after his 30th birthday and baby Bert's arrival, he applied to join Scottish geologist Joseph Thompson in exploring trade routes between Africa's east coast and Lake Victoria. The Royal Geographical Society replied European presence in the expedition would be limited solely to the leader. Assistant Secretary H W Bates wrote nevertheless that had there been a vacancy, then Mr Wragge's "well-known ability and acquirements would have been a strong recommendation to engagement".

In August 1882, when registering his second son's birth, Wragge gave his occupation as scientific worker and, in December 1883, introduced himself to an Adelaide newspaper as a scientific traveller with strong interests in geography and meteorology. Before leaving Britain, he declared in a letter to *The Times* his desire to explore New Guinea, depicted by another correspondent, Port Moresby missionary W G Lawes, as "a great unknown land". If funds were available for such an expedition, he envisaged making a complete scientific and geographical report of the island's resources and climate. In readiness, he obtained references from his Royal Geographical Society mentors Alexander Buchan and Thomas Murray, and Challenger expedition superintendent John Murray.[171]

Wragge wrote a sentimental account for *Good Words* of journeying to South Australia on the Queensland-owned steamer *Maranoa*'s maiden voyage. He sent *Nature* a separate report on his shipboard meteorological work. In 1885, he delivered a paper on these observations to the Royal Society of South Australia. The *Good Words* story entwined nostalgia for England with his yen for Australia's foreignness:

> Wandering albatross bear us company and follow as a convoy into the Australian Bight. And now we realise the distance from *home*. The sun is yonder in the northern sky and shadows to southward tell of the "nether world". The grand old Bear has disappeared and the 'Cross shines forth in everlasting light.

---

171   Reference to East African Expedition in copy of letter from RGS Assistant Secretary H W Bates, October 5, 1883; *The Times*, September 5, 1883, letter from W G Laws; *The Times*, September 14, 1883, Wragge's interest in a New Guinea expedition; *The South Australian Register*, December 7, 1883, p5

Recalling December 6, 1883, when they reached Cape Borda, on the western end of Kangaroo Island, with a view of the distant Mount Lofty hills, he remembered his first sight of this coast, 10 years earlier and having found "decidedly remarkable" those long, lonely lines of blue hills. He showed some sympathy for Nora in his acknowledgement that a couple of hours beyond Kangaroo Island, disembarking at Largs Bay, she would at last be nearing her home town. They took the train from the port into Adelaide where they found greengrocers' shops replete with strawberries, loquats and cherries. The temperature that day in Adelaide reached 100 degrees Fahrenheit in the shade.

CHAPTER 7

SOUTH AUSTRALIA, 1883-85

Most men pretend, more or less, to be able to predict changes in the weather. It is one of man's most man-like failings to hug unto himself the idea he is a born weather prophet — Sydney weekly broadsheet, *Bird of Freedom,* May 11, 1895.

Wragge gave unsolicited advice to South Australia's veteran meteorologist Charles Todd within a week of landing in Adelaide on December 6, 1883. It followed Todd's acknowledgement in the press that he could not predict an end to the heatwave then marinating the city. Todd told the *South Australian Advertiser* he felt confident of forecasting weather only 24 hours ahead. Given his dependence on 9 am weather readings, he saw no sense in issuing forecasts for the next day's paper. "We should never hear the last of it if we were to predict a wet, stormy day, of which there might be every indication, and it turned out beautifully fine," he said. By the way, he suggested, the apparent recurrence in Adelaide of sultry, stormy summers every 11 years or so might be caused by a sunspot cycle.[172]

Wragge replied immediately that Todd should devote more

---
172   *South Australian Advertiser*, Adelaide, December 12, 1883, p5

research to his forecasting problem. Why not invest in two new observation stations, one at sea level and the other in the Adelaide Hills, on the 2334-feet-tall summit of Mount Lofty? Data recorded simultaneously for several months would surely be of scientific interest and value. The next day, Todd responded defensively. Of course, he knew the likely value of observations on Mount Lofty. In fact, high-level stations had been fully discussed by Australasia's meteorologists at conferences in Sydney in 1879 and Melbourne in April 1881. He planned to soon build a station near the top of the mountain.[173] In fact, Todd had greater priorities as the colony's Postmaster-General and coped with a chronically miserly budget for meteorology. In 1884, his weather bureau staff consisted of two observers and a clerk.

On January 1, 1884, Wragge opened a weather station in the backyard of the villa he rented in Gilberton, two miles from the government observatory in the centre of Adelaide. He pledged to strictly follow the British Weather Office regime of 3 am, 9 am, 3 pm, 9 pm and 9.22 pm readings — the latter equivalent to eight minutes past noon in London and synchronised with leading observatories around the world. But he cut himself some slack by using self-registering instruments for the 3 am data. He named the station the Torrens Observatory after the nearby Torrens River.[174]

In coming months, he set up a high-level station on crown land on the summit of Mount Lofty, seven miles directly east, or 19 miles by railway and another five miles on foot. On October 1, 1884,

---

173 *South Australian Advertiser,* December 13, p6 and December 14, p6
174 The name honours Robert Richard Torrens (1814–1884), a South Australian politician famed as a land titles law reformer. Aboriginal people called the stream *Karrawirra Pari* — red gum forest river.

he began recording air pressure on the peak with a clockwork aneroid barograph simultaneously with his sea-level observations. By November, after adding further automatic instruments, he was able to record a mass of barometric, temperature, rainfall and ozone data on weekly excursions to the summit. Nora took charge of the Torrens readings while he was away.

In 1887, while getting a grip on Queensland weather, he found time to write for *Good Words*, recalling life as an amateur meteorologist in South Australia, 1884–86. He told of camping overnight on the mountain and tramping home the next day through Waterfall Gully, "one of the most fertile spots in the world", sucking in gum leaf and bushfire scents, plucking native shrub twigs for identification, and dissecting white ants' nests for the wonder of their honeycombing. Reading this report, you can see escaping from home was a way of life. He did so regularly in Scotland and South Australia with an eager accomplice in Robin Renzo — his daily companion on the Ben.[175] He argued the hills' climate suited Britishers better than Adelaide's plains, with shade temperatures frequently 12 degrees Fahrenheit lower and humidity 22 per cent greater. The South Australian Government gave him a rail pass to Mount Lofty in return for copies of results.[176] In November 1885, the government voted £291 for upkeep of the observatory.

---

[175] Dogs shipped from Europe were held in quarantine for six months, by order of South Australia's Chief Inspector of Sheep, to control contagious hydrophobia. Wragge's much-travelled black Newfoundland died in Brisbane in 1892, according to a report in *The Telegraph*, Brisbane, January 13, 1892, p5.

[176] Wragge, Clement, *Experiences of a Meteorologist in South Australia*, pp 21–24, No 1 in the Pioneers Books Reprints Series, Warrandale, South Australia, 1980, from a three-part series published in *Good Words*, London, 1887, pp 621–26, 685–90, 754–61

In 1879, Todd and the meteorologists of New South Wales, Victoria and New Zealand had envisaged an observatory on Mount Lofty as one of a string of storm-warning stations for the Southern Ocean and the Bass and Tasman Straits. Their resolution at the Sydney conference listed suitable high-level sites as Kiandra (4600 feet) in the New South Wales Snowy Mountains, Tauhara Taupo (4600 feet) and Mount Herbert (4000 feet) in the north and south islands of New Zealand, respectively, Mount Wellington (4000 feet) in Tasmania, Mount Macedon (3600 feet) in Victoria, and Mount Lofty. Studying vertical barometric gradients was not on their agenda.[177] The experienced trio of Todd, the New South Wales Government Astronomer and Meteorologist H C Russell, and Victorian counterpart Robert Ellery had charge of a jumble of weather-watching activities in their colonies. They were old friends. Todd (1826–1910) and Ellery (1827–1908) were both nearing 30 years of employment with their respective governments. Todd trained in astronomy and meteorology as a 15-year-old making computations at the British Royal Academy, Greenwich, and later in the academy's galvanic department. Ellery had qualified as a surgeon in England but preferred astronomy, which he practised as a volunteer at Greenwich before migrating to Victoria in 1852. Russell (1836–1907), a Sydney University physics graduate, joined the Sydney Observatory as a first assistant in 1859 and advanced to government astronomer in 1870, when he established the colony's

---

177   *Meteorological Work in Australia, a Review*, by Sir C. TODD, K.G.M.G., M.A., F.R.S., F.R.A.S., Government Astronomer, Adelaide, S.A., from 'Report of Fifth Meeting of the Australasian Association for the Advancement of Science, Adelaide, 1893', p246, http://www.archive.org/stream/reportofmeeting51893anza/reportofmeeting51893anza_djvu.txt, accessed October 3, 2022

meteorological department.[178] James Hector (1834–1907) was a Scottish-born surgeon and natural history enthusiast who ran New Zealand's meteorological department in his role as manager of the colony's Geological Survey and Colonial Museum. He was New Zealand's first chief scientist.[179]

In his biography, *Behind the Legend – the Many Worlds of Charles Todd*, published in 2017, historian Denis Cryle examined more than a decade of niggles between Todd and Wragge. He concluded the genial, old-school and pragmatically collegiate Todd never came to terms with Wragge's individualism, which ironically paved the way for Australia's national weather service.[180] In 1884, Todd was among South Australia's highest-paid officials, on £950 per annum, a well-respected member of Adelaide's professional middle class, preparing for a sabbatical in Britain after 30 years of public service. He was then still best known as "Telegraph Todd", the can-do construction supervisor of the telegraph line from Adelaide to Port Darwin that, in 1872, put the antipodean colonies in instant touch with Britain. He ruled Adelaide's massive Italianate-style Post and Telegraph Office, also opened in 1872, commanding, as superintendent of electric telegraphs, the web of copper wires strung around the colony since his arrival in 1854.

---

178   *The Maitland Daily Mercury*, February 23, 1907, p3, obituary of H C Russell.
179   Dell, R K, 'Hector, James', *Dictionary of New Zealand Biography*, first published in 1990, Te Ara – the Encyclopedia of New Zealand, https://teara.govt.nz/en/biographies/1h15/hector-james, accessed October 3, 2022
180   Cryle, Denis, *Behind the Legend — the Many Worlds of Charles Todd*, p132 and p218, Melbourne, 2017. Cryle cites an assessment of Wragge's Chief Weather Bureau in Douglas, Kirsty, *Under Southern Skies: Understanding Weather in Colonial Australia, 1860–1901*, Australian Government Bureau of Meteorology, Metarch Papers, No 17, May 2007, pp 1,3

Writing for *Good Words* in 1887, Wragge tacitly acknowledged, in an homage to the city of Adelaide, Todd's work in establishing it as Australia's telegraphic hub — connected with Melbourne since 1858 and Perth since 1877. He likened Todd's headquarters to London's GPO in St Martin's Le Grand, imagining both with thousands of miles of wire radiating in all directions — in Adelaide's case connecting the colonies through a labyrinth of telephones as well as bringing the latest news from around the world. In fact, in 1882, South Australia's telegraph messages sped along 8070 miles of copper wire. Adelaide and Port Adelaide had only 124 telephone subscribers between them that year.[181]

Todd, Ellery and Russell had exchanged and publicised weather observations formally since early 1877, with data informing the weather map that Russell and his staff supplied the *Sydney Morning Herald* for daily publication.[182] Russell published barometric readings from around Australia but refrained from issuing forecasts. Todd also publicised barometric readings from observers on the Overland Telegraph Line, in the press and on a noticeboard at the GPO. By 1884, his duties included issuing the press a report daily at 1 pm, and posting weather charts at the GPO and major commercial outlets based on data from up to 250 stations around the continent. While Western Australia and Queensland contributed observations, neither colony employed a government meteorologist. The West's Surveyor-General Malcolm Fraser kept meteorological records from 1870 to 1883, when promoted to Colonial Secretary. In Queensland, Edmund

---

181 *South Australian Register,* October 24, 1883, 'Supply Debate on Posts and Telegraph Budget', p2

182 *Adelaide Observer,* March 3, 1877, p4

MacDonnell, a public-spirited jewellery firm manager, had been part-time Government Meteorological Observer since 1868, and by 1884 was recording weather readings from at least 130 observing stations around his colony.

On January 24, 1884, the *South Australian Advertiser* published Wragge's first report from the Torrens Observatory in a letter analysing wild weather in the preceding fortnight and predicting more rain. Todd had recorded the same roller-coaster of hot-and-parched, cold-and-sodden days in daily weather reports in the *Advertiser* and *South Australian Register*. He gave his latest synopses of pressure systems but stopped short of making forecasts. Wragge told how the freakish weather began with a dry north-easterly wind that baked Adelaide in maxima of 109.6 Fahrenheit in the shade and 152.6 Fahrenheit in direct sunlight. Next, a south-westerly change brought four days of almost English gloom, followed abruptly by the return of hot, blinding dust-laden north-easterlies, then another south-westerly change yielding three days of rain he thought likely to continue — given low barometric pressure ruling between Victoria and New Zealand. The heavy rains, rolling clouds, gales and squalls from the south-west reminded him of North Atlantic weather. His forecast, he explained, was based on physical laws governing the continued circulation of winds around the north-west segment. More occasional squalls and low temperatures were inevitable while the low-pressure system prevailed in the Tasman Sea.

* * *

Meanwhile, the marital squalls that blew him and Nora home from Scotland in 1883 continued. They were not the kind solved by sunny skies, sea travel or a new baby. Wragge's memoir notes from 1884 to 1887 contain distressed asides such as "the row re the nurse", "my mental agony" and "the old mother at Kensington — the trouble continues". His mother-in-law, Anna Thornton, lived in Kensington, about four miles from Gilberton. Wragge's breezy piece for *Good Words* in 1887, covering his return to Adelaide and setting up the Torrens Observatory, gave only one hint at his home life, a suave observation on finding and hiring elusive domestic staff: "Next to flies the greatest bore in Australian housekeeping is absence of good servants, although wages are nearly double the amount usual in the Old Country." He wrote that well-to-do colonists typically preferred villas of four to eight rooms with running water, an orchard and kitchen garden, costing at least £70 a year in rent if not owned outright. Gilberton House, the Wragges' two-storey villa, was built as a rental property about four years before their occupation. His notes suggest he bought it despite serious financial difficulties caused by losses in the marriage-settlement trust account managed by his "safe-as-a-church" brother-in-law Edward Thornton. This is how he summed up his pickle: "The Ingleby mortgages expire and money loose [sic] — speculators after me."

Their third son, George, was born at Gilberton on March 13, 1884. He was a blond, fair-skinned child, like Bert. Two photographs archived in South Australia's State Library show Nora, Clement and the children, taken at home in 1884 and '85. Gilberton House dominates both, signposted "Torrens Observatory" at the front gate, and with one of Clement's souvenirs at the front door — an Arab canoe picked up in Aden in 1883. On the top veranda in the 1884 picture there's Nora standing tall, flanked by four-year-old Eggie and two-year-old Bert using chairs like stilts. She's looking down confidently, balancing Bert with both hands, while Eggie steadies himself on her right shoulder. It's probably autumn, as the yard is sparse. In the other photo, Wragge and the children have been posed in the front yard. He looks as thin and whiskery as reported in his Ben Nevis days, a skinny disciple in an old suit and jaunty felt hat. Eggie holds his hand and Bert, stocky and independent at three, holds his ground between them and big sister May, who has not managed to hold toddler George still through the photo plate's exposure. Robin Renzo, almost invisible in foreground shadow, is fuzzy too. No sign of Nora, who midway through 1885 was pregnant with their fourth son, Reg, born February 20, 1886. Wragge's memoir note for 1886, "that boudoir and my study and instrument room", sketches his domestic troubles as he understood them.

Roughly two years passed between the first and second of Nora's three Adelaide pregnancies. Wragge fidgeted through those months, his intentions flickering like lightning in a never-quite-consummated storm. His notes show that in early 1884, while setting up his observatory, he also booked to sail back to England then changed his mind when George was born. On April 2, he announced he had trained a competent assistant for his observatory, had begun 3 am readings and intended making weather charts "showing the meteorological conditions of a large part of the globe".[183] Yet his name remained on the passenger list for the barque *Myrtle Holme*'s voyage to England via Cape Horn, leaving on June 5. *The South Australian Weekly Chronicle*

---

183   *The South Australian Advertiser*, April 2, 1884, p4

reported on June 7 that he would soon leave for Perth where he was likely to take charge of Western Australia's reorganised meteorological department. "Mr. Wragge intended proceeding to England at once, but we believe he has abandoned the trip in favour of one to Western Australia," the *Chronicle* reported. I've found no other mention of his having sought work in Perth that year. Western Australian meteorology was run by a Surveyor-General's Department officer until 1896 when Charles Todd's deputy Ernest Cooke became the colony's first meteorologist.

On June 14 in 1884, Wragge advertised for a well-educated youth of good character to learn the duties of assistant at the Torrens Observatory and make himself generally useful. The successful applicant would be treated as a Cadet or Midshipman, entailing a 12-month indenture, for sound scientific training in habits of precision. I don't know if he found one. A year later, William Russell, a sailmaker and amateur astronomer, offered voluntary help, beginning a friendship that stretched into the late 1890s. "William Russell comes into my life," Wragge exclaimed in his 1884–86 memoir notes. Their common interests included astronomy, sailing and rowing. Russell was born in the Orkney Islands in 1842, came to South Australia as a child and, by 1884, had a long-established sail-making, ship-chandling and rigging business in Port Adelaide. He, Wragge and Todd each used 44-inch equatorial telescopes, then the largest in the colony. Wragge wrote little for the press in 1884 and '85 of his astronomical studies, limiting himself to reminding fellow enthusiasts of phenomena such as Jupiter's visibility in the afternoon sky in April 1885. Like other 19th century stargazers — including Todd — Wragge saw life on Mercury and Venus as a strong possibility:

We can easily see that if their atmospheres are tempered by a denser envelope of aqueous vapour than we possess, it is quite possible that people not unlike ourselves may inhabit those worlds.[184]

During autumn 1884, Wragge dug and planted gardens and an orchard around the Gilberton villa. In the next two years, he raised a young vineyard, fledgling orange and olive groves and many tropical and semi-tropical trees in the teeth of hot northerly winds. "So, aided by Nature, I 'made the desert smile,'" he wrote in *Good Words*.

Patchy rain in the first half of 1884 gave settlers hope that nature would revert to Charles Todd's estimation of average rainfall — based on his 30 years of data. Most of the continent had been in drought since 1880. Stock and station agent Elders Wool and Produce Co Ltd estimated, in May 1884, that three parched years had cut Australia's sheep flock by 10 million and the wool clip by 200,000 bales. Eighteen months later, *The Queenslander* offered its country readers this advice:

Year after year the same tale has had to be told, and those who were at first disposed to regard droughts as something out of the common must have begun to get accustomed to them … Indeed, every reader of the short history of Australia must agree … extremely dry seasons are not phenomenal but normal.[185]

* * *

---

184 *South Australian Weekly Chronicle,* May 7, 1884, p14
185 *The Queenslander,* December 26, 1885, p1025

In early 1885, drought returned to South Australia and Wragge's funds were drying up too. In March, he advertised his credentials as a meteorological instruments expert when guest speaker at a free evening lecture for the South Australian Chamber of Manufactures. Todd introduced him as a valuable cooperator in studying the colony's weather and climate, taking his wider fame as a given. This engagement coincided with Wragge trying a new venture at his observatory, "truing" thermometers and barometers for the standard Kew Observatory fee. Adelaide's establishment had, by then, welcomed him into the Royal Society of South Australia and the Bohemian Club. John Langdon Bonython, owner and editor of the *Advertiser*, was among fellow men of letters, musicians and artists in the Bohemian Club.

In April 1885, Todd took a year's long-service leave and sailed for England, leaving his deputy Edward Squire in charge of posts and telegraphs and delegating the observatory to the recently appointed cadet William Cooke. It's unlikely Todd considered the novel alternative of hiring his irritating counterpart at Torrens Observatory, as an anonymous *Advertiser* correspondent suggested he should in May 1885. Later that month, the *Advertiser* acknowledged Wragge's weather nous by seeking his opinion of an old colonist's dire prediction of continuing drought. He replied with a then-orthodox wariness of long-range forecasting and an unorthodox warning on Australia's rainfall:

> If the rainfall of Australia is to be increased it will be by tree-growing — trees with broad leaves, giving a wide condensing medium — and by making the utmost of our means of irrigation by tapping our few rivers and permanent creeks. The valley of

the Murray, for instance, would be as fertile as that of the Nile were irrigation from it adopted instead of letting its waters run to waste as at present.[186]

---

[186] *The South Australian Advertiser*, May 2, 1885, p4 and May 28, 1885, p5

CHAPTER 8

QUEENSLAND CALLING, 1885–86

He is popularly declared not to sleep, but to watch the stars by night and write about them by day, anyhow, day and night certainly seem alike to him ... — Kapunda Herald, December 21, 1886.

Wragge began his Queensland adventures in July 1885, on the rebound from missing two official expeditions to British New Guinea that year. In June, the *Advertiser* reported he would soon leave for Brisbane to discuss his plans for meteorological work in New Guinea and northern Queensland with government officials and the new administrator of British New Guinea, Sir Peter Scratchley. In Wragge's absence, his friend William Russell was to take charge of the Mount Lofty weather station.

New Guinea's lure was summed up in 1885 by veteran Queensland overlander Augustus Gregory as its virginal wilderness, "The only remaining field of any magnitude to tempt the ardour of the present generation of explorers."[187] In April 1883, Queensland had annexed the south-eastern part of the island, known as Papua, prompting Wragge's letter to *The Times*

---

187   *The Capricornian*, December 19, 1885, p26, report on Gregory's inaugural address to the Queensland branch of the Geographical Society of Australasia

of London with his credentials for stocktaking the Empire's latest possession. In 1884, Britain took charge of Papua while Germany formalised its control of the north-east coast and islands. The Netherlands had claimed the western end, adjacent to the Dutch East Indies, since 1660.

In February 1885, Wragge had applied to the Geographical Society of Australasia to work as a meteorologist and collecting naturalist on an expedition to the Aird River near Papua's western border. German zoologist Wilhelm Haacke, a former director of the Adelaide Natural History Museum, was chief scientist for this venture, accompanied by a taxidermist and two collectors. But the Sydney-based society luckily had no call for a weatherman. Diverted to the Fly River district, the survey lasted from June to November 1885, and yielded little beyond condemnation of its leader Captain H C (Henry) Everill, and his accident-prone 70-ton steam launch, *Bonito*.

The second venture originated in England in 1884 with sponsorship from the British Association for the Advancement of Science and the Royal Geographical Society. British botanist and zoologist Henry Ogg Forbes' expedition into the mountains and coastal areas north-east of Port Moresby was in part meant to introduce Sir Peter Scratchley to his new domain. After setbacks, including an accident in Batavia that ruined £800 worth of equipment, Ogg Forbes began trekking from Port Moresby to Mount Owen Stanley on September 25. He was accompanied by Deputy Commissioner Captain Anthony Musgrave, Port Moresby missionary William George Lawes, photographer John William Lindt and 22 Malay carriers. Ironically, Lawes' letter to *The Times* two years earlier on New Guinea's potential prompted Wragge's

Chapter 8 Queensland calling, 1885–86

declaration in the same forum of his willingness to explore the island thoroughly and report on its resources. I've found no evidence he took this informal job application any further, or that Lawes remembered him. Sir Peter briefly joined the survey party in November before being flattened by malaria. When his condition deteriorated, he was carried back to Port Moresby and transferred to a government steamer for treatment in Brisbane, but died at sea on December 2, between Cairns and Townsville. He was aged 50, a veteran of the Crimean and Zulu wars, and had made light of his heart problems when accepting the New Guinea post. His private secretary, Dr Glanville, a fellow veteran African campaigner, told *The Queenslander* that New Guinea's climate was the worst he had experienced anywhere.

* * *

Wragge's first sight of Queensland's tropical eastern coastline was coincidentally aboard the steamer that carried him to Australia in 1883. In mid-July, he joined the *Maranoa*'s Sydney-to-Cooktown run with stopovers at a string of ports, beginning in the Queensland capital, Brisbane, 475 nautical miles north of Sydney. While there, he contacted the colony's governor, Sir Anthony Musgrave, to offer his services as a meteorologist, saying he would be back in a week with a formal proposal.[188] On July 25,

---

188   *The Queenslander*, August 22, 1885, p293, 'Exploration of New Guinea'; *South Australian Register*, January 6, 1886, p5, 'Mr Clement Wragge's Trip to Queensland'. He met the Queensland Governor, Sir Anthony Musgrave, who was the uncle of the same-named New Guinea Deputy Commissioner, Postmaster-General Thomas MacDonald Paterson and Premier Samuel Griffith's under-secretary.

the *Maranoa* reached Cooktown, a further 1000 miles north, and stayed three days.

Captain W W Hampton, who skippered the *Maranoa* on her maiden voyage to Australia in 1883, remembered Wragge's specimen-gathering passion and delegated a crewman to take him to the nearby Turtle Island group of coral cays. Recalling this trip for a South Australian newspaper, Wragge wrote, "I waded through the warm ocean wash with my arms laden up to my chin with corals, beautiful shells, gorgeous starfish and other interesting objects, some of which are seldom seen in museums." In another letter, to a friend in Staffordshire, he likened exploring the coral- and clam-spangled satellite of the Great Barrier Reef to visiting a new planet. He wrote, "I am bankrupt for adjectives to adequately tell you of the prodigality and exquisite forms of tropical marine life." Some of the booty spilled from his grasp as he chased ever more exotic and elusive creatures for his trove in the sloop that took him to the reef.

In July 1770, the crew of Captain James Cook's holed ship *Endeavour* angered local people by taking a dozen turtles from shoals in this locality. Those Englishmen were marooned for nearly eight weeks at the mouth of the river that Cook named after his patched-up barque.[189] On August 22, 1770, Cook reached Torres Strait, 400 nautical miles further north. Here, he claimed for the King of England everything inland from the coastline he'd mapped for the previous four months. He called this new British possession New South Wales. In 1859, the British Government transferred 715,000 square miles of it into a new colony named in

---

189 The traditional owners of this country, the Guugu Yimithirr Bama knew the river as Waalumbaal Birri.

Chapter 8 Queensland calling, 1885–86

honour of Queen Victoria. And a decade later, the Queensland government chose the mouth of the Endeavour River for a settlement at first called Cook's Town — a port for the Palmer River gold diggings, a tramp of 85 miles away.

Back ashore, in July 1885, Wragge got the *Maranoa*'s mate to take him by night to an Aboriginal camp near the settlement. They got there about midnight, having followed campfire sparks through the scrub and sand hills. Wragge recalled being confronted by hairy, naked men with painted faces and hungry women who clutched at the bag of stale bread he brought in exchange for dilly bags and spears. Tin billies and old blankets, bundles of spears and boomerangs lay scattered around shelters made of boughs and branches he called "wurleys". He and the sailor stayed a couple of hours before taking what he described as "trophies of the expedition" back to Cooktown.

Wragge returned to Brisbane on the *Maranoa* and then, on July 30, moved into Gowrie House, a new guesthouse on Spring Hill, not far from the convict-built stone mill that was meant to house the city's meteorological observatory. On August 3, he wrote to Queensland's Premier Griffith offering to take charge of meteorology in the colony and the south coast of New Guinea for a starting salary of £400 per annum. He declared his experience in Britain would enable him to elevate meteorology in Queensland to a level of excellence found in the Old Country and the United States. He would run the entire system in strict accordance with the Royal Meteorological Society — modified to the requirements of the climate. "I would yield to no other country in my plans of reorganisation for which I propose special aptitude," he pledged. Besides, Queensland was ideal for scientific research because of its geographical position. Apart from his salary, costs would include employing an assistant and a boy at the Brisbane headquarters, annual inspections of Queensland's weather stations, and a £300 grant for equipping these stations. But he offered to economise by using his own instruments in the Brisbane observatory. He pressed for an early reply, disclosing he had also approached Sir Peter Scratchley with a proposal for meteorology in New Guinea. [Perhaps in a letter. Sir Peter was then in New Zealand, seeking financial support for his New Guinea assignment.] Wragge made no comment on the work of Queensland's long-serving part-time Government Meteorological Observer Edmund MacDonnell.

* * *

In the next few weeks, Wragge wrote stories for the *Courier* and *The Queenslander*, introducing himself to the government and general public. His pitch for work was audacious and disciplined, based on reliable advice he would be well received. A memoir note, "per Ellery Queensland", perhaps suggests encouragement from Victoria's astronomer Robert Ellery.

In 1885, regardless of the drought and unemployment in western districts, Queensland's political leaders shared a grand vision for their fledgling colony. During the 1880s, the Liberal pragmatist Griffith and his conservative Nationalist opponents Thomas McIllwraith and Boyd Dunlop Morehead borrowed lavishly to advance their bit of the British Empire, with faith that science could help exploit and manage it. Griffith expected the Department of Agriculture, established in 1887, to maximise practical cultivation of the soil. In 1889, the Morehead government chose Professor Edward Shelton, formerly of Kansas State Agricultural College, to head the new department and share his knowledge of efficient US farming methods. The same year the government appointed a British marine biologist William Saville-Kent (1845–1908) as the colony's first Commissioner of North Australian Fisheries, intending to build a profitable fishing industry.

Visited in the early 1890s by Canadian writer Gilbert Parker, Brisbane looked raw and half-finished compared with Adelaide's aura of godly cleanliness. A contemporary essayist, Francis Adams, praised Brisbane's energetic citizenry, the pleasures of its river and bay, and its peerless, sub-tropical winter climate. But he felt the Australian dread of ridicule was hindering true adaptation to the sub-tropical climate, especially their insane adherence to Old Country dress codes.[190] Named after Thomas Brisbane, the Governor of New South Wales who planted it as a penal settlement in 1824, the convict town became capital of the new colony of Queensland in 1859. By 1885, its population had reached 52,128 and Queensland's 344,668 — including Chinese, Polynesians and members of other alien races, according to the colonial census.

In 1886, Griffith identified water conservation as the colony's most crucial issue. The same year, meteorological observer Edmund MacDonnell hesitantly declared the latest drought ended

---

190 Gilbert Parker, *Round the Compass in Australia*, 1892, cited by Judith A Jensen, *Unpacking the Travel Writers' Baggage: Imperial rhetoric in travel literature of Australia 1813–1914*, PhD thesis, James Cook University, https://researchonline.jcu.edu.au/10427/4/04part3.pdf, accessed October 5, 2022; Francis Adams, 'Social Life in Australia', *The Fortnightly Review*, Vol 56, July–December, 1891, p392

and hoped for prosperous seasons ahead. But he conceded in his 1885 report that 18 years of rainfall recordings were insufficient to make any universal rule on weather-cycles.

Wragge received good news two days after the *Courier* published his second essay, a lecture on pitfalls for inexperienced barometer readers. Griffith's under-secretary wrote a letter on August 28, confirming Wragge would be paid £50 expenses to report on the best means of establishing meteorological stations in Queensland, including Cape York Peninsula and Torres Strait. Wragge replied immediately, eager to start, but had to wait a week for his railway pass, a list of the existing weather stations, and a letter of introduction.

Wragge picked up his pass on September 8, then a day later introduced himself to a wary MacDonnell, who baulked at him inspecting the Brisbane first-class station without the Postmaster-General's approval. Wragge was unconcerned, writing later, "I accordingly informed him that, in face of my official credentials, I did not feel justified in waiting over to inspect the Brisbane station, and I pushed forward the inspection of other stations." Two months later, MacDonnell let him check the instruments at his home in Wickham Terrace that was in effect Brisbane's first-class weather station. Wragge rated management of the station as good, despite condemning the use of cage wire to protect thermometers from accidental breakage. He reported to Griffith that MacDonnell was aware the colony's weather stations were disorganised but had felt powerless to remedy the problem. MacDonnell had no control over the observers and had never been authorised to inspect the stations.

Wragge visited 25 stations from Stanthorpe to Thursday Island

Chapter 8 Queensland calling, 1885–86

and west to Normanton during his two-month tour and found only seven in working order. His subsequent report to Griffith reveals observers' grievances at lack of training and remuneration and their frustration at the government's reluctance to replace broken equipment. For example, he found Cooktown's first-class station staffed by a harassed observer who confessed to fudging readings from the broken maximum thermometer and to recording minimum temperatures with a 15-degree Fahrenheit error after letting the alcoholic column volatilise.

Wragge's itinerary was shaped by the timetables of the Australian Steam Navigation Company (ASNC) and Queensland Steam Shipping Co. In early November, he spent 11 days in the Burketown district waiting for the return of the ASNC's steamer *Dugong* on its fortnightly service from Kimberley — present-day Karumba. He considered Burketown ideally placed in the far north-west corner of the Gulf of Carpentaria for establishment of a first-class station to track weather from the Top End. But the fever-blighted settlement did not impress him in any other way, so he accepted an invitation to join the owners of Gregory Downs Station in a ride to their property, 75 miles south.

In those days, Gregory Downs was run by twins Sid and Harry Watson and their younger brother Greg, siblings of the famous Archibald Watson, recently appointed Professor of Anatomy at the University of Adelaide. Wragge enjoyed himself. Inspired by the lush growth of cabbage-tree palms, pandanus and ti-trees beside the Gregory River, he judged the country ripe for sugar cane, rice and other tropical crops. Along the way, he filled his saddlebags with shields and spears, soil samples, specimens of mangrove scrub, freshwater mollusca, and the famous *ceradotus* lungfish. He needed a packhorse to take these treasures back to Burketown. The Aboriginal weapons were, in his words, "procured" from people on the station who also cooperated in his research into what he called inter-racial adaptation to tropical climates. He took the temperatures of an unrecorded number of healthy Indigenous women and found some had readings of up to 100.2 degrees Fahrenheit, compared with 98.6 degrees for healthy Europeans.

Back at Kimberley, waiting for the stranded Thursday Island steamer to be floated off a mudbank, Wragge measured the heads and limbs of Aborigines and sketched their profiles. "My experiments among them will form the subject of a corroboree for a whole generation," he wrote later. Writing to his Stoke-on-Trent friend W H Goss, Wragge reimagined the Gulf's lonely littoral as a lunar seabed and confessed to yearning memories of the Churnet Valley's pine trees and larches. He signed off to check the tide, wondering if in his absence, the skipper had shot any of the alligators rife in the Norman River.[191]

* * *

Wragge finished his report for Griffith in late December, while sailing south from Cooktown. It contained 46 recommendations for the overhaul of the colony's weather service in line with the rules of the Royal Meteorological Society. The director of this service would be responsible alone to the government and have absolute control of the department, including training every post

---

[191] *Staffordshire Advertiser,* March 20, 1886, copy of a letter from Wragge to W H Goss, dated January 25, 1886

or telegraph officer or railway stationmaster taking charge of a weather station. He envisaged the Queensland service would produce daily weather charts for the entire Australian continent, based on data telegrammed from other colonies. Ultimately, the Brisbane office would issue weather forecasts once or twice daily and telegraph them to all chief stations in the colony. While the meteorological systems of other colonies were excellent enough in their way, Queensland could show the scientific world that she was second to none: "That she has, in fact, spirit enough to establish a meteorological organisation of her own, even better than any of them."

Wragge dawdled home from Brisbane. He spent a choppy Christmas Day aboard the south-bound *Maranoa* and waited in Sydney until New Year's Eve for a steamer to Adelaide, arriving January 5, 1886. His Queensland cargo was said to have filled a room of the family villa. In February, a reporter listed among the contents: samples of coastal rocks from Cape York and the Gulf, soil specimens, lichens and other botanical material, skins of birds, an almost-complete collection of Queensland shells and molluscs, and a host of miscellaneous specimens still in sealed boxes. Another room contained Aboriginal artefacts: spears from Cooktown and the Gulf of Carpentaria, necklaces of pearl made from Norman River shells, food bags, reed-and-bean necklaces, boomerangs, north Queensland woomeras and Gregory Downs shields.

Adelaide was mired in a heatwave in early January. Wragge condensed 1886 in his jottings into: "My return to Adelaide. Report to QG." No grizzles at Queensland taking 10 months to employ him, no hurrahs for his fourth son's birth on February 20. As usual, he left a thread in newspapers, beginning on January 6 with a paragraph from Torrens Observatory on the longed-for southerly that slaked the city on the afternoon of his return. Nora was still nearly invisible, solely "the wife of Clement L Wragge, FRGS" in the birth notices he placed in February for their son Reg. In December 1886, in farewell to Mr Wragge, the *Adelaide Observer* acknowledged Nora's invaluable assistance in his scientific labours. Reporting she would take charge of the observatory while he was finding a home in Queensland, the *Observer* noted their presence in Adelaide owed much to her family ties.[192]

Nora's brother and four surviving sisters all lived around Adelaide then, with more than 20 children between them. Bernard, the younger of her eldest sister Margaret Ingleby's two sons, was about the same age as Eggie and Bert, and a decade later joined them in Wragge's Kosciuszko adventure. The Thorntons were respectable folk. Nora's brother Edward and his wife Anna had four sons and a daughter born between 1877 and 1884. Edward was described after his death in 1909, aged 64, as one of the most popular and highly respected solicitors in South Australia. Nora's Canadian-born brother-in-law John Vereker Lloyd was the people's warden at St Paul's Anglican Church, in the city on the corner of Flinders and Pulteney Streets. Vereker Lloyd explored South Australia's Top End as a trooper in the 1860s, married Nora's sister Mary at St Paul's in 1871, became a commission agent and grazier, then later ventured into local government as a member of Mitcham District Council. St Paul's people were High Church and well-to-do. In 1877, Wragge and Nora were wed there by

---

192 *South Australian Weekly Chronicle*, February 27, 1886, p4; *Adelaide Observer*, December 18, 1886, p43

the long-serving Dean of Adelaide, Alexander Russell, and when Russell died suddenly in May 1886, Clement Wragge was listed in the who's-who of South Australian society at his funeral.

In 1886, Wragge opened his observatory to the public twice a week, charging two shillings and sixpence a head for views through the large telescope and an educational talk on meteorology and popular astronomy. Interviewed in February 1886, he seemed to the *Advertiser* the epitome of a gentleman scientist and destined for a useful and honoured career.

Beyond voluntarily recording daily weather data for the British Meteorological Office and Royal Meteorological Society, Wragge was writing a textbook on Australasian ethnology and natural history and preparing to launch an antipodean meteorological society. The book was never finished. But in July 1886, the Meteorological Society of Australasia materialised with him as honorary secretary and Henry Mais, South Australia's ageing engineer-in-chief, the first president. Todd accepted Wragge's invitation to become a trustee of the new organisation, with reservations about the likely duplication of some of his weather-service activities.[193] Membership peaked at 80 subscribers and 15 honorary members, but collapsed when Wragge moved to Brisbane. In November 1889, an Adelaide newspaper referred to the society as defunct.[194]

* * *

---

193 *South Australian Advertiser*, Adelaide, July 31, 1886, p6, copies of correspondence between Todd and Wragge

194 *The South Australian Advertiser*, May 14, 1887, p7; *Evening Journal*, Adelaide, November 29, 1889, p2, reference to "a fellow of the late Meteorological Society of Australasia"

Nearly a year elapsed between Wragge's report on Queensland's under-funded weather service and his appointment to fix it. It was too late in December 1885, when he finished his inspection tour, for any dramatic changes to the coming year's £350 meteorology budget. While waiting, he wrote compulsively for the South Australian press. He began issuing tentative 24- and 48-hour forecasts from his Torrens Observatory, and contributed long articles explaining basic meteorology and astronomy. He shared in a series titled "The Evening Sky" his certainty of a sentient, controlling presence in the cosmos that he doubted took any direct interest in human affairs.

In April 1886, when Evangelical church leaders called for a day of prayer for drought-breaking rain, Wragge advised them in his weather column to "Master the science of meteorology instead of nagging the Great Author of Evolution to suspend the laws of physics for human convenience."[195]

In following decades, Wragge and readers from all over Australia and New Zealand argued over conflicting understandings of God's involvement in human affairs. In July 1886, a columnist in the town of Gawler, north of Adelaide, advised Wragge, "the mighty meteorological savant", to learn more about the Almighty before issuing dogmatic pronouncements. The average bushman and farmer knew more about the weather than Mr Wragge: "Fancy, he [says he] knows exactly how the world was formed, and yet cannot tell us two days beforehand whether we can expect rain!"[196]

* * *

---

195 *South Australian Advertiser*, July 13, 1886, p4
196 *Bunyip*, Gawler, July 9, 1886, p2, 'Current Topics' by Quts Quts

Chapter 8   Queensland calling, 1885–86

On November 18, 1886, Premier Sam Griffith acknowledged Wragge's long wait when arguing for a doubling of the meteorology budget to £713 in 1887. He told the Legislative Assembly he had held over the vote to the 1887 supply debate believing Mr Wragge's proposed changes needed parliament's sanction. Griffith conceded the existing service was practically useless because of chronic neglect. He understood the need for data from all over the continent for weather forecasting and climate studies, but for now he hoped this new departure in Queensland meteorology would at least shed light on general causes underlying Brisbane's weather patterns. Griffith introduced Wragge as a distinguished meteorologist and thoroughly competent scientific man. The vote included provision for an annual salary of £400 and hiring an assistant on £150 per annum. In later years, Wragge said the then Queensland Governor, Sir Anthony Musgrave, had strongly supported his employment. He acknowledged this by including Musgrave among the given names of his fifth son, Lindley, born in 1891.

Griffith also had high hopes for the Department of Agriculture, established in 1887, to educate the colony's numerous cocky farmers. It took shape as a subsidiary of the Lands Department with a budget of £1800 and a small staff led by under-secretary Peter McLean, previously Inspecting Commissioner of Crown Lands. McLean spent the first four months of his tenure touring the southern colonies and New Zealand, gleaning ideas on scientific farming. In August 1889, he recruited Professor Edward Mason Shelton, the director of a Kansas farming experiment station, as Queensland's first Inspector of Agriculture, on an annual salary of £750. At Griffith's suggestion, McLean had asked the United

States Department of Agriculture to find a suitable candidate.[197]

In his final months of waiting, Wragge volunteered to be the meteorologist for a grandiose Antarctic expedition proposed by the British Association and the Victorian branch of the Geographical Society of Australasia. The venture quickly stalled. It had been estimated to cost £150,000 — equivalent to A$30 million today — and hinged on funding of £10,000 by each of the Australasian colonies. Griffith rejected it as unaffordable well before reports of Wragge's offer. Given the Australian colonies' growing debts, Griffith doubted the time was right, regardless of likely discoveries.[198]

But a decade later, the world's last unexplored continent still tantalised Wragge. In May 1897, before setting up his weather station on Mount Kosciuszko, he declared in a letter to *The Queenslander* his willingness to join the Royal Geographical Society of England's long-anticipated Antarctic expedition. He intended using weather data from far southern latitudes for seasonal and long-range forecasting.

Between 1901 and 1904, Captain Robert Falcon Scott led the Royal Geographical Society's reconnaissance of Antarctica known as the *Discovery* Expedition. The party had only one scientist with meteorological pretensions, Belgian-born Louis Charles Bernacchi, who had trained in astronomy at the Melbourne Observatory.[199] Bernacchi had previously served as a meteorologist on Norwegian adventurer Carsten Borchgrevink's privately funded foray from

---

197 *The Queenslander*, August 13, 1887, p260, and December 10, 1887, p952
198 *The Age*, June 9, 1886, p5; *Argus*, August 27, 1886, p6
199 Louis Bernacchi, Wikipedia, https://en.wikipedia.org/wiki/Louis_Bernacchi, accessed October 5, 2022

1898 to 1900. While in Brisbane in 1897, seeking backing from philanthropists, Borchgrevink unwisely cautioned Wragge that at 45 years old he was too old and unfit for his team. Wragge replied at once that he had no wish to join the Norwegian's escapade. His ambition was then confined to the British project, assuming the Queensland Government approved.[200]

---

200 *The Queenslander*, May 29, 1897, p16

CHAPTER 9

THE BIG PICTURE, 1889–92

Despite the meteorological systems of the other colonies which are excellent enough in their way, Queensland proves to the scientific world that she will be second to none; that she has, in fact, spirit enough to establish a meteorological organisation of her own, *even better than any of them.* — Clement L Wragge, Brisbane, December 19, 1885[201]

I doubt Wragge ever briefed Samuel Griffith on his Big Picture for Queensland. In November 1886, Griffith recommended him to parliament as a competent scientific man capable of reforming the colony's ramshackle meteorological service. He believed Wragge would accurately observe, record and analyse the colony's rainfall and climate — in other words, characteristic natural conditions. He expected Wragge's insights would assist Queensland's development. Moving for a significant increase in the meteorological budget, from £250 to £713, Griffith said he had taken a "very great deal of interest in the matter".[202]

---

201 Wragge, Clement L, 'Meteorological Inspection and Proposals for a New Meteorological Organisation, Part III', 'Recommendations for a new Meteorological Organisation', Gowrie House, Brisbane, December 19, 1885 (Queensland Colonial Secretary's Office, 9653, received December 21, 1885)
202 Queensland Parliamentary Debates, Legislative Assembly, 1886, Fourth Session, Ninth Parliament, Parliamentary Debates (Hansard), November 18, 1886, Supply Debate, pp 1788–89

Wragge prefaced his 46 recommendations by pledging a weather service superior to the other colonies, without spelling out his wish to supersede them. But he boasted in his first annual report that the Brisbane office had by December 31, 1887, become Australasia's Chief Weather Bureau, "As a matter of fact ... by reason of Queensland having taken the entire initiative in issuing Australasian and inter-colonial forecasts." He reported that before moving to Brisbane, he arranged interchange of weather telegrams with counterparts in Adelaide, Melbourne and Sydney, and had later added the Northern Territory, Tasmania and New Zealand to the network. From January 1, 1887, he had begun building a new weather service on the bedrock of British Meteorological Society and London Meteorological Office rules and principles. He had taken the initiative in issuing forecasts for Australia and New Zealand similar in principle to those framed at London and Washington for Britain and the United States, respectively. This amounted to establishing a Chief Weather Bureau, "Whether recognised as such inter-colonially or otherwise."[203] In 1894, he wrote that his ambition from the outset of his Queensland appointment had been: "To make Brisbane the Washington of Australasia in the intercolonial interests of Queensland, or die in the attempt."[204]

* * *

---

[203] Queensland Legislative Assembly, Votes and Proceedings, 1888, First Session, Tenth Parliament, Vol III, pp 1013–14, 'Preliminary Report of the Government Meteorologist for the Year 1887'; QLA V&P, 1889, Vol IV, pp 1073–75, 'Meteorological Report for 1887'

[204] *Clarence and Richmond Examiner*, Grafton, New South Wales, October 6, 1894, quoting from Wragge's article, 'How I Make My Weather', published in the Australasian edition of *The Review of Reviews*, September number, 1894

Chapter 9 The Big Picture, 1889–92

Wragge began work in Brisbane on New Year's Day 1887, and was anointed a few weeks later by Queensland's summer rain god. Between January 20 and 24, a nasty, persistent low-pressure system flayed the colony's south-east corner, causing massive flooding and at least 28 deaths. Brisbane's 18.3 inches (465 mm) of rain on January 22 remains the city's highest-ever 24-hour tally. Wragge monitored it from the convict-era windmill on Spring Hill that his predecessor Edmund MacDonnell had tolerated as a weather station. He augmented MacDonnell's existing equipment with some of his own — dry- and wet-bulb thermometers, a special barometer from his Ben Nevis project, and an electric humidity meter from Adelaide.[205]

On January 13, Wragge forecast stormy weather in northern districts with the chance of heavy rain. During the next week, he deduced in observations of plunging air pressure that a cyclone was advancing on Brisbane, west-south-west across the Coral Sea. On January 20, the storm hit the Noosa coastline with gale-force winds and passed just north of Brisbane on January 21. It interrupted telegraph communications with almost every port in the colony. Griffith and his wife postponed travelling to Sydney by steamer. Nine people drowned and 60 or so families were washed from their homes by the flooded Logan River, south of Brisbane.[206]

---

[205] *Brisbane Courier,* January 17, 1887, p5. Introducing Queensland's new meteorological department, Wragge described his Board of Trade Principle, Kew pattern standard barometer as "probably the best in existence". He had bought it for his Fort William base station and intended to use it to verify all official meteorological instruments In Queensland.

[206] *The Brisbane Courier,* January 22, 1887, p5, *The Telegraph,* Brisbane, January 22, 1887, p4, Callaghan, Jeff, 'Case Study: Brisbane Cyclone, 1887', Hardenup Queensland, https://hardenup.org/umbraco/customContent/media/646_Brisbane_Cyclone_1887.pdf, accessed September 25, 2022

Wragge quickly established himself as Queensland's new but much-experienced weather man. *The Queensland Times,* one of the colony's oldest newspapers, praised him on January 22 as "the new broom". He impressed the daily *Brisbane Courier* as an enthusiastic true lover of science, dedicated to establishing in Queensland a meteorological service surpassing any other in Australia and equal to Britain's.

Interviewed a few days before the cyclone, Wragge declared the other colonial meteorologists must, for the sake of absolute uniformity, strictly follow him in adhering to the rules of the Royal Meteorological Society of England. His aim in forming the Meteorological Society of Australasia in 1886 had been to achieve uniformity in observations in all colonies, to enable valid comparison and discussion of issues of weather and climate.[207]

* * *

In October 1888, the then Postmaster-General, Walter Horatio Wilson, acknowledged Wragge's pre-eminence "at the head of his profession in the colonies" and increased his pay from £400 to £500 per annum. The £1600 vote for the meteorological branch in the McIlwraith Government's 1888–89 budget covered his salary and those of his three assistants. His deputy Archibald Anderson had joined the branch in January 1887, and junior assistants Edgar Lambert Fowles and Inigo Jones during 1888. Anderson, 25, transferred from the Post and Telegraph Department's accounts branch where he was renowned for feats of mental arithmetic.

---

[207] *The Queensland Times,* January 22, 1887, p5; *Brisbane Courier,* January 17, p5, 'The New Government Meteorological Department'

Jones and Fowles were aged 15 and 16, respectively — Jones was the son of Wragge's rowing friend Owen Jones, a Brisbane civil engineer.[208]

During 1888, Brisbane's two daily newspapers validated Wragge's self-appointed status by publishing his reports under the heading "Meteorology of Australasia — Chief Weather Bureau". Inter-colonial forecasts from Brisbane appeared in Adelaide papers from July 1888, then in Sydney, Melbourne and Launceston in the next few months, all under his name. This annoyed his peers, who tried in vain at a conference in Melbourne in September 1888, to get him to do the gentlemanly thing and desist. That was the gist of a resolution from Western Australian Surveyor-General Sir John Forrest asking each of the colonial weather bureaus to stick to providing forecasts for their own patches. Wragge declined, arguing it was his public duty to share his forecasts. The other directors made no public comment after the Intercolonial Meteorological Conference at Melbourne from September 12 to 15. Their case, interpreted by the *Argus* on the final day, was that the public gained nothing from publication of conflicting forecasts for the same locality. Hence, Queensland's recruit, if a true man of science, must confine prognostications "within the narrowest of limits".

But Wragge received strong support from Brisbane newspapers, *The Queenslander* and *The Brisbane Courier* and Sydney's *Daily Telegraph*. A columnist in *The Queenslander*

---

208 Queensland Parliamentary Debates, Legislative Assembly, 1888, First Session, Tenth Parliament, Vol LV, Parliamentary Debates (Hansard), October 30, 1888, Supply Debate, p977; Evans, Raymond, *A History of Queensland*, Melbourne, 2007, p116. Thomas McIlwraith's "Australia for Australians" Nationalist party had trounced Griffiths' Liberals in June, 1888.

accused his critics of jealousy and ridiculed their effort to muzzle "our Clement". The *Brisbane Courier* argued his weather wisdom must be published wherever of use, and so should the prognostications of his meteorological brethren. A year later, the *Daily Telegraph* asserted it would be impossible to estimate the extent of Australia's indebtedness to the meteorological skill of Queensland's weatherman. By 1893, the Chief Weather Bureau's forecasts were regularly published alongside state meteorological bureau-generated forecasts in newspapers across New South Wales, Victoria and South Australia.

* * *

Wragge stayed with Nora and the children in Adelaide only twice in the two years before they joined him in Brisbane — for a month midway through 1887 and a week in September 1888. He was in Adelaide on business in July 1887 when she gave birth to their second daughter, Violet.[209] In 1888, he returned to Adelaide after the Intercolonial Meteorological Conference in Melbourne from September 11 to 15. In his memoir notes, Wragge wrote, "N takes no interest," in reference to buying land in Brisbane in 1888 and building the 12-room bungalow he named "Capemba" — an Aboriginal word for the place of water. The airy house with river views, three miles west of his city office, cost him £3500, entailing an £800 mortgage loan.

Nora arrived with their six children in January 1889. She

---

[209] Leonore (Violet) Eulalie Clementine Adelaide Wragge, born at Walkerville, July 6, 1887, birth reported in *South Australian Register*, Thursday July 7, 1887, p4

was pregnant again. Six weeks later, Wragge left for western Queensland, announcing plans to place barometers in new locations as far away as the South Australian border. But in late March, he came home unexpectedly. *The Western Champion*, quoting *The Charleville Times,* reported he was ill from drinking muddy water in droughty country between Thargomindah and Adavale. His deputy Archie Anderson told the Brisbane press he was recovering from diarrhoea and also needed to supervise the printing of his annual report.[210]

But there are scraps of a different story in his memoir notes — of his having received a letter in Charleville from "Muirhead" summoning him to Petty's Hotel in Sydney, and the upshot: "the disclosure and the embezzlement". This is likely to have been from Adelaide lawyer Charles Mortimer Muirhead. *The Sydney Morning Herald*'s railway passenger lists published April 16, 1889, show C L Wragge bound for Sydney via Tenterfield, and Mr and Mrs C M Muirhead, of Adelaide, also in transit to Sydney from Albury.

Muirhead, a sailing friend, had a long-standing arrangement to pass on his South Australian forecasts to the Adelaide press. From 1888, he wired them from Brisbane, defying Todd's efforts to halt their publication. Petty's Family Hotel, in York Street, Sydney, was a renowned rendezvous for wealthy squatters, founded decades earlier by Thomas Petty, an enterprising valet of squatter and Battle of Waterloo veteran Colonel Henry Dumaresq.[211] Decades later,

---

[210] *The Telegraph,* Brisbane, February 13, 1889, p5; *The Queenslander,* March 30, 1889, p581; *The Western Champion,* Blackall, April 9, 1889, p2

[211] *Australian Postal History and Social Philately,* 'Petty's Hotel: A Sydney Site with a Long Provenance', http://www.auspostalhistory.com/articles/15.php, accessed September 25, 2022; *Sydney Morning Herald,* April 2, 1950, p27 and July 12, 1950, p2

Wragge wrote that Edward Thornton had confessed to having embezzled £1200 of the £2000 trust fund and that Nora had insisted he must not be prosecuted. "She would allow no prosecution and shielded him in every possible way," he recalled in 1921. "The cruel words she used to me — most cruel — greatly embittered me."

A cryptic "Nora's flight to Adelaide" in his memoir notes suggests Nora visited South Australia briefly in 1889, after the birth of their third daughter, baptised Annie Maria Marguerite in November. By then, Wragge was away on a three-month patrol of inland weather stations. He set up four new observation posts in central Queensland, one of them at Avon Downs station near Clermont. *The Daily Northern Argus,* in Rockhampton, praised the participation of railway stationmasters, telegraph officers and the manager of Avon Downs in these "new and most important duties".[212] Wragge returned exhausted just after Christmas and was grounded for two months with Gulf malarial fever. Anderson took charge during massive coastal rains in January and February 1890, and another big Brisbane flood on March 10. A fortnight later, a cyclone flayed Townsville, killing two people, blowing 30 houses off their blocks and causing damage estimated at £10,000.[213] Wragge returned to work later in March and opened a first-order observatory at the Western Downs town of Mitchell and stations at seven other south-west centres.

* * *

---

212 *The Daily Northern Argus*, Rockhampton, p3
213 *The Telegraph*, Brisbane, March 14, 1890, p4; *Brisbane Courier*, March 26, 1890, p5

Wragge was rumoured to have "carried away" two Aboriginal children on his travels in 1889. The boy and girl were reportedly seen with him at Heley's Hotel in Normanton, where he waited for a Brisbane-bound steamer to reach the nearby port of Karumba. The Normanton correspondent of Rockhampton's *Daily Northern Argus* newspaper assumed he intended to put the children to work at Capemba:

> Mr Foundling Wragge, as he was named, stuck to his proteges, though why he was allowed to carry them away from their very respectable and ancient families puzzles us not a little. I don't know Mrs Wragge, but if she has been unused to black foundlings, I fancy the temperature will go up four or five thousand in the Wragge establishment.[214]

In 1894, a *Sydney Mail* writer pen-named "Australian Native" recalled having been waited on in Wragge's household by a young Aboriginal woman, "a native of Rockingham Bay".[215] Five years later, Archibald Meston, the recently appointed Protector of Aboriginals in Queensland's southern division, noted Wragge's employment of an Aboriginal woman named Maud Jeffries — "a half-caste girl from Carandotta". Historian Victoria Haskins, of the University of Newcastle in New South Wales, believes Wragge removed Maud — better known as "Mutta" — from western Queensland about 1896. She had crippled feet and was of a similar age to the younger Wragge children, having been born about 1887 or '88. Haskins has found that, about 1899, Meston took Maud from Wragge and placed her with a Brisbane woolbuyer's

---

214 *Daily Northern Argus*, Rockhampton, January 7, 1890, p3
215 *Sydney Mail*, September 8, 1894, p491

family. Wragge wrote "Mutta" twice in his notes, first connected with 1896 events, then among his "Queensland agonies". Between October 1895 and July 1896, he checked rainfall observations at Carandotta Station, 170 miles south-west of Cloncurry, while touring the colony's vast north and north-west.

* * *

Wragge left few clues to his home life in 1890 and '91, beyond conceding he had often escaped to the garden. In July 1890, he boasted having grown bananas equal to Queensland's best on his deep-trenched, rubbly hillside. His memoir notes suggest a servant and governess helped with the seven children. Their family was not especially large for the 1890s. His contemporaries, merchant and politician Robert Philp (1851–1922) and Archibald Meston (1851–1924) both had seven children too. And his long absences were fairly typical of other scientists and engineers in Queensland's public service. Government geologist Robert Logan Jack (1845–1921) spent many months in the field during his 20-year career, based first in Townsville, then Brisbane. The colony's first instructor in agriculture Edward Mason Shelton (1846–1928) and fisheries commissioner William Saville-Kent were also frequently at large along the coastline and littoral north and north-west of Queensland's poorly sited capital city.

* * *

In early April 1890, Queensland's then-Postmaster-General Charles Powers ordered an investigation of extreme weather in

Chapter 9   The Big Picture, 1889–92

the just-finished wet season. Brisbane received a near-record 39 inches of rain for the first quarter of 1890, in contrast with 6.2 inches in the first three months of 1889. Wragge responded with a preliminary report suggesting solar disturbances as a possible cause of cyclonic storms, floods and epidemics around the world in early 1890. In Australia, the north-west monsoon had, at times, engulfed almost the whole continent, including the arid inland. As requested, Wragge made sketch charts showing where principal towns had been flooded. He lacked relevant overseas data to attempt any long-range prediction, but anticipated improving his daily forecasts once he set up new observatories at Cape York, the Gulf and the colony's north-west.[216]

Wragge worked in Brisbane from April to early December 1890. Typically, he spent mornings to late afternoons in the city, either at the weather station on Spring Hill or in his General Post Office cranny in nearby Queen Street. At 5 pm daily, the bureau issued forecasts for all Australian colonies, New Zealand and surrounding waters. Usually by then, he had caught a train to Taringa, three miles up-river and had dashed up the hillside to Capemba. He was a renowned power walker, notoriously hard to keep up with. His famously named junior assistant Inigo Jones joked that, some mornings, weather bureau staff could hear the boss's hobnailed boots clattering all the way from Brisbane's Central Station. Jones, distantly related to the 16th century English architect Inigo Jones, later became a long-range weather forecaster. In 1898, a journalist recalled Wragge having dashed recklessly over hills and down gullies while leading him on a short-cut to Capemba from Taringa station. "The dance that Weather Man led

---

216   *The Telegraph*, Brisbane, April 25, 1890, p3

us shall live in my memory forever," the reporter from *Freeman's Journal* wrote.[217]

Wragge's few mentions of his marriage while in Brisbane suggest he is unlikely to have rushed home for Nora's sake. In old age, he recalled his life in Brisbane after her arrival as a time of intense unhappiness that had driven him to excesses. He wrote: "My home became a torture. I would go to my office with things apparently brighter and return to find it a pandemonium, so that no man could stand it." He wrote that their arguments over servants had continued in Brisbane, with a consequent churn. Nevertheless, Capemba was his sanctuary. In particular, he loved the shady bush garden which, in 1890, he mattocked and hoed down the lee side of his two acres of ridge-top in the new suburb of Taringa. The bushland was intact before his arrival. In May 1889, the *Courier* reported the Indooroopilly Divisional Board's regret at having mistakenly felled a tree on his block, believing it to have been on public land. They pacified him by cutting up the debris for firewood.

In later years, Wragge told reporters he had drained and trenched the whole scrubby hillside while coaxing apparently barren schistose soil into a ferny maze. Terraced stands of palms, conifers and bananas thrived through the wet early years of

---

[217] *Freeman's Journal*, Sydney, July 16, 1896, p6; *Queensland Geographical Journal*, No 40, Vol LIV, Inigo Jones, 'The life and Work of Clement Lindley Wragge', pp 48–49; Steele, John, 'Inigo Owen Jones (1872–1954)', *Australian Dictionary of Biography*, National Centre of Biography, Australian National University, http://adb.anu.edu.au/biography/jones-inigo-owen-539/text11915, accessed April 17, 2019, published first in hardcopy 1983; *Royal Historical Society of Queensland Journal*, January 1, 1952, Inigo Jones, 'My Seventy-Seven Years in Queensland', p688

his occupation, with rainfall supplemented by water from nine 1000-gallon tanks and three deep wells. In 1901, when trying to sell Capemba, he said his reclamation of the ridge had been prompted by strong memories of the meandering pleasure gardens of Alton Towers. Writing to Queensland Home Secretary Justin Foxton, he confessed to having often dug the garden in the witching hours "as a relief from office care and the tension of the forecasts". He likened himself to the then ageing British Opposition Leader William Gladstone, who was renowned for felling oak trees to stay fit, and planting seedlings in their place.[218]

Two other favourite spots were the astronomical observatory Wragge erected in May 1890 and his den off the front veranda. The observatory had a revolving roof, enabling a 360-degree view of the night sky through his four-and-a-half-inch equatorial telescope. He had, by then, also set up a home weather station, in case of after-hours emergencies, connected by wire to the GPO.[219] In early 1893, a Sydney reporter invited to Capemba wrote the den was like a ship's cabin — an eight-foot by ten-foot chamber with many charts suspended from a ceiling rack, a cedar trunk covered in sextants and other nautical devices, tiers of drawers containing thousands of rocks and shells, and photographs of exotic women met on his travels.

Typically, on his Queensland treks, Wragge carried rain gauges, thermometers and barometers, as well as specimen-collection jars, a camera, photographic dry plates and journals

---

218 *The Australian Star*, Sydney, May 26, 1894, p7; Clement Wragge to Colonel Justin Foxton, Home Secretary, August 8, 1901; William Ewart Gladstone (1809–98), Wikipedia, https://en.wikipedia.org/wiki/William_Ewart_Gladstone, accessed October 6, 2022

219 *Darling Downs Gazette*, June 7, 1890, p2

for his other scientific interests. The reporter from the *Australian Town and Country Journal* did not mention the Aboriginal skull that 14-year-old Inigo Jones saw in 1887 among souvenirs in Wragge's temporary den in the old windmill on Spring Hill. In 1894, journalists from the *Sydney Mail* and *The Australian Star* both reported skulls on display in the dining room at Capemba.[220]

---

[220] *Sydney Mail and New South Wales Advertiser*, September 8, 1894, p491; *The Australian Star*, Sydney, May 26, 1894, p7; *Queensland Geographical Journal*, No 40, Vol LIV, Inigo Jones, 'The Life and Work of Clement Lindley Wragge', p50

CHAPTER 10

SUNSPOTS, 1889–90

I am well aware that the sun is the great source of our weather system and that an outbreak of spots is followed by magnetic disturbances on earth, but the connection, if any, between sun spots and rainfall cannot in the present state of science be demonstrated...
— Western squatter, *Daily Telegraph*, Sydney, December 26, 1889.

Between April and December 1890, Wragge worked on overdue reports, gave public lectures and, when asked, commented cautiously on seasonal forecasting, a contentious topic in colonial meteorology. The Sydney and Brisbane press had widely reported the fate of popular young New South Wales sunspot-cycle theorist Charles Egeson, sacked from the Sydney Observatory in October 1890. His dismissal for insubordination and neglect of duty followed three years of niggling that stemmed, at first, from his venturing into daily forecasting in 1887 while observatory director Henry Russell was on leave in Europe. On July 1, 1887, Egeson launched an unsigned column called Our Weather Glass in the *Evening News*, a Sydney afternoon daily. Egeson, the observatory's weather map compiler, wrote later he had been approached expressly by the newspaper to produce the colony's first daily forecasts. When

Russell returned in late November 1887, he roasted his protégé and briefly banned the column before giving in to the newspaper's editor and continuing it himself.

In March 1889, Egeson further riled his mentor by publishing a private study of sunspot meteorology titled *Egeson's Weather System of Sunspot Causality: Being Original Researches in Solar and Terrestrial Meteorology*. His ideas were well received by a reviewer in *The Sydney Morning Herald*, especially his assertion that studying sunspot cycles might enable weather to be forecast years ahead. Egeson's ingenious research looked most relevant to Australia, "where so much depends on rainfall and the seasons".[221] The arrival of drought-breaking rain over New South Wales, as predicted, seemed to validate his method. Word spread to Queensland of his generally good news for pastoralists — plentiful rain through 1890, dry times in 1891–92 and a deluge in 1893. In April 1889, Rockhampton's daily newspaper, the *Morning Bulletin*, ranked him as Mr Wragge's equal in weather wisdom, astonished he had forecast the break-up of the latest drought:

> Mr Wragge, our meteorological observer, has come on so prominently before the public recently and has created such a favourable impression by his predictions, that Queenslanders will be loath to believe there is another weatherwise man in Australia ... [now] people are beginning to rub their eyes and wonder who Mr Egeson is.[222]

---

221 Egeson, C, *Egeson's Weather System of Sunspot Causality: Being Original Researches in Solar and Terrestrial Meteorology*, Sydney, 1889, National Library of Australia digital copy, https://nla.gov.au/nla.obj-2558311020/view?partId=nla.obj-2558321668#page/n8/mode/1up, accessed October 24, 2022; *Sydney Morning Herald*, March 5, 1889, p6, 'Publications received'

222 *Morning Bulletin*, Rockhampton, April 18, 1889, p5

Chapter 10 Sunspots, 1889–90

In his pamphlet, Egeson argued the procession of droughts and floods chronicled in New South Wales since 1788 had been driven by waxing and waning sunspots, in cycles varying from 31 years to 35 years. This theory rested on his correlation of decades of sunspot counts with patchy colonial weather records from the First Fleet onwards and also Aboriginal oral history. Essentially, he believed — following the prevailing wisdom in India — that waves of maximum solar activity, on average every 33 years, produced waves of copious rainfall, whereas extended periods of minimum sunspots caused droughts. He observed in New South Wales that the arrival of general rain was most delayed in times of minimal sunspots. He did not explain exactly how solar disturbances caused the cyclical fluctuations he observed in records of thunderstorms, winds and atmospheric pressure.

Egeson's seasonal forecasts assumed sunspot cycles would continue to periodically inhibit and enhance rainfall in New South Wales. His projections were based on the pattern of past drought and flood years. For example, his prediction of good rainfall in 1889 was based on similar conditions in 1858 and 1825. Using the same method, he warned a severe drought like that of 1827–29 was likely in the coming decade, probably beginning in 1894–95. Old hands remembered the 1827–29 drought as the colony's worst. Explorer Charles Sturt had called it fearful. Egeson's seasonal forecasts were essentially broad-brush affairs, with tolerances of one or two years for fluctuations in the start and finish of mean maximum and minimum solar activity. Hence, he watched the actual weather unfold and fine-tuned his "probabilities" accordingly. On October 2, 1889, he warned in a letter to Sydney's dailies that the latest sunspot cycle was two years in advance of the model he had used to forecast the beginning of the severe drought in 1894–95. He explained the most recent sunspot maximum was 2.2 years later than its mean position and advised farmers and graziers to prepare for a big drought, starting in six to nine months. He concluded:

> The evidence concerning our present position, on the eve of a most inauspicious climate change is to me overwhelming, and I give the warning with firm conviction it will be verified.[223]

Instead, within days, Wragge, Russell, Todd and Ellery all rejected his forecasts. On October 3, Wragge told Brisbane's *Telegraph*, the prediction of three years' drought was "decidedly premature in the present state of meteorological and solar physical science". Meanwhile, farmers and graziers across eastern Australia were reportedly in turmoil over whether to sell off stock, because of Egeson's recent renown. Russell intervened, telling New South Wales Minister of Public Instruction Joseph Carruthers that his maverick map-compiler's predictions were unscientific and fanciful. Carruthers subsequently banned Egeson from publishing any more forecasts.

An unpleasant truce ensued for the next 12 months during which – to the relief of pastoralists, politicians and perhaps Russell – the revised drought warning proved premature. New South Wales received above-average rainfall in 1890, as Egeson had at first forecast, and his detractors ridiculed him for the anti-climax of "Egeson's drought". In July 1890, he defied the forecasting embargo by sending the Sydney press a supplement to his weather system, leading to Russell complaining to Carruthers of his "spirit

---

[223] *The Sydney Morning Herald,* October 2, 1889, p4

of insubordination". This resulted in Egeson's suspension from official duties, and he was eventually dismissed in October 1890. The Federation Drought began in New South Wales in 1897.[224]

* * *

Meteorologists more experienced than Egeson had, since the 1860s, searched the sun's freckled face to explain cycles of flood and famine. Today, the spots that 19th century weather-watchers counted and correlated with wet and dry years are understood as electromagnetic eddies in the sun's 5500-degree-Celsius photosphere. Astrophysicists have found the spots to be relatively cooler — at 3520 degrees Celsius — and hence darker than surrounding plasma and argue they periodically modulate the intensity of solar radiation reaching Earth.

In 1801, British astronomer and musician William Herschel made a tentative connection between sunspot lulls and cereal crop failures. In a paper to the Royal Society, he linked spikes in wheat prices to times of low sunspot activity and suggested fewer spots meant too-little sunlight for healthy cropping. Herschel believed sunspots were shafts of light flickering in and out of view through gaps in the sun's cloudy photosphere. While lacking sufficient data to quantify any pattern in sunspot occurrence, he hoped his study would spark discussion on the sun's unexamined role in world affairs.[225] In 1816, a spate of sunspots, grotesquely magnified by

---

224 *South Australian Register*, October 2, 1889, p5; *Sydney Morning Herald*, October 24, 1889, p4

225 *Geophysical Research Letters*, Vol 40, 4171–76, Jeffrey J Love, 'On the Significance of Herschel's Sunspot Correlation', https://agupubs.onlinelibrary.wiley.com/doi/pdf/10.1002/grl.50846, accessed October 2, 2022

volcanic dust from a massive eruption in the Dutch East Indies, looked to anxious Londoners like omens of Armageddon. *The Times* was unfazed, though, concluding, "Generally speaking, the physical state of our little world is incomparably more stable and steady than its moral state." Herschel had, by then, suggested another hypothesis, that the body of the sun was opaque and dark and that the black spots observed there at intervals were merely the summits of very elevated mountains, visible between openings in solar clouds.[226]

Across the next 60 years, astronomers reimagined the sun using chemistry, physics and pioneering spectroscopy and astrophotography. In 1845, English physicist and chemist Michael Faraday (1791–1867) succeeded in magnetising a ray of sunlight; then, 20 years later, Scottish physicist James Clerk Maxwell (1831–1879) proved mathematically that light was electromagnetic energy.[227] By then, German scientists Gustav Kirchhoff (1824–1887) and Robert Bunsen (1811–1899) had unlocked the sun's chemistry, which had been manifest but long unrecognised in sunlight's spectrum of colours — also known as spectral signatures. Using Kirchhoff's invention, the spectroscope, they identified hydrogen, sodium, calcium, iron and other elements in the sun's atmosphere. In August 1869, English astronomer Joseph Norman Lockyer summed up the latest knowledge of Earth's nearest star,

---

226   *The Times*, July 26, p3
227   Michael Faraday, Wikipedia, https://en.wikipedia.org/wiki/Michael_Faraday, accessed October 8, 2022, citing Day, Peter,1999, *The Philosopher's Tree: A Selection of Michael Faraday's Writings*. CRC Press, p125; *The Royal Society Journal*, December 31, 1865, VIII, 'A dynamical theory of the electromagnetic field', James Clerk Maxwell, Abstract, https://royalsocietypublishing.org/doi/abs/10.1098/rstl.1865.0008,   accessed October 2, 2022

speaking at the 39th annual congress of the British Association for the Advancement of Science, in Exeter:

> We must look upon him as a specimen of millions of other stars, more remote, which together formed what might be called the skeleton of the universe, the scaffolding of which doubtless supported systems like our own and other habitable worlds past, present and to be.

In 1868, Lockyer — then a War Office clerk in London — identified a strange yellow line in the solar spectrum as the signature of an unknown element which he named helium. His analysis corroborated French astronomer Jules Janssen's coincidental discovery of helium's signature a couple of months earlier, while observing that year's solar eclipse in India. Lockyer, founding editor of the science journal *Nature* in 1869, predicted that year's findings would be soon left far behind, "paling into ineffectual fire beside future beautiful and brilliant pages in the book of nature".[228] In 1871, Scottish physicist Sir William Thomson predicted refinements in spectrum analysis would enable direct investigation of the physiology — in other words, "laws of being" — of the sun and stars, and glimpses of their evolutionary history. Thomson observed that, for at least a decade, dedicated volunteers from all nations had searched the universe for the distinctive prismatic colours of the firmament. Rarely in the history of science had there been such a flood of discoveries, underpinned by tireless scientific observation as well as enthusiastic perseverance leavened by penetrative genius:

> To non-scientific imagination, accurate and minute measurement seems less lofty and dignified work than looking

---

228 *Essex and Plymouth Gazette Daily Telegrams*, Exeter, August 24, 1869, p4

for something new. But nearly all the grandest discoveries of science have been but the rewards of accurate measurement and patient long-continued labour in the minute sifting of numerical results.[229]

* * *

Nineteenth century weather-cycle theorists trusted in the enthusiastic perseverance of generations of amateur weather-watchers and sunspot counters. In 1890, German geographer Eduard Bruckner (1862–1927) proposed a 35-year weather cycle based on historical rainfall records retrieved from 804 observation posts scattered over every continent, augmented by chronicles of weather dating back to 1000 AD. Comparing water level records for lakes in Europe, the Middle East, North America and Australia, he found similar times of maximum shrinkage and replenishment, and inferred an extra-terrestrial influence — namely the sun. His analysis of 53 years of sunspot counts — made between 1832 and 1885 — and older records of aurorae and magnetic storms, suggested a recurring 35-year pattern in global weather, hypothetically shaped by the 11.1-year cycle of maximum and minimum sunspots that, by 1890, astronomers widely accepted. Incidentally, Egeson's proposition of a cycle of around 33 years was published in Sydney several months before Bruckner's findings

---

229 *Bell's Weekly Messenger*, London, August 5, 1871, p7, reporting Thomson's address, as incoming president, to the biennial meeting of the British Association, in Edinburgh

appeared in Vienna.²³⁰ Egeson's work is recognised by present-day physicians and physicists in the Bruckner-Egeson-Lockyer climatic cycle of 30 to 40 years, known as the BEL cycle. Some historians and scholars of economics and human physiology argue "paratridecadal" cycles such as the BEL can be seen in the occurrence of military and political events and disease patterns.²³¹

Swiss astronomer Rudolf Wolf identified the 11.1-year cycle in the early 1850s, using a 17-year daily sunspot count by German pharmacist Heinrich Samuel Schwabe, supplemented by records from 1750 onwards. Ironically, Schwabe's disciplined sun-watch failed to disclose his desideratum, the transit of the phantom planet Vulcan, theoretically in orbit between the sun and Mercury. In 1843, he wound up the project and reported his subsidiary finding that the sunspots seemed to have followed a 10-year cycle of maximum and minimum occurrence. In 1848, Wolf, then-director of the Bern Observatory, revised Schwabe's counts and tentatively extended the cycle to 11.1 years, a period endorsed in 1851 by naturalist Alexander von Humboldt in his natural history compendium, *Kosmos*. In 1864, Wolf officially published his proof of the cycle, as new director of the Zurich Observatory.²³²

Mathematician Charles Meldrum, a native of north-east

---

230 Bruckner, Eduard, *The Sources and Consequences of Climate and Variability in Historical Times*, Vienna, 1890

231 Halberg, F, Cornelissen, G, Bernhardt, K-H, Sampson, M, Schwartzkopff, O, and Sontag, D, 'Egeson's (George's) transtridecadal weather cycling and sunspots', *History of Geo and Space Sciences*, 1, 49–61, 2010; www.hist-geo-space-sci.net/1/49/2010/doi:10.5194/hgss-1-49-2010, accessed October 2, 2022

232 Bean, Michael, 'Heinrich Samuel Schwabe, 1789–1875', *Journal of the British Astronomical Association*, Historical Section, Vol 85, 1975, pp 532–33, https://articles.adsabs.harvard.edu//full/1975JBAA...85..532B/0000533.000.html, accessed October 24, 2022

Scotland, is recognised as the first meteorologist to test Herschel's hunch of a link between sunspots and weather extremes. He did so during his nearly 50 years on the Indian Ocean island of Mauritius, beginning in 1848, when posted to teach maths in the capital, Port Louis, and continuing from 1862 as Government Meteorological Observer. In 1872, he reported a strong correspondence between the 11.1-year sunspot cycle and occurrences of cyclones in the Indian Ocean south of the equator. Situated 700 miles east of Madagascar, Mauritius typically experienced clusters of cyclonic wet seasons and occasional droughts. Meldrum used records of cyclones he had observed in his first 24 years on the island and shipping logs from some of the many India- and South Africa-bound vessels that called at Port Louis in pre-Suez Canal days. He also found tentative evidence of a general law of sunspot-influenced rainfall periodicity in the Southern Hemisphere, using 22 years of weather data from Adelaide and 12 years from Brisbane.[233]

Joseph Norman Lockyer reported Meldrum's findings encouragingly in *Nature* in an editorial scolding British meteorologists, state-endowed and amateur alike, for their contentment with simply amassing data. The 11.1-year sunspot period as applied by Meldrum was a crucial cycle for meteorologists — in his opinion as monumental as the discovery of the 18-year-11-day Saros cycle had been for astronomy more than 2000 years earlier.

---

[233] Michaud, Edley, 'Meteorologist's profile — Charles Meldrum', *Weather* (2000) 55, pp 15–17, https://rmets.onlinelibrary.wiley.com/doi/pdf/10.1002/j.1477-8696.2000.tb04013.x, accessed October 2, 2022; *The Queenslander*, April 5, 1873, p6, transcript of Meldrum's paper on the supposed periodicity of rainfall

Lockyer wrote he had been told while visiting India for the 1871 eclipse that everybody in Ceylon recognised a cycle of about 13 years in the intensity of the monsoon. He reminded readers to remember that in meteorology, as in astronomy, cycles are clues to enlightenment. He said meteorologists of the future must obtain an accurate knowledge of the "currents of sun and ... earth" by approaching their trade as a physical science, not a mere collection of weather statistics.[234]

Between 1876 and 1878, drought and famine in southern India, with a death toll of more than five million people, forced further investigation of the subcontinent's cyclical weather patterns. In 1877, Lockyer and the Indian Statistical Department director William Wilson Hunter (1840–1900) identified in six decades of data from southern India an 11-year rainfall cycle coinciding with the 11.1-year sunspot cycle. But Hunter, a Glasgow-born historian and statistician, was wary of long-range forecasting and argued the need for a solar observatory for further in-depth study.[235] In 1876, British meteorologist Edmund Douglas Archibald suggested floods and famines in northern India followed a 33-year sunspot cycle. He argued Calcutta's winter rainfall records showed a "distinct periodicity" with *maximum* rainfall occurring at times of *minimum* sun spots, and minimum rainfall coinciding with maximum solar disturbances. Reporting Archibald's theory, *Nature* noted winter rainfall in Sydney had followed a similar

---

234   Lockyer, J N, *Nature*, 'The Meteorology of the Future', December 12, 1872, pp 98–100, https://www.nature.com/articles/007098a0, accessed October 8, 2022

235   *Leeds Mercury,* March 29, 1877, p3; *Pall Mall Gazette,* November 10, 1877, p12

pattern to Calcutta's, and that both cities occupied similar latitudes south and north of the equator.[236]

Sydney was tormented by drought too, from mid-1875 onwards. In October 1876, H C Russell predicted more of the same for 1877, if the cycle he'd found in the city's patchy 89 years of weather records held true. Undoubtedly, the sun was the chief source of changing weather in this 19-year cycle, probably in conjunction with meteoric rings that sometimes crossed between the sun and Earth. He was unconvinced by theories linking sunspot maxima and minima to weather extremes. In a letter to *The Sydney Morning Herald* in August 1877, he cited inconsistencies in a study purporting to show low rainfall recordings in the Australian colonies in periods of minimum sunspots, which he found actually only partially true for Queensland and South Australia and contradicted by figures from New South Wales, Victoria and South Australia.

More than a decade later, Russell and his colonial peers viewed India as a special case and remained wary of weather cycles. Sydney's *Daily Telegraph* published a letter in December 1889, from an anonymous squatter, "burnt and browned by the sun for over 40 Australian summers", questioning if a clockwork-like pattern for seasonal weather forecasting could ever be found amidst "the terrible forces eternally at work on the sun's surface".[237]

---

236 *Nature*, August 2, 1877, p267, https://www.nature.com/articles/016267a0.pdf, accessed October 8, 2022
237 *Australian Town and Country Journal*, Sydney, January 1, 1876, p1, and October 21, 1876, p17; *The Sydney Morning Herald*, May 5, 1877, p8; *Daily Telegraph*, Sydney, December 26, 1889, p5

CHAPTER 11

NEW CALEDONIA, MUNICH AND BRISBANE, 1890-91

The greater our experience of New Caledonia, the more does one regret that our friends the French have made it a convict dépôt — the more do we wish it was British. They freely admit that they have not the peculiar colonizing instincts of the Anglo-Saxon ... — Clement Wragge, introduction to *The Romance of the South Seas*, Part One, "La Nouvelle" or The Prison of the Pacific.[238]

In mid-December 1890, Wragge sailed from Sydney to New Caledonia, leaving Nora pregnant with their eighth child, and Archie Anderson again in charge of the bureau. He was on a working holiday, intending to visit Samoa, Fiji and the New Hebrides (Vanuatu) after installing barometers and practising his French in Noumea. He left aboard the A.U.S.N. line's steamer *Waroonga*, savouring four days at sea for socialising, sightseeing and scientific experiments. The latter included taking the temperature of a cabin boy from Solomon Islands for

---

238 Wragge, Clement, *The Romance of the South Seas*, London, 1906, p53, https://archive.org/details/romanceofsouthseoowragiala/page/n19/mode/2up, accessed October 8, 2022

comparison with temperatures of Aboriginal women recorded while in Queensland's Gulf country. The boy's reading, 99.1 degrees Fahrenheit, was lower than the Gulf women's 100.1 to 100.2 degrees. He reported later that all subjects had been in tiptop condition. Recalling the trip, he wrote he had been bent on seeing all he could, "making the most of the privilege of human existence". He concluded, "What a charm, what a keen fascination, has travel to the intelligent mind."

He was particularly intent on seeing for himself the French colony's convict camp on Ile Nou and the notorious Nouvelle prison. During 1890, two escapers from the prison camp were guillotined after being extradited from Brisbane. Hundreds of the thousands of prisoners sent by France to New Caledonia from 1864 had fled to Queensland — so many that the Sydney press fretted their numbers in Queensland and New South Wales had seriously swelled the criminal class.[239] In October 1890, New Caledonia's principal warder identified another four men in Brisbane's jail as truants. Wragge returned to Queensland in late January 1891, aflame with what he saw. He made little of the *Telegraph's* praise for his shouldering of the light of meteorology through the South Seas, stressing instead, when interviewed by the *Brisbane Courier*, France's need for enlightenment in humane treatment of its convicts. He deplored the callousness of a guard whom he saw kill a sparrow found concealed in a solitary confinement

---

[239] Most of the 30,000 men and women sent to New Caledonia between 1864 and 1897 were common law prisoners convicted of crimes such as theft and murder. Men aged over 16 years and sentenced to more than eight years were forbidden to return to France. Paterson, Lorraine, 'New Caledonia The Penal Colony', in *Convict Voyages, A Global History of Convicts and Penal Colonies*, School of History, University of Leicester, UK, 2018, http://convictvoyages.org/expert-essays/new-caledonia, accessed October 8, 2022

prisoner's shirt. He also disapproved of the guard's confiscation of the convict's pipe and shred of tobacco — "the lonely one's greatest solace". Feeling helpless, Wragge had pocketed the tiny bird's broken body and later preserved its half-feathered wings. He concluded in dismay that on Ile Nou hope was trampled daily:

> I cannot help thinking from what I saw that a more humane and scientific system of treatment based on psychological principles would do more to reclaim these hardened criminals ... so austere is the treatment of some of these prisoners that the smallest consolation is denied them.

In following weeks, *The South Australian Register* and *Adelaide Observer* lifted the story of the sparrow's cruel death. Country papers in Queensland and New South Wales ran it too. Wragge spoke out indignantly about the French penal system again in 1892. Writing for the *Illustrated Sydney News*, he listed instances of prejudice against the ex-convicts known as *libérés* who were still officially under surveillance in New Caledonia. Surely, they had paid the penalty of their misdeeds and had a right to make a fresh start?[240] Memories of the guillotining chamber and convicts' cemetery on Ile Nou bothered him. In 1906, he rewrote and republished the Sydney articles in his book, *The Romance of the South Seas*, published that year in England. He pitied equally the permanently shackled fourth-class lifers on Ile Nou and their fifth-class brothers deemed intractable and awaiting decapitation. Surely, those prisoners who had transgressed through awful perversions of the mind were fit more for the lunatic asylum than the guillotine? France had stopped exiling felons in 1897 but

---

240 *Illustrated Sydney News*, June 18, 1892, p14

still subjected them to relentless, demoralising punishment. This practice precluded any hope of rehabilitation. "It does not bring out the best in man. *Au contraire*, it degrades him and makes him worse," he wrote.

France recorded executions by guillotine of 27 of the 30,000 men and women sent to New Caledonia between 1864 and 1897. Queensland courts sent 56 prisoners to the gallows in roughly the same period, from 1860 to 1899. There's an element of British superiority in Wragge's distaste for the guillotine. He was intrigued and appalled by his Ile Nou guide's revelling in details such as the swish of the three-cornered weighted knife and the thud of the felon's luckless head. Ile Nou stained his imagination. He wrote he would never forget the contrast between the death chamber and the adjacent sublime, blue Pacific Ocean. Oh, if only Britain was in charge of New Caledonia:

> The more I see of this place, the more do I wish our estimable friends, the French, had not made it a convicts' rendezvous ... It is impossible to resist the impression that naturally it is the attaché of Australia which, if subject to the factors of energy and solid pluck which operates in our colonies, would prove the brightest jewel in southern waters, and a grand sanatorium withal.

* * *

In May 1890, Wragge had told *The Queenslander* he was working on sharpening his daily forecasts, especially those for northern Queensland and the northern half of Australia, generally. He intended extending his weather-watching network to stations in Singapore and Java and producing meteorological charts for the Australasian zone from 100 to 180 degrees east of Greenwich — roughly from the Cocos (Keeling) Islands in the Indian Ocean to Fiji in the Pacific. His region stretched from the fifth parallel north of the equator to 50 degrees south, in the sub-Antarctic waters of the Southern Ocean. Conceding his plans hinged on getting the overseas data wired to Brisbane, via Darwin and Adelaide, free or at cut rates, Wragge was confident the owner of the Eastern Extension Telegraph Company, Sir John Pender, would be sympathetic, "considering the many interests affected by a more accurate knowledge of coming weather in the tropics".[241]

Sir John's company, forerunner to the 21st century British multinational Cable and Wireless Communications, monopolised international telegraph traffic in the 19th century. The company later allowed free cipher code transmissions of daily observations from Singapore and Batavia, passed on by Todd, as a public service. The Brisbane bureau also received monsoon-season data from Port Darwin and the first-order stations that Wragge set up on his 1889 expedition at Rockhampton, Mackay, Townsville, Cooktown and Thursday Island. In all, between September and late December 1889, Wragge established 19 new meteorological and climatological stations in the north. In March 1890, after recovering from Gulf fever, he extended the network in the colony's south-west, opening a first-order observatory at Mitchell, two second-order stations and five rain-gauge stations. Data from the new Queensland stations was telegraphed to the other colonies through a long-standing reciprocal agreement.

---

241 *The Queenslander*, May 3, 1890, p821

Wragge had lobbied since 1888 for Queensland to help establish weather stations in New Caledonia, New Guinea and Tahiti. His colleagues at that year's third Intercolonial Meteorological Conference, in Melbourne, agreed observation posts were needed on the eastern side of the Coral Sea. Early in 1890, believing pastoralist Boyd Morehead's National Party government was too broke to help, he asked the Australasian United Steam Navigation Co (A.U.S.N. Co) to assist in setting up a station in New Caledonia. The shipping line agreed to delegate Samuel Johnston — its agent in the capital, Noumea — to take charge and send data to Queensland via its Pacific Islands fleet.

But in August 1890, Morehead and his ministry quit, effectively ousted by their former National Party leader, Thomas McIllwraith, who supported Opposition Leader Samuel Griffith in a no-confidence vote. The Morehead government had begun the financial year with a £147,000 deficit and threatened to fix it with an unpopular realty and personal property tax. Griffith formed a coalition with McIlwraith, his nemesis in the colony's most recent election in June 1888. "It is as if the representatives of France and Germany were to fall upon each other's necks and swear a 'blood brotherhood'," a Melbourne observer wrote.[242]

Griffith knew the need for tracking Coral Sea lows like those that had recently flayed the Queensland coast. In 1889, while Opposition Leader, he backed a modest rise in the Chief Bureau's annual budget to £1620 and also acknowledged a strong case for telegraph contact with New Caledonia. A year later, as Premier, he increased the

---

242 Queensland Parliamentary Debates, Legislative Assembly, Second Session, Tenth Parliament, Vol LIX, 1889, November 6, 1889, pp 2678–79; *The Telegraph*, Brisbane, August 13, 1890, p2; *The Telegraph*, Brisbane, August 13, 1890, p2, citing report in *The Argus*, Melbourne, August 7, 1890

meteorological budget by £50 and through the new Postmaster-General, Theodore Unmack, granted Wragge six weeks to set up a weather station in Noumea and also visit the New Hebrides and Fiji. On November 21, 1890, speaking in the meteorological branch supply debate, Griffith acknowledged all the great circular storms along Queensland's coast had come from the direction of New Caledonia and praised Wragge's Coral Sea initiatives.[243]

* * *

In February 1891, Wragge returned home from his South Pacific excursion, and left for Germany in July. He was among 80 directors of the world's meteorological services notified of the International Meteorological Committee's gathering in Munich from August 26 to September 2. The Griffith government gave him six months' leave to attend, but not as its official representative, since he had not been formally invited. Charles Todd chose British Meteorological Office director Robert Scott as his proxy and upset Wragge by disputing — through Scott — the Brisbane weather office's self-appointed status as Australia's chief bureau. Wragge's sole mention of the conference in his memoir notes is: "Munich! How I fought for my Qld office — Scott."

Letters between Todd, Ellery and Russell show that, by 1891, Wragge had upset each of them. In December 1890, Russell scolded *The Sydney Morning Herald* for praising Queensland's push for new Pacific island weather stations and overlooking his own past

---

243 *The Herald*, Melbourne, September 14, 1888, p3; *The Queenslander*, November 22, 1890, p992 and February 7, 1891, p273; *The Telegraph*, Brisbane, November 28, 1890, p4

efforts and those of his mentor George Robarts Smalley (1822–70). Russell wrote he had tried without luck to get data from New Caledonia, had recruited an observer on Lord Howe Island and received regular reports from the New Hebrides island of Tanna. Smalley, New South Wales astronomer from 1864 to 1870, had sent equipment to Fiji but was let down by his observers.[244] Russell was sceptical of the *Herald*'s affirmation of Wragge's forecasting successes, cautioning that meteorology as then practised was still an inexact science.[245] Neither Russell, Ellery nor New Zealand's astronomer Sir James Hector took part in the Munich gathering. Wragge later argued in his 1891 annual report the best outcome for Queensland had been the conference's endorsement of his proposed South Pacific weather network. It was taking shape with the special stations he established in January 1891 — in the New Caledonian capital, Noumea, on Aneityum in the New Hebrides, and on Norfolk Island — and he hoped to add others in the Solomon Islands, Samoa and Tahiti.[246]

In his five months in Brisbane in 1891, before sailing for Europe, Wragge parried some more with Todd, set up three coastal weather stations and joined the brand-new Queensland Theosophical Society. He continued loading Capemba with curiosities, having returned from New Caledonia with a herbarium, numerous corals

---

[244] Russell complained in December 1890, in a letter to Robert Scott on scratchy Pacific Islands observations, "... the reason is very obvious — the Climate is so enervating that Europeans get too lazy for meteorological observations. See Sydney Observatory collection, Powerhouse Museum archive, http://www.sydneyobservatory.com.au/tag/wragge/ accessed October 8, 2022.

[245] *The Sydney Morning Herald*, December 4, 1890, p4, and December 11, 1890, p9

[246] Clement L Wragge, *Meteorological Report for 1888, 1889, 1890 and 1891*, Chief Weather Bureau, Brisbane, June 1, 1892, p4

and shells, and specimens of scorpions, centipedes, butterflies, beetles and other insects. He was rumoured to have kept a pet carpet snake in his study to keep his children out.[247] Nora and the children are absent from his 1891–92 jottings. A note, "Mrs Burton comes in," probably refers to the Mrs Burton he introduced to a Sydney reporter in 1893 as one of his childhood nurses, a fellow Staffordshire native then living in Brisbane. She worked for him at Farley in 1878 and probably helped Nora after the birth of their fifth son, Lindley, on July 10, 1892. Sarah Burton (nee Alcock), born in Cheadle in 1843, was 10 years older than Nora and Clement, and she had eight surviving children of her own — the youngest were twin girls, born in 1885.

On July 2, 1891, Wragge left by train for Adelaide where he boarded a steamer for Europe. A few days earlier, the *Brisbane Courier* had reported his nearly immediate six months' leave, giving no specific reason for the delayed approval except the other colonies had turned down Queensland's idea of sharing the cost of sending him to Munich. The colonial weather services all had ample notice of the conference. British Meteorological Office director Robert Scott kept in touch on the International Meteorological Committee's wish to foster wider international knowledge-sharing among weather services. As he left, Wragge told the *Courier* he anticipated debating weighty meteorological science, which for him meant the case for uniform observation standards. He would also visit major European observatories and

---

247 *The Brisbane Courier*, February 3, 1891, p 7; *The Telegraph*, Brisbane, February 3, 1891; *Australian Town and Country Journal*, April 15, 1893, p19; Newman, B M, 'News Cuttings and Biographical Notes', John Oxley Library, State Library of Queensland, M761, quoted in Dunstan, Keith, *Ratbags*, Melbourne, 1979, p56

weather centres. Today, the gathering is remembered as the world's first mass meeting of weather service directors, a forerunner of the World Meteorological Organisation, established in 1950. The 80 participants agreed on procedural changes that lasted until World War I — the election of an executive bureau funded by the weather services of Britain, France and Germany, and provision for occasional international conferences of meteorological service directors. The British press judged the meeting's most important scientific outcome was adoption of a cloud classification system proposed by Scottish meteorologist Ralph Abercromby and Swedish scientist Professor Hugo Hildebrand Hildebrandsson, famed for having identified and named cumulus clouds.

The talks ended on September 2 with acknowledgement of Europe's need for rapid transmission of weather reports from North America, Iceland, the Faroes and Azores, and also the impossibility of funding such a service. Afterwards, in Paris, French Central Weather Bureau director Eleuthere Mascart showed Wragge the electronic link between the top of the recently completed Eiffel Tower and his office beside the Seine. Wragge also inspected observatories in Utrecht, Hamburg, Berlin and Vienna, but declined an invitation from the chief bureau of the United States weather service in Washington, DC, as he could not find suitable passages across the Atlantic. Instead, he visited his old friend William Henry Goss in Stoke-on-Trent.

* * *

At the beginning of the conference, in a meet-and-greet at a restaurant with a view of the snowy mountains beyond Munich, the world's leading meteorologists toasted him as "the man with the largest family". He confessed to *The Queenslander* on his return to Brisbane to having drunk many litres of beer towards the good health of various causes, ending with a swig for prolific fatherhood. "It is believed that on that occasion Mr. Wragge himself should have remained seated," *The Queenslander*'s reporter remarked.[248] On January 4, 1892, Wragge was welcomed home by his whole family, corralled by Archie Anderson.[249] The tribe that Anderson brought to meet the Royal Mail steamer *Jelunga* at the South Brisbane Wharf comprised Nora, then nearly 40 years old, their three daughters and five sons. It was Clement's first sight of his youngest boy, Lindley, born on July 10, 1891, just after he left for Munich.

* * *

Wragge had arrived in South Australia in 1883 as the famed weather-watcher of Ben Nevis, the True-Grit Brit of his *Good Words* tales. Farewelling him in December 1886, an Adelaide journalist wondered if the public really recognised they had in their midst a scientific man of worldwide celebrity. *South Australian Register* reporter William Sowden knew him well for his late-night arrivals at the paper carrying meteorological notes for the next morning's edition, accompanied by his big black dog. Writing under a pen-name for the *Kapunda Herald*, Sowden judged him an eccentric philosopher, "as simple and unaffected as he is wise". William Sowden (1858 – 1943), who began his newspaper career as a printer's devil on the *Castlemaine Representative* in Victoria,

---

248 *The Queenslander*, January 16, 1892, p123
249 *The Week*, Brisbane, January 15, 1892, p6

edited the *South Australian Register* from 1899 to 1922, and was knighted in 1918 for numerous community-service activities. Like Wragge, he was a field naturalist and supported forest conservation. Sowden had joined the *Register* in 1881 and became chief author of leading articles in 1892.[250] During Wragge's time as Queensland Government Meteorologist, the Adelaide press depicted him more often as an ambitious upstart than an irrepressible scientist. In early 1893, a *Register* editorial, probably by Sowden, called him the unruliest member of Australasia's family of meteorologists and weather prophets. The *Register* argued his deadlock with Todd proved the need for a chief federal bureau, but not one necessarily of his making and certainly not located in Brisbane.

The Brisbane press consistently defended Wragge's reasoning in opposing Todd and the others and adopted him as their own. Wragge's defiance at the Melbourne meteorological conference in September 1888, hastened his discovery by Sydney and Melbourne journalists. A columnist in *The Sydney Morning Herald* introduced him in October 1888 as Queensland's irrepressible and storm-loving prophet, with a burden for all of eastern and southern Australia. The writer, barrister and yachtsman Alexander Oliver, concluded, "I suppose we shall get accustomed to him in time as a sort of weather Jeremiah and continue to get on with business."[251]

---

250 *Kapunda Herald,* December 21, 1886, 'Scratchings in the City' column, by A Pencil; *Advertiser,* Adelaide, October 11, 1943, p3; Carl Bridge, 'Sowden, Sir William John (1858–1943)', *Australian Dictionary of Biography,* National Centre of Biography, Australian National University, http://adb.anu.edu.au/biography/sowden-sir-william-john-8593/text15005, published first in hardcopy 1990, accessed June 11, 2019

251 *The Sydney Morning Herald,* October 21, 1888, p7, 'As You Like It'; Woods, Gregory D, *A History of Criminal Law in New South Wales: The Colonial Period, 1788–1900,* Sydney, 2002, p423, identifies Alexander Oliver as the author of 'As You Like It'

In Melbourne, *The Australasian* settled on dubbing him "Mr (in) Clement Wragge, the Queensland Ellery", when chiding his poaching of brother meteorologists' territory.[252]

By 1891, Wragge's forecasting stocks were high in all the east-coast colonies. In December 1890, stories in *The Australian Star* and *The Daily Telegraph* depicted him as a dashing scientist with a nearly ethereal passion for meteorology. He gave interviews to both Sydney papers as he left for Noumea. The *Star* rated him an enthusiast of the first water — meaning the purest quality — whose zeal defied political boundary lines. *The Daily Telegraph* concurred, depicting him as an enthusiast among enthusiasts. The *Star* asserted his uncanny forecasting extended well beyond Queensland, encompassing the entire continent, Tasmania and New Zealand. He was not merely a local scientist, since his South Seas plans embodied a cyclone-tracking strategy to benefit coastal areas of both Queensland and northern New South Wales.[253]

---

252 *The Australasian*, September 8, 1888, p32
253 *The Sydney Morning Herald*, October 21, 1888, p7; *The Daily Telegraph*, Sydney, December 10, 1890, p5; *The Australian Star*, Sydney, December 11, 1890, p2

Clement Lindley Wragge's father, Midlands lawyer Clement Ingleby Wragge, c. 1851.

His mother, Anna Maria Downing, c. 1851.

His paternal grandmother, Emma Wragge (nee Ingleby)

Clement, aged 12 in 1864, when a student of Alleyne's School, Uttoxeter.

A clean-faced 22-year-old, San Francisco, 1874, at the beginning of his rail tour of the USA.

Midshipman Wragge with pipe and whiskers, before sailing to South Australia in 1875.

Wragge described himself as dressed as a "Bedouin Arabian Sheik" in this photograph from his Cook's Tour of the Holy Land in 1873.

Ben Nevis Observatory Sketches: 1. The lake, halfway up. 2. Wragge's hut. 3. Thermometer screen and cage. 4. Rain gauge, barometer cairn and screen. 5. Wragge's Fortin's Principle barometer. 6. Wragge and dog Renzo. 7. General view looking ENE. All sketches from *The Illustrated London News,* June 14, 1883, p36.

Wragge and children at the front fence of their home in Walkerville, Adelaide, 1884. His weather station can be seen in the backyard, left.
Source: State Library of South Australia.

Nora Wragge, then aged in her mid-40s, in Brisbane in 1897. Baptised Nora Thornton in 1852, her given names were registered as Leonore Eulaliecia Edith Florence de'Eresby in 1877 on her marriage certificate. Picture courtesy Mrs Myrtle Wragge.

Wragge, aged 29, and Renzo, pictured in Edinburgh, 1881.

Wragge, third left, second row, among his peers at the first international conference of directors of meteorological services, Munich, August 26, 1891. He is directly behind the first civilian director of the US Weather Bureau, Mark Walrod Harrington, seated second left, front row.
Source: British National Meteorological Library and Archive.

Helpers tow pedestrians in a boat through the intersection of Edward and Queen Streets, Brisbane, during the 1893 flood.
Source: John Oxley Library, State Library of Queensland.

Wragge sets out in July 1885 to open an observatory on Tasmania's highest peak, Mount Wellington, 4170 feet high. He and his helpers overnighted in a hut built at his request by sailors from *HMS Porpoise*, of the British Navy's Australian Squadron. *The Sydney Mail* reported he brought his autoharp for a singalong. "Mr Wragge is an enthusiast. Darkness and light, weariness and ease, weather fair and weather foul are all alike to him," the paper observed on July 27, 1895. Picture: *The Graphic*, London, October 12, 1895, accessed from the British Newspaper Archive.

Charles Egeson, left, stands beside fellow Sydney Observatory staff in 1887, when he was acting meteorologist in the absence of Charles Russell. Henry Ambrose Hunt, first director of Australia's Bureau of Meteorology, is seated third left. Source: Museum of Applied Arts and Sciences, Sydney.

Clement Wragge, left, his deputy Archibald Anderson, second left, and other staff members, standing beside thermometer screens at the Brisbane Meteorological Observatory on Spring Hill, Brisbane, c. 1900. Source: State Library of Queensland.

Wragge in his garden at Taringa, with an unidentified dog, from a photograph in *The Queenslander*, January 25, 1902. Source: State Library of Queensland.

Wragge, right, poses with six Steiger Vortex cannons before the guns were railed to Charleville on August, 1902. The other men are probably William Harvey and son James, engineers of Margaret Street, Brisbane. In October 1902, news that two of the cannons had burst during firing prompted Harvey and son to issue a statement defending the guns' heavy-duty construction.
Source: State Library of Queensland.

Wragge, second right, stands in a fleecy coat at his campsite near the summit of Mount Kosciuszko in December 1897. Among the crew who set up the weather station were overseer Captain Charles Iliff, far left, guide James Spencer with pipe, third left, and observer Bernard Ingleby wearing a beanie, fourth left. Charles H Kerry, *Wragge's Camp, Snowy Mountains, New South Wales*, c.1900, National Library of Australia, Tyrrell Collection.

Two unidentified observers at work at the snowed-in observatory, one standing on the roof, beside a lightning conductor, the other checking thermometers. Charles H Kerry, *Observatory – winter*, album of photographs of the Snowy Mountains, Mitchell Library, State Library of NSW.

Rupert (Bert) Wragge with sons Keith and Raymond, c. 1941. Rupert served with the 1st AIF in the Middle East and Europe from 1914 to 1918, and finished as a Company Sergeant Major. He served in the Voluntary Defence Corps in Brisbane during World War II. Early in World War II, Keith joined the Royal Australian Navy and Raymond joined the Australian Army as a gunner. Raymond survived three years as a prisoner of war after the fall of Singapore in February, 1942.

Trooper Clement Egerton Wragge, 1914. Source: John Oxley Library, State Library of Queensland.

The memorial to Clement Lionel Egerton Wragge, unveiled in Oakamoor's Anglican Church of the Holy Trinity in 1917.

Clement, Edris and baby Kismet, c. 1901.

Kismet with bicycle, c.1910.

Dr Emily Brainerd Ryder.
Source: State Library of New South Wales, Mitchell Library, Scott Collection.

"The grand atmospheric extravaganza played by the myrmidons of the old Wind-god King Aeolus himself who wields the baton ..."
Cartoon from *The Worker*, Brisbane, June 11, 1898, p6, National Library of Australia.

CHAPTER 12

THE MYSTICAL POTTER, 1891–92

Spring, Summer, Autumn, Winter too, the Sun is always ruling.
In spite of all the Sun can do, the World is slowly cooling...
— From *Thomson's Seasons*, undated doggerel from *Punch*, quoted by *The Salisbury Times and South Wilts Gazette*, January 29, 1887.

"Dearly as I love my scientific profession, I was not born to live on isobars, isothermal lines and vapour tensions," Wragge wrote in 1892, recalling his working holiday in New Caledonia. His homily in the *Illustrated Sydney News* continued, "Each man's aim should be happiness, and happy we will be so far as such an abstract entity is realisable in this world."[254] In short, he had been ready for a break after four years of exacting meteorology. He confessed having carried a kit of percussive bones and triangle in hope of some fun. One memorable night, he and a sailor from a recently wrecked, Glasgow-bound barque had belted out sea shanties while travelling by horse and buggy 18 miles from Noumea to the village of Paita. He confided he had offered the recuperating sailor a lift because they both got a kick out of blue-water music. After a nip at

---

254 *Illustrated Sydney News*, June 18, 1892, 'Science and Travel in the Western Pacific', p15

a wayside inn, they had trotted on, their refrain *We're All Bound to Go,* with bones and triangle accompaniment, filling the midnight air. "Why shouldn't we be jolly?" he reflected. "Let him cavil who likes." He prized the advice of his late friend Scottish physician Sir Robert Christison[255] to remain a boy as long as he could. Christison, a fellow Scottish Meteorological Society member, told him to resist ever thinking he was getting old. "Age comes soon enough, and the greatest mistake a man can make is to imagine he is getting old," he'd said, as they scuttled through Edinburgh one winter's morning in 1881. They had become friends after Wragge's feats on Ben Nevis. Christison, then 84 years old and renowned for his athleticism, died unexpectedly in January 1882 of complications from a cold. Incidentally, a story is shared by the New Zealand Wragges of a séance in the early 1950s when Clement's spirit was said to have arrived like a small tornado, singing sea shanties and knocking the medium off his feet.

* * *

Although 20 years Wragge's senior, the renowned potter William Henry Goss was more like his brother and confidante than a father figure. Both of Wragge's uncles had died before his return to Britain in September 1891 — William in 1888, aged 71, and George in 1889, aged 78. Goss shared with Staffordshire

---

[255] Sir Robert Christison (1797–1882), a former Professor of Medicine and Therapeutics at the University of Edinburgh, had controversial views on disparate causes. He campaigned against the university's admission of women as medical students and opposed capital punishment by hanging, advocating instead execution by lethal injection. Robert Christison, Wikipedia, Wikimedia Foundation, updated July 15, 2021, https://en.wikipedia.org/wiki/Robert_Christison, accessed October 8, 2022.

Chapter 12 The Mystical Potter, 1891–92

and London newspapers the letters that Wragge sent him from South Australia between 1884 and 1886, and subsequently from Queensland. William Henry Goss (1833–1906) was, according to his obituary in the *Liverpool Courier*, a student of the hidden wisdom of the world, with a reverence for every form of life and a belief in the worship of God "through the laws of the universe and the primeval righteousness of things". His daughter, Eva Adeline Goss, described him as a broad-minded True Churchman who no longer attended services of worship. He liked quiet and empty churches better than Sunday's bustle. His fame as an artistic potter stemmed from the jewelled porcelain and heraldic china that he and his three sons made in Stoke-on-Trent. He invented the enamels used for colouring heraldic arms in their line of small white-glazed vases and pots known as Gossware.

In October 1891, Goss wrote to a friend recounting long conversations with a visitor from Brisbane, "almost straight from the French penal settlement in New Caledonia". Describing his guest as an astronomer and world traveller, he was struck by his deep fear of annihilation and hope of reincarnation. He recounted his friend's distress at comments by astronomer Dr William Huggins in a speech to the British Association in Cardiff two months earlier. Huggins, then-British Association president, had summed up the previous decade's research on what he described as the sun's approaching extinction — estimated by physicist William Thomson to be 10 million years away. Press reports suggested Huggins took for granted the inevitable end to which evolution "in its apparently uncompensated progress" was carrying all creation. The possibility that collisions between dark suns might create new nebulae would at most only delay that eternal oblivion. Goss recalled his visitor's anxiety over this statement and his near despair that Huggins had meant ultimate annihilation, in other words the death of suns and all things within human knowledge.

In the letter that I believe records Wragge's visit, Goss stated his faith in God's goodness evident in all creation. Eva Goss included the letter in *Fragments from the Life and Writings of William Henry Goss*, published in 1907, a year after his death. It was from his correspondence with Wilhelmina Harriet Anderson, a sister of the famed, late Major-General Charles George ("Chinese") Gordon. Goss met his Brisbane visitor's harassed questioning with his own questions and answers. Could the universe have been called into existence for ultimate annihilation? Certainly not. Surely all creation was progressing from chaos to perfection, from some incomprehensible fall to ultimate complete deliverance? Goss wrote that after discussing the "devilisms" of New Caledonia, his visitor had concurred with him in viewing death as the gate of life, "To those who have availed themselves of their opportunities in the progressive march of this life's journey."

Happily, his visitor had agreed with his view that the soul of even the most debased criminal could be redeemed through reincarnation: "He has a pretty good notion of some folks needing to be born again for he has travelled nearly all over the world and seen much of men and manners." Goss believed those souls falling short would be born again into the toil and trial of life's education onward to the Gate of Higher Life. He continued:

> This interpretation of being "born again" seemed to strike my friend and he exclaimed, "Reincarnate?" And I replied, "Yes, mercifully; instead of going to universal punishment."

Chapter 12 The Mystical Potter, 1891–92

Goss described his friend as a traveller acquainted with Buddhism who shared his dismay at the priests who used the Buddha's image as a begging idol. Wragge recalled in a *Brisbane Courier* story published in February 1892, his stopover in Ceylon six months earlier and having visited the Buddhist temple at Kelaniya, outside Colombo, following the example of the founders of Theosophy, Madame Helena Blavatsky and Colonel Henry Steel Olcott. He wrote of having found comfort in the "Buddhist creed" and its close resemblance to Christian teachings. Olcott, a 59-year-old ex-US Civil War colonel, spoke on Buddhist practice in Brisbane in April 1891, before chartering the Queensland Theosophical Society. A reporter described him as "eloquent of speech and finely venerable of appearance".[256] Wragge was a founding member along with Archibald Meston and Queensland District Court Judge George William Paul. Olcott also promoted Buddhism and Theosophy in Melbourne, at a public meeting chaired by Alfred Deakin, then a Victorian Cabinet minister and later Australia's second Prime Minister. A Spiritualist, Deakin had visited Buddhist temples in Ceylon and India during a fact-finding trip on irrigation schemes in 1890–91.

Many years later, Wragge asserted Spiritualists and Theosophists were on a common human quest to know the reason for life and answer the questions: "Great God, who am I? What am I here for?" Writing in 1919, as leader of a Spiritualist church in Auckland, he argued all thinking men worldwide yearned for Theosophy, which he defined as the wisdom of the infinite. In 1891, the *Brisbane Courier*, commenting on Colonel Olcott's visit, termed Theosophy a product of religious restlessness.

---

256 *The Telegraph*, Brisbane, March 30, 1891, p4

The movement's popularity reflected rejection of the throttling influence of traditional dogma and also a reaction to "the blank and deadening issues of materialism". But the *Courier* rated the substance of the Secret Doctrine of Theosophy, as expounded by its author, Madame Blavatsky, obscure and incomprehensible and less likely than Christianity or Buddhism to further human welfare.[257]

Goss ended his letter to General Gordon's sister with a homily on the joys of friendship and family. He and his wife, Georgina, had by then been married for 37 years and had raised seven children. He found evidence of a beneficent creator in the way marriage generally ameliorated the raw imperative of Darwinian evolution, or the "Go-forth-and-multiply" dictum of Genesis. In his view, man and woman could not be the authors of their own emotions and instincts, but were truly reliant on the creator's fatherly beneficence for "the reciprocal joys of husband and wife".[258]

* * *

---

[257] *Auckland Star*, February 3, 1919, p9, letter to the editor; *Brisbane Courier*, April 11, 1919, p2

[258] Goss, Eva Adeline, *Fragments from the Life and Writings of William Henry Goss*, Stoke-on-Trent, Hanley, 1907, includes extracts from obituaries in *Liverpool Courier*, January 9, 1906 and *Daily Telegraph*, January 5, 1906, pp 11–12. The letter from Goss to Wilhelmina Harriet Anderson dated October 25, 1891, was first published as Letter XIX in the 'Series of Letters to a Lady', *Review of Modern Science and Modern Thought*, Stoke on Trent, 1895, pp 148–151. The Hath Trust's online version of the 1895 *Review of Science and Modern Thought* is the digital copy of a book held by Columbia University and includes a fragment of a letter of dedication from Goss to a Colonel Bolland, referring in passing to Clement Wragge's travels, http://www.ebooksread.com/dl2.shtml?id=163143&ext=txt&f=reviewofmodernscoogossiala&a_id=67326, accessed October 10, 2022

Chapter 12   The Mystical Potter, 1891–92

Wragge arrived back in Brisbane in early January 1892, and three months later Nora left for South Australia with six of the children. She stayed away for two years. Her exit and eventual return, in March 1895, are recorded in shipping notices, corroborated in his jottings. She had just turned 40 and the last of their children, baptised the previous October as Lindley Herbert Musgrave Egerton Wragge, was nine months old. Wragge noted his two eldest sons, Eggie and Bert, had remained at Capemba while the rest of his family — "the others" — were in Adelaide. In March 1892, a few weeks before their departure, he had advertised their former home at St Stephen's Terrace, Walkerville, for sale or long-term lease. Nora probably stayed within reach of her mother, Annie Thornton, sister Margaret Ingleby and brother Edward in the historic suburb of Glenelg. Their extended family included Margaret's daughters, Mary and Mabel, both trainee nurses in their early twenties. Situated six miles seaward from Adelaide on St Vincent Gulf, the old port of Glenelg supported a well-established Church of England school, which some of the children possibly attended. May, the eldest, was fourteen years old in 1892, George eight and Reg six.

Eggie was aged eleven and Bert nine when Nora left. Mrs Burton probably looked after them while Wragge was on fieldwork for two months in 1892, three months in 1893, and four weeks in 1894. In July 1894, *The Telegraph* reported the boys had sung sea shanties during a soiree at Capemba for a visiting Melbourne theatre company. At 11 pm, Wragge got them out of bed and onto the Chinese lantern-lit front veranda to farewell the Brough and Boucicault Comedy Company with *We're All Bound to Go* and *Goodbye, Lady*. He joined in on an autoharp, to wild applause.[259]

---

259   *The Telegraph*, Brisbane, July 18, 1894, p6, July 23, 1894, p5

Newspaper mentions of Eggie in 1894 and '95 suggest he was a precocious eldest son. An *Australian Star* journalist who visited Capemba in May 1894, praised his grasp of weather laws, and his initiative in alerting schoolmates to Epsilon, one of Wragge's early Greek alphabet cyclones. In October 1895, his accidental wounding of a friend while duck hunting in the Brisbane River's Indooroopilly Pocket was widely reported. Theodore Koch, 16, survived being hit in the chest with 30 pellets of duck shot from Eggie's double-barrelled muzzle-loader. Identifying him as "Young Clement Wragge" — the given name his family never used — *The Telegraph* alleged that after taking a pot shot, Eggie, while turning to Koch, had somehow discharged the other barrel. Toowong doctor Wilton Love, who extracted the pellets and staunched the bleeding, pronounced Koch lucky his lungs were intact.[260]

Wragge lobbied in 1892 for a trip to the Chicago World's Fair in 1893. He was strongly supported by William Sheffield Paul, a Central Queensland MP convinced the world would benefit from his attendance. Wragge had announced, in June 1892, his appointment to an advisory council for the meteorological congress being organised for the World's Columbian Exposition in Chicago. The exposition celebrated the 400th anniversary of landfall in the New World by Christopher Columbus in 1492. William Paul assured Postmaster-General Theodore Unmack his advocacy was based on Mr Wragge's great abilities and that he had not, as Unmack scoffed, been swayed by "the ravings of a civil servant".[261] But Unmack discounted any reflected glory for

---

260 *The Australian Star*, Sydney, May 26, 1894, p7; *The Telegraph*, Brisbane, October 21, 1895, p4

261 *Morning Bulletin*, Rockhampton, April 22, 1893, p7

Queensland and ruled, in July 1892, the colony could not afford another absence by its highly regarded meteorologist. Besides, he had other plans for him. In an official minute on October 23, 1892, he asked Wragge to get on with publication of reliable seasonal forecasts using relevant data from India and China and to consider calling an inter-colonial conference on the subject. In his opinion, seasonal forecasts would be of much greater value than the present publication of daily forecasts, weather charts and hurricane warnings.

In 1952, Wragge's old pupil Inigo Jones recalled Unmack had been excited by a German report on Eduard Bruckner's study of cyclical fluctuations in Caspian Sea water levels. Bruckner's book, *Climate Change Since 1700*, published in 1890, correlated records of maxima and minima sunspot activity with evidence of a 35-cyclical pattern of wet and dry years.[262] But Jones — whose own long-range forecasting practice was grounded in Bruckner's work — overlooked, when recalling his mentor's seasonal forecasting genesis, the contentious influence of freelance meteorologist Edmund Douglas Archibald (1851–1913).[263] On October 1, 1892, Archibald, an ex-Bengal Education Department mathematics lecturer and a three-day weather forecaster for *The Times* of London, wrote to *The Queenslander* rating Wragge's daily forecasts as relatively useless and arguing the colony's need for reliable seasonal forecasts. India, the nursery of all that had been

---

262 Jones, I, *My Seventy-Seven Years in Queensland*, from an address delivered at the monthly meeting of the Historical Society of Queensland, April 24, 1952, pp 694–95

263 *Quarterly Journal of the Royal Meteorological Society*, January 1914, obituary of Mr E Douglas Archibald, https://rmets.onlinelibrary.wiley.com/doi/abs/10.1002/qj.49704016909, accessed October 10, 2022

achieved by Englishmen in meteorology, already had a specialised long-range forecasting department — why not Queensland? Archibald said he had been told often during a two-year tour of Australia and New Zealand that daily forecasts were really only useful for coastal mariners. Agriculturists craved weekly, monthly and seasonal predictions. He recommended steps to enable seasonal forecasting in Queensland, including an adaptation of the cyclical theory he had proposed with physicist Professor Samuel Alexander Hill, of Allahabad University, and confirmed by pioneering Indian Meteorological Department director Henry Francis Blandford. On October 24, reporting Unmack's directive, Brisbane's *Telegraph* newspaper noted Wragge had already advised the minister he lacked the time, data and manpower to proceed with seasonal forecasting.

In November 1892, Wragge left on a tour of the colony's remote north and north-west, loaded with barometers for new observatories on the Gulf of Carpentaria and planning to resume his study of Aboriginal body temperatures: "In accordance with instructions issued for the late Congo expedition ... laid down by the Geographical and other learned societies."[264] His fame was growing through the Gulf country. The *North Queensland Register* reported in January 1893, his visit to a pub on the Croydon–Normanton tramline. Told after his departure that she had just pulled a drink for Inclement Wragge, the publican was said to have exclaimed, "Oh, by the Holy Moses, if I had known that, I would have got him to tell my fortune."[265]

---

264   *The Queenslander*, November 19, 1892, p994
265   *North Queensland Register*, Townsville, January 25, 1893, p7

Chapter 12   The Mystical Potter, 1891–92

In late January 1893, after 10 weeks away, Wragge headed for home in the storm-scarred steamer *Buninyong*, which anchored on the rising Brisbane River on February 2. He disembarked near Kangaroo Point, on the eve of a month of consecutive floods in the Brisbane River, described by a Sydney newspaper as one of the greatest natural visitations known to white Australia.[266]

---

266   *The Sydney Morning Herald*, February 13, 1893, p5

CHAPTER 13

THE BUNINYONG HURRICANE, CYCLONE BETA AND BEYOND, 1893-97

It's a good job for Clement L he did not make his appearance before the nineteenth century, or he would have stood an excellent show of having been hanged ... for possessing qualities foreign to the human race generally; in fact, there are some in this enlightened age who regard him with something akin to fear and give him an unchristian blessing when they remark his latest triumph. — *The Worker*, official journal of the Associated Workers of Queensland, Brisbane, November, 1898.

Surviving the "Buninyong Hurricane" energised Wragge. In the midst of the storm, on January 31, 1893, he wrote in his journal: "The longing has often been great to personally observe the conditions attaching to the centre of a tropical hurricane. That wish has to-day been met, and sufficiently so."

The deluge lasted 15 hours as the *Buninyong* steamer threaded south through the jumble of islands and reefs south-east of Mackay. On January 31, Wragge, who had boarded in Townsville two days earlier, watched the glass drop from 29.091 to 28.64 between 2 am and 1 pm. By then, the wind was gusting to 100 miles per hour. The south-eastward swell seethed with billows "awful but truly grand", as he and two other travellers reached the saloon through blinding rain for lunch. In his log later published in the *Brisbane Courier*, he recalled waves bashing the ship's hull like dynamite blasts. He praised skipper Thomas Richardson's navigation through the rocky islands north of Keppel Bay with a smashed compass, no starboard anchor, helm hard a-port and barely a ship's-length visibility. Wragge had been through bad weather in past voyages, but his experience aboard the *Buninyong* capped the lot and "beat Ben Nevis hollow".[267]

Richardson called it the worst weather of his life. One of his old crewmen declared it nastier than a storm he survived in the Bay of Biscay in 1866 when the overloaded Melbourne-bound *SS London* sank and hundreds drowned.[268] Interviewed 20 years later, Richardson believed the storm's legend as the Buninyong Hurricane was entirely because of Wragge's presence. "Are you a believer in Clement Wragge?" a reporter asked him. "Indeed, I am," he replied. "I have proved over and over again the value of his forecasts, and I have profited by his warnings. There is scarcely a shipmaster I know who does not hold similar views." Richardson praised Wragge for having issued shipping warnings some days beforehand. But in fact, his deputy Archie Anderson put out an alert on January 22, warning of "somewhat suspicious" falling barometric readings on the northern coastline. Anderson telegraphed all weather stations south of Mackay to hoist storm

---

267 *The Brisbane Courier*, February 22, 1893, p7

268 *The Brisbane Courier*, February 22, 1893, p7; Wikipedia, *SS London* (1864), https://en.wikipedia.org/wiki/SS_London_(1864), accessed October 10, 2022

signals — a drill the bureau had recently introduced with help from the Brisbane Portmaster, Captain Thomas Almond.[269]

* * *

Brisbane's winding river swamped the city three times that February, as remnants of the hurricane swirled into a monsoonal siege. Wragge's staff at the General Post Office measured 40 inches of rain that month. An estimated 7000 of Brisbane's 100,000 residents fled their homes. The 35 dead included seven Ipswich miners who drowned when a tributary flooded their colliery. Two bridges collapsed and the botanic gardens were ruined. In Toowong, down the ridge from Capemba, the torrent uprooted and crushed many of the wooden houses rebuilt since the 1890 deluge. An evacuee wrote in Sydney's *Daily Telegraph*, after the first of the floods on February 4 and 5, that the disaster was unrivalled in suddenness and the terrible extent of its destruction.

Inigo Jones, who resigned from the Chief Bureau in 1892, observed the downpour at his parents' farm on the headwaters 50 miles north of Brisbane. His measurement for February 3 — 35.7 inches in 24 hours — is still recognised as an Australian record for any recording station. The first flood peaked on February 4 at a near-record 27 feet 5 inches above mean low-tide level, measured on the post office gauge. Total damage was said to be £2 million. The second, on February 12 after further heavy rain, was relatively

---

269 *The Mail*, Adelaide, October 11, 1913, p8. Thomas Richardson was, by then, a senior shipping pilot at Semaphore, Adelaide. *The Brisbane Courier*, January 23, 1893, p5, Archibald Anderson, assistant Chief Weather Bureau, reported storm signals had been lately inaugurated by Captain Almond and Mr Wragge.

harmless at seven feet above low tide level. The third, on February 19, peaked at 26 feet 6 inches. In 2020, a retired Queensland meteorologist, Jeff Callaghan, noted the three successive tropical cyclones that caused the floods had been among 24 destructive weather systems that flayed Australia's sub-tropical coast between 1883 and 1898. Callaghan, former head of the Bureau of Meteorology's severe weather section, found 19 of these events occurred during positive phases of the Southern Oscillation Index, in other words in La Niña years. Callaghan told me that the wild weather of the 1890s was surely the worst in Australia's recorded history and a challenge for Wragge in his first appointment. More than 200 lives were lost in southern Queensland and the northern rivers districts of New South Wales.[270]

I've found no evidence Wragge correlated the colony's rainy years with any hypothetical weather cycle as requested by Theodore Unmack the previous October. There's hardly anything on the hurricane and floods in his memoir notes. In early March 1893, Wragge told the *Sydney Morning Herald* he regretted his forecasting during the wild weather had been hampered by telegraph failures and pledged to continue enlarging Queensland's network of observers. For weeks, his sole insight into fluctuating barometric pressure came from the government weather station on Wickham Terrace and his private observatory at Capemba. The paucity of offshore observations was his greatest frustration. He said he had recently asked skippers of Australia-bound vessels

---

270 Callaghan, Jeff, 'Extraordinary sequence of severe weather events in the late-nineteenth century', *Journal of Southern Hemisphere Earth Systems Science*, 2020, 70, pp 253, 268, 269, https://doi.org/10.1071/ES19041, accessed October 10, 2022; *Daily Telegraph*, Sydney, February 8, 1893, p5; Personal correspondence, Jell Callaghan, September 22, 2022

Chapter 13 The Buninyong Hurricane, Cyclone Beta and beyond, 1893–97

to share their meteorological logs to help his reconstruction of weather patterns.

The *Herald*'s Brisbane correspondent introduced him to Sydney readers as a kind of Renaissance Man weather prophet — wiry and red-bearded with deep and penetrating eyes. "He is determined to succeed, his enthusiasm is genuine," the reporter said of his mission to perfect storm-warning meteorology. The *Herald*'s man, who called unannounced at Capemba, found Mr Wragge unguarded and unpretentious. Before he could introduce himself, he had offered him a smoke. "We lit up and, as men will do, smoked in silence for a minute or two," the reporter wrote. He continued:

> I was wondering how to begin and Mr Wragge as yet had not the slightest idea who I was or what I wanted. "We take a great interest in your predictions in Sydney, Mr Wragge," I started ... His thin face lighted up [realising] ... I was only a newspaper man and not one of the local grumblers who had been wearing the life out of him in the previous few weeks.[271]

Six weeks later, an *Australian Town and Country Journal* writer acknowledged Wragge as "Australia's meteorologist" — either in deference to his self-belief, or as a statement of fact. It was well known that Todd, Russell and Ellery were principally astronomers and that New Zealand's Sir James Hector was a geologist. Pen-named "The Tourist", Wragge's interrogator was possibly John Plummer, a prolific, veteran journalist and

---

271   *The Sydney Morning Herald,* March 1, 1893, p3, reprinted in *New Zealand Times,* April 19, 1893, p4. Wragge complained in his weather notes, published in *The Brisbane Courier* on February 18, 1893, p5, "We have received no information whatsoever from our valuable observatories in the southern colonies ... by reason of telegraphic information."

one-time Thomas Cook & Son guidebook writer. There's a tone of mild amazement in the writer's account of four hours at Capemba after tramping with his host along Toowong's rutty streets and up the Taringa ridge. Plummer was then 62 years old and a staffer on the *Australian Town and Country Journal*. He had grafted his way from poverty in London to gentility in New South Wales, and, like The Tourist, seems not to have shared Wragge's Old Country homesickness. After reporting his host's helter-skelter precis of sundry world travels, The Tourist wrote disbelievingly, "On referring to the mother country, he says he must pause, as he loves it so, and longs to be in it once again."[272]

* * *

In the two years between the floods and Nora's return with the children, in March 1895, Wragge revisited New Caledonia to open a second weather station, invested the annual saga of the Wet with a cast of cyclonic sprites, and plotted mountain-top adventures in Tasmania and New South Wales. He also began giving popular lectures on meteorology and writing occasional homilies for the Brisbane press.

The first of these articles, published in July 1893, cast God as the "Great Master of Evolution and the Great Ruler of the Kosmos". He questioned how any scientist could sustain atheism

---

[272] *Australian Town and Country Journal*, Sydney, April 15, 1893, pp 19–20; Stewart, Ken, 'Plummer, John (1831–1914)', *Australian Dictionary of Biography*, National Centre of Biography, Australian National University, http://adb.anu.edu.au/biography/plummer-john-8066/text14075, accessed October 10, 2022, published first in hardcopy, 1988

in view of the certainties of the fixed laws of Nature.[273] He saw Nora once in that time, in June 1894, when in Adelaide on what the press called urgent private business. His memoir notes record the visit but not its reason. He also caught up with his friend the amateur meteorologist William Russell, who had agreed to send monthly readings from his seaside observatory at Semaphore. The Orkney-born Russell was, by then, a co-trustee of the contentious marriage settlement. They met soon after the recent near-collapse of the Queensland National Bank and with public service retrenchments looming. Sections of regional Queensland viewed their Weather Prophet as a luxury in straitened times. In July 1893, the Townsville-based *North Queensland Register* expressed surprise that Wragge's Chief Bureau had weathered the colony's recession:

> As we can't stop the cyclones, floods etc that Mr Wragge is constantly supplying, the citizens are little better off for his information. A weather prophet is a luxury which Queensland can scarcely afford and the Meteorological Department might well close, pending the close of the century.[274]

In July 1894, the government — led again by Sir Thomas McIlwraith — slashed salaries of senior public servants to save £40,000 to £50,000. Wragge lost his £50 house rent allowance, in effect cutting his salary from £400 to £350. By then, Theodore Unmack had returned to commerce, having lost the seat of Toowong to Labor candidate Matthew Reid in Queensland's election in May 1893. His successors as Postmaster-General,

---

273 *Brisbane Courier*, July 31, 1893, p7
274 *North Queensland Register*, July 5, 1893, p7

solicitors Walter Horatio Wilson (1893–94) and Andrew Joseph Thynne (1894–97), lacked Unmack's zeal for seasonal forecasting.

* * *

In February 1894, Wragge introduced *Brisbane Courier* readers to Cyclones Beta and Gamma, the first of his named Coral Sea disturbances. Beta petered out near Lord Howe Island, and Gamma dispersed harmlessly north of Fiji and east of the New Hebrides. He detected both using barometric data from the observatory he set up at Gomen, New Caledonia, in October 1893. Completion that month of a French-built undersea telegraph cable from New Caledonia to southern Queensland ensured his access to new data for Coral Sea forecasting. The Greek names, he explained in his *Courier* column, were in anticipation of needing a systematic way of recording future tropical disturbances. In April, he began using Hebrew alphabet letters to identify storms of Antarctic origin, known as V-shaped depressions. He detected them with observations from Cape Leeuwin — Western Australia's most south-westerly point. The Western Australian government opened the Cape Leeuwin observatory in February 1894, before building the promontory's famous lighthouse, finished in 1896. The weather station was connected to the transcontinental East-West Telegraph Line, which, by a reciprocal arrangement, also brought Wragge data from the Perth, Albany, Ashburton and Eucla observatories. In June 1893, Western Australia's Chief Harbourmaster had asked him to wire his forecasts for the colony and surrounding oceans, which he did free of charge.²⁷⁵ The closest weather station west of Cape Leeuwin was Cape Town, 5000 nautical miles distant, hence Wragge's occasional advocacy for a station in the Kerguelen Islands, 2500 miles closer to Fremantle. The fanciful names he gave to the storms that annually raked the Southern Hemisphere's mid-latitudes alerted readers to Antarctica's poorly understood impact on Australia's weather.

In a study of 1891's weather, the New South Wales Astronomer Henry Chamberlain Russell found wide variations in the tracks of that year's 42 east-bound high-pressure systems and associated V-shaped lows. The highs were generally closer to the equator in winter and further south in summer, ranging from 27 to 38 degrees south. In June 1891, abnormally low-tracking systems coincided with three-times-above-average rainfall in Sydney.

In 1896, Russell's acolyte Henry Ambrose Hunt analysed the transformation of V-shaped depressions — also called narrow tongues of low pressure between adjacent highs — into cyclonic storms, driven by violent polar winds.²⁷⁶ More than a century later, in 2017, Australia's Bureau of Meteorology hypothesised that the drought in southern states that year might have been linked with a tendency for Southern Hemisphere mid-latitude weather systems to circulate further south of the continent than the previous norm. The bureau's research suggested global warming had pushed winter weather south and diminished the influence of cold fronts

---

275   *The Queenslander*, June 24, 1893, p1193
276   Russell, H C, 'Moving Anticyclones in the Southern Hemisphere', *Quarterly Journal of the Royal Meteorological Society*, Vol XIX, No 85, Jan 1893, reprinted in Abercomby, R, (ed) *Three Essays on Australian Weather*, Sydney, 1896, pp 1–4; Hunt, Henry A H, 'Types of Australian Weather', in *Three Essays on Australian Weather*, Sydney, 1896, p87

and low-pressure systems compared with the climate of the mid-20th century.[277]

By mid-1895, Wragge had exhausted all the letters of the Hebrew and Ethiopian alphabets on Southern Ocean disturbances and began mining the Illyrian alphabet — an ancient Eastern European hybrid with a mixture of Latin and Greek letters. In 1897, writing by invitation in the literary journal *Antipodean*, he said he had an immense stock of names for future use.[278] Before his innovation, natural disasters were sometimes remembered by a nickname. For example, the "Royal Charter Storm" commemorated the steam clipper wrecked in the Irish Sea in 1859 with a death toll of 400 that spurred Captain Robert FitzRoy's reform of British meteorology.

* * *

On March 7, 1895, Nora and the children returned to Brisbane, having coincidentally travelled on the *Buninyong* between Adelaide and Melbourne. They reoccupied, by then, a shrine-like dwelling. Wragge told a Scottish reporter in 1896 that his compulsive hoarding of Aboriginal artefacts began while waiting for transport between remote weather stations. His collection would someday be of great value to anthropological science, since Queensland's first people were dying out, he

---

277 'Subtropical Ridge Leaves us High and Dry this June', Australian Government Bureau of Meteorology, issued July 2017, http://www.bom.gov.au/climate/updates/articles/a025.shtml, accessed October 10, 2022

278 *Antipodean*, No 3, Christmas 1897, p87, https://nla.gov.au/nla.obj-52817616/view?partId=nla.obj-96779214#page/n170/mode/1up, accessed October 10, 2022

asserted. [279] In 1898, replying to critics of his long absences on fieldwork, he wrote that he busied himself while his horses were spelling by meeting and studying Aboriginal people. His research included photographing them, measuring their heads, taking their temperatures, recording vocabularies, and also as many proper names as he could muster to give to his tropical storms.[280] In his 1898 weather almanac, quoted by *The Bulletin* in Sydney, he gave the following bald precis of his dealings with Aboriginal people: "To secure an offending 'blackfellow' and thereby teach him a lesson, tie the thumbs and little fingers together, the hands being back to back behind him."[281]

In September 1894, a visiting *Sydney Mail* journalist found Capemba's verandas, dining room, drawing room, hallways and corridors packed with Aboriginal, South Sea Islands, Middle Eastern and European artefacts. Three of the outer walls were shaded by deep verandas protecting what the writer called "a remarkable collection of Australian Aborigines' weapons of offense". The front door was framed by a *chevaux de fries* of spears and the dining room bristled with weapons, dilly bags, fish spears and yam sticks. [282]

Three months earlier, another Sydney paper, *The Australian Star*, judged his weapon collection as one of the finest in the colony. While the *Sydney Mail* presented him as a True White Man living a fine bachelor's life, the *Star* acknowledged he had eight children — five boys and three girls — and that he was proud

---

279 *Glasgow Herald,* November 23, 1896, p8
280 *Brisbane Courier,* February 22, 1898, p2
281 *The Bulletin,* Sydney, April 16, 1898, p11
282 *Sydney Mail,* September 8, 1894, p491

of his eldest boy's knowledge of meteorology. Wragge told the *Star* he abhorred conventionalities and that in studying natural phenomena he trusted in nature's God.[283]

Wragge's gregarious portrait in the *Sydney Mail* corroborates Brisbane reports of his doings, beginning in July 1894 with the previously mentioned Brough and Boucicault Comedy Company farewell. In December 1894, *The Queenslander* confided he had responded with his usual hospitality to a surprise party at Capemba organised by a number of gentlemen. In February 1895, a fortnight before Nora's return, he was reported to have guided a group of cricketers from England, New South Wales and Queensland through his garden and museum. Later that month, the *Telegraph* reported his presence at a wine party in Brisbane's Centennial Hall, thrown by the Sydney bookmaker Humphrey Oxenham. The Postmaster-General Andrew Thynne was present too, as chairman of the city's Tattersall's Club. Oxenham, who was celebrating his 40th birthday and also the recent opening of a Brisbane branch of his business, praised Wragge's forecasting renown in the southern colonies.[284]

The flow of visitors continued after Nora's return. In early April 1895, Wragge escorted members of the Brisbane Natural History Society through his gardens, showed off his collections and served them afternoon tea. They stayed until the evening was far advanced in order to view the solar system through his equatorial telescope, the *Courier* reported.[285]

---

283 *The Australian Star,* May 26, 1894, p7

284 *Queensland Times,* March 5, 1895, p3, citing *The Telegraph,* Brisbane, February 25, 1895

285 *Brisbane Courier,* April 9, 1895, p4

Chapter 13   The Buninyong Hurricane, Cyclone Beta and beyond, 1893–97

* * *

Between May 1895 and February 1897, Wragge opened an observatory on Mount Wellington — Tasmania's highest peak — spent eight months inspecting 30 or so north and north-western Queensland weather stations, visited Europe, Britain and the United States, and befriended Edris on the voyage from Hawaii to Sydney in January 1897. He was away from Brisbane for 17 months in all, entrusting the Chief Bureau's reputation to his experienced assistants. Archie Anderson had been his right-hand man since 1887, Edgar Lambert Fowles joined him in 1888, and both his juniors George Grant Bond and Percival George O'Mahoney in 1892.

From 1898, Wragge broadened his naming of tropical and Southern Ocean disturbances to include Aboriginal and Biblical words and random historical and contemporary personalities such as 14th century English theologian John Wycliffe and the Boer commander General Louis Botha. In his absence in February 1897, Anderson applied the last of the Greek letters — Omega — to a tropical disturbance that skidded harmlessly down the Queensland coast. Wragge was away in January 1896, when his deputy named and did his best to give early warning of Cyclone Sigma, which drenched Townsville's floodplain with 20 inches of rain in 24 hours and caused 17 drowning deaths. His critics said his wet-season absences were unfair to Anderson.

In January 1896, Wragge had been away re-establishing a burnt-out weather station at Boulia in western Queensland and, in January 1897, he was at sea, nearing the end of six months of leave for an international meteorological conference in Paris. But he was on duty in early March 1899, to unwittingly give Queensland's most deadly cyclone a feminine Tahitian name, Mahina. "Mothers will agree that no infant daughter can bear a softer or prettier name," he wrote in the *Courier* on March 7, unaware at least 300 people died on March 4 and 5 when Mahina flayed the pearling fleet at remote Princess Charlotte Bay, on Cape York, nearly 1100 miles north.

* * *

In early February 1897, Wragge returned excitedly to Australia from his latest European conference, with, he boasted, international affirmation for his dream of opening a weather station on Mount Kosciuszko.[286] A *Daily Telegraph* journalist, who met his ship in Sydney, wrote he had insisted on an interview, although it was just after midnight: "Mr Wragge sat up in his bunk in dishabille ... and chatted on the successes of his tour," the *Telegraph* reported in a late edition. Wragge said he would call at the Sydney Observatory to set straight his New South Wales counterpart H C Russell on his view Australia had no need for any new high-level meteorological data. This meeting, confirmed by Wragge five weeks later in a letter to *The Sydney Morning Herald*, reinforced their differences. Russell judged existing High Country readings from Kiandra (4600 feet above sea level) were adequate for his purposes and rubbished Wragge's Kosciuszko scheme as costly and pointless. He wanted no part of it and ruled

---

[286] NAA: J1, Q400/1/22, Wragge, circular seeking sponsors for "a tentative high-level station on the summit of Mount Kosciuszko", Brisbane, April 30, 1897, "My proposal has been warmly supported by my eminent colleagues in Europe and by the foremost physicists of the day, including Lord Kelvin and Professor Tait."

out involving any of New South Wales' sea-level observatories in the project.

This forced Wragge to set up independent, low-level stations in Merimbula on the New South Wales south coast and, for a brief time, in Sydney and the Victorian town of Sale.[287] His parrying with Russell over Kosciuszko continued in letters to the Sydney press that spilled into exchanges over other hobbyhorses — Wragge's pride in the Ben Nevis observatory and Russell's prediction of widespread rain in 1897 and '98 based on his controversial 19-year weather cycle. In addition to Russell's official forecasts, the *Daily Telegraph* ran a daily precis of the Brisbane bureau's predictions for New South Wales. In early April, that newspaper scolded Wragge for calling Russell's cyclical theory a failure because of continuing drought conditions:

> Controversy between meteorologists is, no doubt, valuable to the science, and thus universally beneficial ... but such discussions can best be carried on discreetly. To endeavour to conduct them, or incidentally make a point at the expense of some disputant, in connection with forecasts, is to deprive those forecasts of the uncontroversial, business-like manner that is essential to them.[288]

Disquiet over the squabble was echoed in a letter to *The Worker* newspaper in Brisbane — the official journal of the Associated Workers of Queensland. An anonymous correspondent

---

287 *The Daily Telegraph*, Sydney, February 9, 1897, p6; *The Sydney Morning Herald*, March 18, 1897, p7; National Archives of Australia: J1, Q400/1/22, Wragge, memo to Under-Secretary and Superintendent of Post and Telegraph Department, Brisbane, September 6, 1897

288 *The Daily Telegraph*, Sydney, April 8, 1897, p4

condemned Wragge's sneers at southern scientists and argued the government must rein him in. "It would appear he can do as he darn well pleases." After all, the Brisbane office seemed to function perfectly well during his absences — travelling the length and breadth of the colony collecting curios for his private museum and attending old world conferences with a tour through America thrown in.[289]

* * *

In mid-May 1897, Brisbane church leaders and the rabbi of the city's synagogue held special meetings to pray for rain. Brisbane Ministers' Union members said their gathering had been spurred by despair over widespread drought through eastern Australia. The *Brisbane Courier* reported participants at Albert Street Wesleyan Church on May 17 had, before pleading for help, begged forgiveness for ungratefulness of God's past favours. In the preceding two months, government-sanctioned days of humiliation and prayer for rain were also held in New South Wales, Victoria and South Australia. Today, historians see the dry spell that began in January 1896 — with heatwaves, dust storms and bushfires — as the first innings of the notorious Federation Drought. In April 1897, *The Age* in Melbourne reported nearly all back-country selectors were carting water. The Lachlan River was said to have stopped flowing in the Condobolin district of New South Wales.[290]

---

289 *The Worker*, Brisbane, May 8, 1897, p7
290 *The Daily Telegraph*, Sydney, April 10, 1897, p9; *The Argus*, Melbourne, April 27, 1897, p4; *The Daily Telegraph*, Sydney, May 1, 1897, p1; Brisbane Courier, May 11, 1897, p4; *The Age*, April 8, 1897, p6

Wragge scorned praying for rain. In a postscript to his newspaper forecast on the day of the Brisbane gatherings, he wrote the worst drought-affected regions were "subject to a rigid law in meteorological physics, operating unalterably", and relief was hence unlikely. In subsequent reports through May and June, he described seasonal conditions as most unusual — akin to what might be expected in late spring and early summer months. It was evident to him that some abnormal influence had operated over south-eastern Asia and the neighbourhood of the Malay Archipelago to bring about this change. In 1898, *The Bulletin* quoted his view on "puerile supplications":

> Since we cannot in meteorology give ... [checkmate] in three moves to the Master Weather Man, we look on with resignation, and endeavour to learn yet more at His hands, instead of indulging in puerile supplications, which could be of no more avail than if we asked Him to make the earth rotate the other way.[291]

In July 1897, perhaps prompted by newspaper chatter on "Acts of God", Wragge began writing articles for the Brisbane press to inspire educated weather-watching.[292] His author of the science and poetry of meteorology — the "Eternal Dynamo" — was the antithesis of the Brisbane Ministers' Union's "cajolable Jehovah". He seems to have written the first of his Meteorological Pictures after a religious journal scolded his lack of faith in rain prayers and gall in inventing an "Infinite Dynamo". Reporting these criticisms, a Brisbane journalist commented drily on Wragge's "high falutin talent" for finding scientific poems between the loops

---

291 *The Bulletin*, Sydney, November 12, 1898, p13
292 *Brisbane Courier*, July 14, 1897, p7

and isobars of his weather charts. The unnamed Brisbane stringer for the biweekly *Western Star and Roma Advertiser* wrote: "As a good general rule, I would say that the more these things point to coming rain, the more poetically will they strike the feelings of Australians."[293]

In the first of his Meteorological Pictures, Wragge declared anyone with an eye to the sky could learn the signs of Australia's procession of highs and lows. For example, on a bright, sunny morning in Brisbane, factory chimney smoke tearing away in a long black wreath from the west was a sure sign of the transit of Antarctic depression over the Tasman Sea. And Brisbane's westerlies heralded an approaching high-pressure system, its south-eastern edge colliding with the north-western edge of an Antarctic low with inevitable condensation of vapours. He then launched into how this knowledge carried him into an almost out-of-body realm, imagining snow-capped Kosciuszko:

> The snow driving and tearing over the Australian Alps, under the steep atmospheric slopes; steamers rounding Gabo to westward labouring in the teeth of strong head winds, with their sea-sick passengers, and note the rain and slush of the Melbourne streets with their turbulent gutters.[294]

In August 1897, he was grounded by a recurrence of dengue fever and what the *Courier* called congestion of the liver. But he pressed on. In the next of his weather lessons, he took readers on anti-clockwise flight over the continent, viewing the balmy green waters of Torres Strait, the saltbush plains of South Australia, the

---

293 *Western Star and Roma Advertiser*, Roma, July 17, 1897, Brisbane Notes, p3
294 Brisbane Courier, July 15, 1897, p6

sleet-lashed summit of Tasmania's Mount Wellington and back to Brisbane, "well satisfied with [this] experience of the grandeur of the atmosphere". He declared the certainties of atmospheric physics as proof of a governing Eternal Dynamo.

Brisbane's afternoon daily, the *Telegraph*, indulged him, conceding his poetic gift, but asking why readers had to endure forecasts invariably 24 hours out of date. "Why can't you get your forecasts to us by 3 pm for immediate publication?" the paper pleaded.[295] *The Queenslander* advised him to restrain his gush or risk the scorn of southerners: "Already the newspapers down that way are beginning to make fun of his work, and nothing is more dangerous to a public man than public ridicule."[296] A month later, in Melbourne, the *Australasian* flayed his forecasts as generally only useful for sailors, not the ordinary people of Australia: "When he gets the chance, Mr Wragge goes off into hysterics ... this is the inane stuff that is put before the Brisbane public as scientific meteorology."[297]

In December 1897, the *North Queensland Register* parodied his alert published in Brisbane a fortnight earlier for the stultifying impact on inland Australia of an approaching monsoon he'd named Ithiel — an Old Testament name for "the words of God". He forecast intense heat in western Queensland, with sandflies and mosquitoes west from the Diamantina rivalling the plagues of Egypt, while Barcoo rot would make travellers long for the more genial conditions of the Pacific coast.[298] The Barber, a *Register*

---

295 *Telegraph*, Brisbane, October 12, 1897, p4
296 *The Queenslander*, October 30, 1897, p834
297 *Australasian*, Melbourne, November 20, 1897, p36
298 *Brisbane Courier*, November 17, 1897, p3

columnist, contended that, in fact, Wragge was no better at tipping the weather than the average Aboriginal rainman. The writer lampooned Wragge's forecasts with asides from an imaginary rainman dubbed "King Billy":

Wragge: Look out for certain drought, Ithiel gives us warning.
King Billy: Big ants have all come out, my word, rain heavy in the morning.
Wragge: Excessive heat anticipated soon, be careful of a fire.
King Billy: One white ring around the moon say white man awful liar.[299]

In April 1898, Wragge announced in the *Courier* he would in future spare the public the poetry of isobars and confine himself to meteorology pure and simple. *The Bulletin* in Sydney reported he had given in to press criticism, and speculated how long he would conduct himself like a good, staid, respectable weather prophet, "plain, scientific, dull and uninteresting". A couple of months earlier, *The Bulletin* had defended him against Queensland editors' distaste for his use of the royal *we*. "Surely, Mr Wragge had equal entitlement to this quirk as the owner of the *Banner of Bugaboo* or the *Clarion of Dead Dog*. He was, after all, a most plural sort of person — prophet, poet and meteorologist combined."[300]

Meanwhile, news of his enigmatically named weather systems

---

299 *North Queensland Register*, December 1, 1897, p5. In recollections published in the *Chronicle*, Adelaide, September 29, 1938, p53, the North Queensland journalist Frank Reid recalled Wragge had often mentioned in his lectures a famous Aboriginal rainmaker, King Billy of the Kalkadoon tribe in the state's north-west.

300 *The Bulletin*, Sydney, January 29, 1898, p21

reached North America. In May 1898, a newsagency report, published in US regional papers, introduced the Coral Sea cyclone he'd named Sana, "the storm empress of the Pacific", rushing westward from New Caledonia "to perform her marvellous evolutions and dance her wildest step".[301] That month, only weeks after reining himself in, he began writing a 10-part series titled "Meteorologica" in monthly instalments for *The Worker*. He introduced the first as a gift to his "brother bushmen of the Never Never":

> We have no political tenets — ours is the platform of science. All Nature speaks to us of the Infinite Dynamo of Whom you and ourselves form a part — every zephyr, every sand ripple, every list in the trees, every rolling "roly poly bush" has its own tale to tell ... and these and many others we shall as mates investigate ...[302]

He felt an affinity with other bushmen when on his outback rounds, on the wallaby in the spinifex, humping barometers and thermometers between weather stations. In following weeks, he would simply share his knowledge of the weather. *The Worker* published the first instalment on May 23, 1898, and the last — written from the summit of Kosciuszko — on October 22. He explained in the latter his hope that simultaneous high-level and sea-level readings would improve the Chief Weather Bureau's forecasts. He finished with a homily for all Children of the Infinite "with thankful hearts impelled by the life-currents of grand Kosciuszko":

> If we strive for the good, we shall be good, and all the absurd misunderstandings and humbugs of life will melt away leaving

---

301 *Brainerd Dispatch*, May 20, 1898, p1
302 *Worker*, Brisbane, May 23, 1898, 'Meteorologica I', p11

the refined gold of our natures pure and as unalloyed as the crystals of God.³⁰³

In November 1898, Sydney's weekly *Australian Town and Country Journal* praised his democratic impulse in writing for *The Worker*. The *Bulletin*'s main rival in the Australian colonies and New Zealand, the *Journal*, quoted Wragge's invitation to bushmen, bullockies, swagmen and stockmen to cultivate the capacity for intellectual enjoyment inherent in every person in the dawning free nation of Australia. "Think how it would hearten our poor bushmen and all workers generally to be addressed in this strain by a Government meteorologist," the *Journal*'s columnist Rus declared.³⁰⁴

* * *

Wragge's memoir notes on 1895–97 end with "the agony of return to Q" and, in capital letters, "KOSCIUSZKO!" In essence, his meteorological endeavours through those three years culminated in December 1897 with his first barometric observations on Mount Kosciuszko. In official instructions to his first crew of volunteers, he predicted their endeavours would be chronicled "in the science annals of the history of Australasia".³⁰⁵ His drive to replicate his

---

303 *The Worker*, Brisbane, October 22, 1898, p2

304 *Australian Town and Country Journal,* Sydney, November 26, 1898, p34, quoting from *The Worker*, Brisbane, September 17, 1898, p10, 'Meteorologica VIII'

305 National Archives of Australia, NAA: J1, Q400/1/22, 'Establishment of Mt Kosciuszko Observatory by Mr C L Wragge, Government Meteorologist, Official Instructions to Captain Iliff, Mr Bernard Ingleby and Mr Basil de Burgh Newth', December 9, 1897

Chapter 13   The Buninyong Hurricane, Cyclone Beta and beyond, 1893–97

Ben Nevis triumph entwined his vision for a truly Australia-wide weather forecasting service with a relentless, gnawing need to prove himself once more.

CHAPTER 14

BRITAIN, 1900

Mr Wragge is a tall, thin man, and no amount of labour daunts him. He seems to be in communication with all points of the compass. Hundreds of years ago, he would have been accounted a great magician. His weather prognostications are regarded by all seamen as thoroughly reliable and his accuracy is marvellous. — Nat Gould, *Town and Bush: Stray Notes on Australia*, 1896, page 47.

Trooper Allan Cameron knew he'd met a famous man while boarding the Highlands train in Glasgow one October morning in 1900, but not his exact legend. "We shook hands as if we were old friends," he wrote home of the cheerful traveller who gave his name as Clement Wragge. Cameron was a wounded Boer War soldier from Glen Innes, on the Great Dividing Range in New South Wales. Writing to his parents, he assumed Mr Wragge had once been that colony's astronomer.[306] In those days, Wragge was probably better known around the Northern Tablelands than the veteran, Sydney-based astronomer Henry Chamberlain Russell. Their sometimes-contradictory weather forecasts were published in tandem by many country newspapers. In 1898, Wragge was

---
306 *Glen Innes Examiner*, January 18, 1901, p 3 and February 26, 1901, p3

described as nearly six feet tall, lean and blue-eyed with a head of sparse ginger hair, stringy whiskers and typically dressed like a stockman on holidays. He favoured tight old suits for cover from the elements and gigantic Blucher boots for power-walking.[307] In person, he could never have been mistaken for the balding, full-bearded, avuncular and chunky Russell.

Trooper Cameron was a casualty of the South African War. Sent to Britain after being wounded in action with the First New South Wales Mounted Rifles, he craved seeing his father's birthplace in the Western Highlands of Scotland. He and Wragge were passengers for the Lochaber town of Fort William, 100 miles north of Glasgow at the foot of Ben Nevis. During their five-hour journey, Wragge shared all of his excitement in vistas of peaks, lochs and heaths. They shuttled between the carriage windows for panoramic views. "We passed along the banks of Loch Tulloch, and round the foot of the original Ben Lomond ... [and other] wild, grand-looking mountains," Cameron recalled, continuing:

> After stopping at several small stations, we came in sight of Ben Nevis looming through the clouds. Wragge went into ecstasies at the sight of it, as he had been stationed there at the observatory for a long while and had not seen it for years.

On the platform at Fort William, the Scottish migrant's son saw that if their arrival was in any way a homecoming, it was, despite his ancestry, less so for him than his ebullient guide. "Mr

---

[307] Harald Ingemann Jensen diary, 29742, John Oxley Library, State Library of Queensland, from 'Chapter 3, The Kosciuszko Observatory'; *Auckland Star*, December 18, 1926, p8 and March 28, 1935, p6

Wragge seems to be acquainted with lots of people here, and introduced me to a number of them," he wrote. A few nights later he accepted Wragge's invitation to a lantern-slide show at the town hall. Subject matter included the secrets of weather forecasting, Australia's grand Snowy Ranges, and recent astrophotography of the abysses of space. Trooper Cameron wrote home that Mr Wragge had introduced him unexpectedly, and his being Australian and direct from the South African Front had yielded him brief fame. He later credited his recovery from enteric fever to the hospitable people of Lochaber.[308]

* * *

Wragge's jaunt to Fort William in October 1900, was his third trip back to the Scottish Highlands since his adventures on Ben Nevis between 1881 and 1883. On his return to Britain in early November 1891, he hiked up Ben Nevis with Colin Livingston, the Fort William school teacher who, 10 years earlier, had guided his first ascent. It was a sunny, blue-sky late autumn day. A wall of snow shrouded the ruin of his hut, derelict since the opening of the official weather station in 1883. He recognised some of his equipment being used in the main observatory. "Mr Wragge greatly enjoyed his visit to old familiar scenes," a local newspaper reported. "[He] could hardly be induced to leave the summit, where he would wish to spend one night at least, did the time at his disposal make it possible ... darkness was setting in before he could be induced to leave."[309] That excursion was pure rest and

---

308 *The Scotsman*, October 18, 1900, p4
309 *Inverness Courier*, November 6, 1891, p7

recreation after the meteorological conference in Munich that secured his six months' leave.

In 1900, Wragge made his third excursion to the Scottish Highlands after attending the Paris Exposition Universal, a world fair celebrating major achievements of the 19th century. Again, he and Livingston ascended the Ben together, this time in unusually mild weather. He also met Angus Rankin, his meteorological assistant in 1882 who had since succeeded Robert Ormond as superintendent. Ormond was, by then, secretary of the Scottish Meteorological Society.[310] The Paris exposition's international meteorological conference, from September 10 to 16, gave Wragge a platform to promote his Ben Nevis-inspired Kosciuszko project in a paper based on data from the mountain-top as well as Merimbula on the New South Wales south coast. He reported that delegates had snatched up printed copies of his Kosciuszko records and charts, and Scottish Meteorological Society secretary Dr Alex Buchan and other UK scientists had insisted the project must continue.[311] Before returning to England, Wragge joined conference delegates on an inspection of the Dynamical Meteorological University at Trappes, near Versailles, owned and managed by Teisserenc de Bort (1855–1913), a champion of high-level observatories. Wragge glimpsed the future in de Bort's test flights of helium balloons equipped with self-recording instruments for monitoring atmospheric pressure. At £100 per balloon, plus a finder's reward, the cost looked prohibitive. He told the *Brisbane Courier*'s London correspondent that de Bort seemed to have money to burn. By contrast, Wragge had almost

---

310 *The Glasgow Herald*, August 13, 1894, p4; *The Oban Times*, November 28, 1896
311 *The Advertiser*, Adelaide, February 14, 1901, p7

no funds for research. For example, his study of ocean currents hinged on collecting beer bottles, stuffing them with RSVPs in seven languages and, full of hope, tossing sealed bottles overboard from various maritime co-ordinates.

In 1896 and again in 1900, Wragge sought affirmation from other champions of high-level observatories. Critics argued uncertainties of weather forecasting had not been improved by data from Ben Nevis, nor by the twin observatories he had controversially opened in May 1895 on Tasmania's 4140-feet-high Mount Wellington and at sea level in Hobart. Henry Russell and Victoria's Government astronomer Pietro Bararcchi both ridiculed his tandem observatories on Mount Kosciuszko and on the coast at Merimbula. Furthermore, the British government seemed unconvinced of the value of data from the Ben Nevis station. In 1898, it faced closure for want of government funding.[312] Before leaving for Europe in June 1900, Wragge wrote to *The Sydney Morning Herald*, pledging to keep the soon-to-be federated colonies abreast with "the ever-growing and eminently practical science of meteorology". Organised by the International Meteorological Committee, the Paris conference drew representatives from more than 30 countries.[313] But the record of Wragge's official business in France and the UK suggests Mount Kosciuszko loomed over all other interests. He needed to reassure the New South Government its then annual grant of £400 for running expenses of the Kosciuszko and Merimbula observatories was a wise investment.

---

312 *The Times*, London, August 8, 1898, p7
313 *Science*, NS Vol XII, No 308, November 23, 1900, Rotch, A Lawrence, 'The International Congresses of Meteorology and Aeronautics at Paris', p797

On June 25, 1900, Wragge had left Sydney on a circuitous voyage to Europe via New Zealand, Cape Horn, Montevideo and Tenerife,[314] announcing he would study in transit the Southern Ocean's notorious low-pressure systems he called Antarctic V-shaped disturbances. As usual, he packed a swag of empty beer bottles for tracing ocean currents. He planned to gradually lob them overboard trusting some would be returned from far away. Each contained a note written in eight languages giving his return address and details of latitude and longitude at the time of sealing and despatching.

Unmentioned anywhere other than the passenger list for the steamer *Mokoia,* bound for New Zealand, was the presence aboard of his de facto wife.[315] He and Edris had lived together in Brisbane for about a year since his official separation from Nora in May 1899. The *Paparoa* reached England in mid-August. Edris, who was six months pregnant, joined her elder sister, Ida Blackwell, in Tonbridge, Kent.

* * *

On September 28, 1900, back in England from Paris, Wragge wrote to his old rowing friend Queensland Postmaster-General James Drake, urgently seeking an extra six weeks of leave, on half-pay if necessary. He explained the dates of proposed speaking engagements in November and December were later than expected and would delay his return. Drake gave his approval,

---

314 *Evening News,* Sydney, June 26, 1900, p2; *Hawkes Bay Herald,* July 13, 1900, p2, listed "Mr Clement Wragge, the Queensland weather prophet" among the *Paparoa*'s passengers.

315 *Evening News,* Sydney, June 26, 1900, p2; the *Mokoia*'s passenger list

supported by Queensland's Agent-General, Sir Horace Tozer, who booked him to address the Imperial Institute in London on November 26. Drake had previously readily given him six months' leave for the Paris conference. "Colonial life to an Englishman without occasional breaks does not tend to mental development," Wragge wrote on May 5 to Drake, whose department managed Queensland's weather bureau. "I feel that *alone* out here. Except for the conferences, I am liable to become rusty from the lack of personal contact with my colleagues and the great scientific minds of the day." In the same letter, he wrote he hoped to lead the Federal Bureau of Meteorology that he assumed would be set up once the Commonwealth of Australia was formally established on January 1, 1901. He revealed he had sent applications for the as-yet unadvertised position to all colonial premiers, except Queensland's Robert Philp. He wrote: "You *know* that I *can* and *will*, if elected, give Australia a meteorological service equalling if not surpassing that of England and the United States — All I want are appreciation and encouragement."

* * *

In late November 1900, Sir Horace Tozer introduced him to the Imperial Institute in London as a popular and able scientist whose work extended over the whole of Australasia. After his London engagement, *The Echo* newspaper called him "the very policeman of the weather" whose name in Australia was synonymous with "prediction fulfilled".[316] *The Times* reported his pride in observers'

---

316 *The Times,* London, November 28, 1908, p4, *The Echo,* London, December 12, 1900 — reprinted in the *Goulburn Herald,* New South Wales, January 23, 1901, p1

accurate rainfall records, in a report that overlooked Queensland's disastrous lack of rain since 1897. The drought was still Australia's biggest weather story in 1900, especially for British pastoral industry investors. The Scottish Australian Investment Company Ltd put stock losses on its Queensland properties in 1899–1900 at 276,244 sheep, 27,417 cattle and 614 horses. Birdsville, in the far west, recorded 2½ inches of rain for the year to late October, after 4 inches the year before. Even the kangaroos were dying of thirst and hunger.[317] In April 1899, Wragge wrote of having seen, during a tour of south-west Queensland, death and prostration everywhere: "Never have we witnessed such a depressing sight … One run we could name, not 50 miles from Cunnamulla, dead indeed is all the stock, and even has the mulga — that grand standby in times of drought — been all used up … the musty smell of death pervades the air and the buzzing flies are all attention."[318]

In 1900, the Queensland Government was mired in drought-related debt and looking for ways to save the colony's farmers and graziers. Since becoming Premier in December 1899, Glasgow-born businessman Robert Philp had waived payment of pastoral rents to encourage battlers to stay on their leased country.[319]

In October 1900, the Pioneer River Farmers' Association in Mackay asked Lands and Agriculture Minister James Chataway to test a controversial rainmaking method. Canvassed earlier that year in the *Queensland Agricultural Journal* and *Sydney*

---

317 *Dundee Courier*, November 9, 1900, p2, for losses by Scottish Australian Investment Company; *Reading Mercury*, October 27, 1900, for Birdsville rainfall; French, Jackie, *Let the Land Speak*, Sydney, 2013, p277, for comment on kangaroos

318 *The Brisbane Courier*, April 26, 1899, p6

319 Percy, Harry C, *Memoirs of Sir Robert Philp, 1851–1922*, pp 207–08

Mail, the practice entailed bombarding clouds with cannon shot. Orchardists in Northern Italy and Austria reported rain as a side effect of firing at menacing clouds to break up hailstorms. Chataway, who was also Mackay's MP and owner of the *Mackay Mercury* newspaper, referred the question to his department, which in turn delegated it to Wragge. Queensland's Chief Secretary's office telegrammed Agent-General Tozer on December 7, 1900, asking if the roaming government meteorologist could use his remaining leave "to obtain information relative to methods for controlling rainfall used in North Italy [and] Styria [in Austria]".

Official correspondence shows Wragge complied after securing an extra fortnight's leave and payment in advance of £40 for expenses. He agreed to go to Austria when he had finished his December lecturing engagements, and also to postpone by a fortnight his return to Australia by mail steamer. After quizzing landholders in the Austrian province of Styria about *Das Wetterschiessen*, or "weather-shooting", Wragge recommended a Queensland trial. But he cautioned it had been mid-winter in Austria and had hence not viewed the procedure.[320]

The Bulletin in Sydney was unconvinced. In February 1901, the weekly newspaper argued Wragge had naively accepted as fact a rainmaking superstition held only in the backblocks of Europe. *The Bulletin*, also known as "The Bushman's Bible", flayed him as a second-rater who camouflaged his ordinariness with loud self-promotion. His high-level weather station on Mount Kosciuszko was little more than a fad, the revival of an idea that had fallen

---

320 National Archives of Australia, Chief Weather Bureau Brisbane, 'Report on the Steiger Vortex', by Clement L Wragge, June, 1901, Chief Secretary's Office 07218, Department of Agriculture 06908

into comparative disrepute in America and Europe. Now he was grandstanding with yet another gimmick.[321]

Edris gave birth on November 18, 1900, to their son, Kismet Kent Wragge — her first child and Wragge's ninth. On January 8, 1901, she and the baby were reunited with him in Marseilles on the Australia-bound steamer *Oruba*. Four weeks later, they reached Fremantle slightly late having been delayed by a cyclone in the Indian Ocean. Interviewed by the *West Australian* before disembarking, Wragge trusted his pioneering work in inter-colonial forecasting would be recognised in planning for the proposed Federal Weather Bureau. If he had anything to do with the new service, he would make things hum in the West, "[Or]my name is not what it is."[322]

The *Oruba* reached Adelaide on February 12. But he disembarked too late for that night's public reception at Adelaide Town Hall for Edmund Barton, Australia's first Prime Minister. James Drake, soon to become a Queensland Senator and Federal Postmaster-General, was also present. He was well aware of his friend's ambitions, but their subsequent meeting in Sydney on February 18, at Wragge's request, was inconclusive. The Commonwealth Constitution Act empowered Australia's Parliament to "make laws ... with respect to astronomical and meteorological observations", but welding together the remnants of existing colonial weather services looked likely to be time-consuming and complicated.

* * *

---

321   *The Bulletin*, Sydney, February 2, 1901, p15
322   *The West Australian*, February 9, 1901, pp 5–6

Wragge's separation from Nora in May 1899, formalised a long estrangement. Press accounts of his official duties, confirmed by government reports, show he had barely been home in the 18 months before returning from the Paris conference in February 1897. He had previously spent eight months from October 1895 to July 1896 inspecting weather stations on Cape York and in the far west of Queensland, then left for Europe via the United States a month later. His notes on 1897–98 refer to "the agony of return to Queensland". He and Nora had married in Adelaide in September 1877, just before his 25th birthday and a few months after hers. Their two decades-long partnership formally ended with him agreeing to pay maintenance of £20 per month from his annual salary of £500. They decided against divorce, believing the publicity could harm their three daughters and five sons.

Writing in 1921, Wragge gave conflicting dates for when he first met Edris but it seems to have been in early 1897, aboard the steamer *Zealandia* while crossing the Pacific from Hawaii to Australia.[323] He loved sea travel. The lottery of the passenger list suited his gregarious nature. His shipboard ritual allowed time for reading, conversation, music and smoking, as well as weather-watching and star-gazing.[324]

In February 1897, the *Zealandia's* list of passengers arriving in Sydney included a Miss Horne and Dr Emily Ryder, both of

---

323   Wragge's notes, headed 'My Life', written in 1921, place "meeting Ed" [Edris – his nickname for Louisa] alongside his return from the 1896 Paris meteorological conference, via America and Honolulu, Auckland War Memorial Museum Library, MS 1213, Box 6(a) 'My Life', p21. The draft of a letter to his solicitor, also written in 1921, 'Partial Abstract of the Tangle of my Life', puts the date at 1898, during travels abroad.
324   *The Queenslander*, February 27, 1898, p413

whom had boarded in Honolulu. A US-born physician, Dr Ryder was renowned in India, Britain and Australia for her fight to lift the age of consent for Hindu child brides. During two months in Hawaii, she had set up a circle of the Little Wives of India Mission.[325]

Edris had been Dr Ryder's private secretary since 1892, when hired in Bombay. Between 1893 and '95, they promoted their cause throughout the Australian colonies and New Zealand. Typically, they wore saris in their presentations to church and missionary societies and temperance groups.[326] Described by an Australian reporter in 1893 as "a young Hindoo lady", Edris was a native of Karachi, the younger daughter of a recently deceased official of the Indian Civil Service. Her father, Edwin Horne, died six months before her birth in 1864. Four years later, her mother, Mary Mansfield Horne (nee Malvery), married another civil servant, William Kelly. Edris's father and stepfather were both of Eurasian ancestry. Coincidentally, she and Wragge spent their formative years with ageing grandmothers. She resigned from Dr Ryder's employment around October 1898, after helping form a Brisbane branch of the Medical and Legal Mission for the Relief of the Child Wives of India. Dr Ryder, then aged 55, retired to the US a few years later. In early 1900, refuting a gossip columnist's take on his relationship with Edris, Wragge said she had quit the child brides campaign because she needed a rest after

---

325 *The South Australian Register*, Adelaide, November 25, 1896, p5 and March 8, 1897, p5; *Evening News*, Sydney, Tuesday, February 9, 1897, p4; Brainard, Lucy Abigail, *The Genealogy of the Brainerd-Brainard Family in America (1649–1908)*, Hartford Press, 1908

326 *The Express and Telegraph*, Adelaide, Saturday, April 13, 1893, p3; *The Advertiser*, Adelaide, Saturday June 10, 1893, p3; *The West Australian*, August 11, 1894, p4

nearly eight years of travelling. He described her as his housekeeper and assistant in literary and scientific work.

* * *

In the three-and-a-half years between meeting Edris in early 1897 and his extended leave from June 1900, Wragge dovetailed a maze of private projects and public duties. In 1897, he launched his Mount Kosciuszko and Merimbula observatories and sustained them by political lobbying, fund-raising lectures, articles in his just-launched annual *Wragge* almanac, and inspiring the zeal of voluntary observers. He also supervised hundreds of weather observers around the colony during frequent tours by train, steamer and on horseback.

In 1898, he opened Capemba for visits by the Brisbane Ladies' Cycling Club and a series of public astronomical lectures. Between October 1899 and February 1900, when scraping for cash, he sold off hundreds of Aboriginal artefacts to the state museums of South Australia and Queensland. The former paid £100 for a collection that included wooden swords and painted shields from tropical north Queensland; the latter paid an undisclosed sum for 577 objects, such as necklaces, baskets and string bags as well spears and shields. His personal records show he paid Nora maintenance of £20 a month for three years after their separation.[327]

---

327 *Adelaide Observer*, June 24, 1899, p29 and October 28, 1899, p44; McGregor, Russell, *Objects of Possession: Abstract Transactions in the Wet Tropics of North Queensland, 1870–2013*, James Cook University, http://www.jcucollections.org/?page_id=924, accessed on October 10, 2022; email from Peter Volk, Queensland Museum, September 26, 2013; *The Age*, Melbourne, January 10, 1900, p7

In May 1900, Drake readily approved his leave for the Paris conference and rejected Archie Anderson's protest over his likely extra workload. Wragge gathered his boxes of lantern slides, beer bottles, reports on the Mount Kosciuszko project, Queensland weather reports and records, then left Brisbane for Sydney by train on June 16, 1900. He broke his journey at Newcastle, to give a lecture at the invitation of his friend, the Reverend Donald Fraser, MA (1864–1940), a minister of the recently formed grass-roots, social justice-driven Australian Church. Fraser had emigrated from Scotland in 1893 and remembered Wragge from his Ben Nevis days. During his talk, Wragge endorsed the Australian Church's liberal theology on evolution. Fraser later retrained in medicine and practised for many years in Sydney as a pioneering psychiatrist.[328] Before leaving Sydney, Wragge wrote to *The Sydney Morning Herald* thanking the editor for past support and looking forward to six months of respite from the strain of official routine:

> Happy in the reflection that in Australasia I have not laboured in vain, for proudly do I carry to my dear native England written thanks of the many from the Murchison to Cape York, and the Leeuwin to Noumea, Hobart, and the Bluff.

He conceded a handful of detractors would probably wish him good riddance, but he believed his friends far outnumbered enemies, and that, anyway, "The latter are — with but one or

---

[328] *Newcastle Morning Herald*, June 25, 1900, p4. In 1904, Donald Fraser retired as minister of Newcastle's Australian Church and moved to Sydney, where he had begun practising as a psychotherapist: *Sydney Morning Herald*, obituary, September 4, 1940, p16

two solitary exceptions — of that honest and noble type that one cannot fail to respect, and to whom in fact I am ever ready to extend the hand of good-fellowship and brotherly love."[329]

---

[329] *The Sydney Morning Herald*, June 21, 1900, p9

CHAPTER 15

FEDERATION FIASCO, 1901

A peaceful spot is Piper's Flat. The folk that live around —
They keep themselves by keeping sheep and turning up the ground.
But the climate is erratic; and the consequences are
The struggle with the elements is everlasting war.
We plough, and sow, and harrow — then sit down and pray for rain;
And then we all get flooded out and have to start again ...
— From *How McDougal Topped the Score,* by Thomas E Spencer (1845–1911), first published in *The Bulletin*, Sydney, March 12, 1898.

In February 1901, Archie Anderson was peeved and grumbly after eight months running the Chief Weather Bureau on deputy's wages. He had no grace left for homecomings. Fourteen years as Wragge's right-hand man had worn him down. In the early years, he used to formally welcome the boss back from his adventures. For example, in early January 1892, he joined Nora and the children at the South Brisbane wharf when he came home from his first European trip as Government Meteorologist. But 10 years on, Anderson delegated the bureau's long-serving draftsman, Edgar Fowles, to meet the Sydney train carrying their famous leader. Anderson had fallen out badly with Wragge in mid-June

1900, just before the jaunt to Britain and Europe from which he was finally returning.

Had he read between the lines of the troubled letter that caused their breach, Wragge might have cut his offsider some slack. Anderson's complaint to the Post and Telegraph under-secretary amounted to a plea for collegiate management of the bureau. He argued Wragge should have checked with him before packing the bureau's scrapbook of weather reports to show off at the fabulous Paris Exposition Universelle Internationale. Precious public property, the so-called *Courier* book, held years of official climatological data from the Australian colonies. Anderson explained his concern at losing access to the records:

> I am absolutely certain that during the ensuing six months, it will put this office, and particularly myself, at a great disadvantage and many subjects may turn up by request from interested people in Australia for verification or otherwise.[330]

He also questioned Wragge's jealous hold of the official letters and papers he kept locked away from other staff members. He asserted that the Queensland Public Service Board chairman Captain William Townley disapproved of this habit and had ruled during a recent visit that all staff needed access to official documents. Anderson also alleged Wragge had been careless with other government property such as the ocean-current papers and beautiful cloud photos he gave away at the 1896 Paris conference.

Wragge replied in aggrieved haste while he and Edris packed

---

330 National Archives of Australia, C. 04983, 'Archibald Anderson to Under-Secretary and Superintendent, Post and Telegraph Department, Brisbane', June 14, 1900; Wragge's reply hand-written on reverse side of letter.

for England. A scrawled draft of his reply — held by the National Archives of Australia — seems disproportionately vehement, considering his deputy's plaintive tone and relatively mundane concerns. In summary, he accused Anderson of atrocious and ungentlemanly insinuations. Such miserable spite was unworthy of any student of science. Of course, he needed the scrapbooks to show Paris delegates the breadth and depth of the bureau's service to Australasia. And he had never treated official letters as private property. Besides, the papers kept under lock and key were so old and useless they would not be missed if burnt. As for the notes found in ocean-current bottles, he planned to cast overboard many hundreds more on his coming trip. And the cloud photos had been gifts from several of the colony's weather observers, grateful he had passed their prints on to the world-renowned cloud scientist, Professor Hugo Hildebrand Hildebrandsson of Uppsala University in Stockholm. Wragge lambasted Anderson at length for his quibbling, fault-finding and ungratefulness:

> Instead of hampering him in any way, I have for the last 13 years and a half done my *very* utmost to help him ... to instil into his mind how absolutely he should guard the interests of the office in every way and how loyalty to the department and to his immediate chief are paramount virtues ... He would be better occupied ... in cultivating the virtue of gratitude in return for all I have taught him ...[331]

Wragge began his leave on June 18, 1900, craving in a farewell letter to the press his respite from the strain of official routine, and leaving Anderson once again in charge of the bureau. James Drake

---

[331] Ibid

had allowed him a generous six months of leave to attend the Paris Exposition. His sole obligation was presenting his papers on the meteorology and floods of Queensland, as well as a comparison of data from the Mount Kosciuszko observatory and the sea-level observatory at Merimbula. That was to have left more than three months of spare time in England with Edris before resuming work in Brisbane on December 18. Sea travel between Britain and Australia took about four weeks then. Wragge and New South Wales Chief Commissioner of Railways Charles Oliver were the only antipodean participants in the Paris world fair conferences. Oliver's six months' leave entailed inspecting tramways in North America before attending his international railroads congress. He took his wife and three daughters with him.[332]

Edris's pregnancy is not mentioned in Wragge's official letters, nor in his memoir notes. It's probable he organised his leave to maximise her privacy. On September 28, 1900, he applied for a six-week extension on half-pay, officially to give a series of important lectures about Queensland to British learned societies. He conceded possible concern at how the Brisbane bureau would handle his prolonged absence, but there's no record of input from Anderson on this question. Drake approved the additional leave based on Wragge's assurance, in a letter to the under-secretary of the Post and Telegraph Department, that the routine of the weather office would not suffer in the interim.

The arrival of the baby on November 18 should have left ample for time for Wragge to resume work in Brisbane in late January 1901. But there were complications that, frustratingly for Anderson, further delayed his return by another four weeks. The rainmaking

---

332 *The Sydney Morning Herald*, January 11, 1900, p10

investigations in Austria and Italy took a week, then an Indian Ocean cyclone held up the steamer *Oruba* carrying him, Edris and Kismet back to Australia. Finally, the Queensland Government gave him a week's break in Sydney to sort out provisioning of the Mount Kosciuszko observatory.

Anderson was tired and depressed when Wragge finally returned to work more than two months late, on March 1, 1901. Married with two young sons, he surely resented having to unexpectedly work the long hours over the Christmas and New Year holidays.[333] On April 2, 1901, Anderson lodged a request for long-service leave, pointing out he had taken only four weeks of holidays in the 14 years since they opened the bureau in January 1887. His health was shot and he feared worse to come unless he was given leave immediately:

> During the many absences on your account, I have been in charge not to say unofficially and at many hazards to my frail frame, it is my desire by virtue of the statutes to ask for a very extended leave of absence to take place at a very early date, otherwise physically I may suffer considerably.[334]

Three years earlier, the popular weekly *The Queenslander* touched on Anderson's health in a profile that was prompted by an outcry in the press over Wragge's almost annual absences from Brisbane in the wet season. The newspaper praised his accurate

---

333 *The Brisbane Courier*, February 22, 1898, p2, comparing their personalities, Wragge summed up Anderson as by nature a domestic man: "He has taken honours in home life, and is, withal, an excellent officer in the office, but he is hardly as yet a bushman."

334 Queensland State Archives, Inwards Correspondence, Chief Sec Dept, 1901, 'A.A. Anderson to Government Meteorologist', April 2, 1901, letter 03840

forecasting of moderate flooding in Brisbane on January 13, 1898, while Wragge was in the Snowy Mountains. Anderson had also run the bureau conscientiously through the 1890, 1893 and 1896 floods, and issued early warnings as Cyclone Sigma swept towards Townsville in January 1896. In short, he had a nose for "the worst moods of the weather fiend".

The writer was dismayed at the physical toll on Anderson since his recruitment by Wragge in January 1887. His demanding vocation had withered his smallish, once somewhat fleshy frame and left him no time for playing violin in the Liedertafel Orchestra. He had hung up his fiddle, swapped copper wire for cat gut, and learnt to read isobars as instinctively as bars of sheet music. "Meteorology very absorbing work; leaves little room for recreation ... but victim does not grudge time; feels is doing big duty to country," *The Queenslander*'s writer concluded.[335] Some years later, Harald Jensen, a member of Wragge's Kosciuszko crew in 1898, recalled Anderson had been absorbed in his work, but at least allowed himself time for a cold beer in warm weather. Jensen recalled him donning his hat and vanishing from the office for a few minutes to quench his thirst, then reappearing looking contented.[336] Anderson was granted five months of leave from April to October 1901, including four unexpected weeks recovering from a broken wrist. Coincidentally, Wragge needed extended sick leave in May and July, when incapacitated by an ulcerated right eye and what Edgar Fowles, the bureau's first assistant, termed nervous excitement and prostration.

---

[335] *The Queenslander*, January 22, 1898, p165

[336] Jensen, Harald Ingemann, *Reminiscences of a Geologist* [unpublished manuscript], circa 1957, State Library of Queensland, Harald Ingemann Jensen Records, 1905–65, Call No OM69-29, p18

* * *

Month by month, Wragge's Federation dreams were dying. The Barton Government, which first met in Melbourne in May 1901, ranked nation-building and social cohesion well ahead of weather-watching and climate science. Early laws secured customs-and-excise revenue, introduced the White Australia policy, and merged colonial post-and-telegraph departments into a national communications service. James Drake became Australia's first Postmaster-General, having been invited to join Barton's Cabinet in early February 1901, after the death of fellow Queensland Protectionist Sir Robert Dickson. Ironically, Drake's tasks included reminding former Queensland government colleagues that the Commonwealth was not obliged to take over their state's weather bureau, regardless of its national pretensions.

In Queensland, meteorology was a branch of the Post and Telegraph Department, not, as in some of the other colonies, a stand-alone agency or the province of the government astronomer. In effect, Queensland's wealthy post-and-telegraph service had sustained the Chief Weather Bureau's ambitious reach by carrying the cost of weather telegrams for the previous 14 years. In 1897, Dickson — who preceded Drake as the colony's Postmaster-General — estimated the annual cost of meteorological department telegrams at £25,000. In his first newspaper weather report of 1901, published on March 2, Wragge praised Australia's colonial post-and-telegraph services for their many years of accurately transmitting his data. He acknowledged without comment his department's takeover by Queensland's Chief Secretary's Department. Brisbane's daily papers, the *Courier* and *Telegraph*, were both alarmed. *The Courier* pictured Wragge in limbo, "hung like Mahomet's coffin between heaven and earth", and blamed the state government's complacency. The *Telegraph* saw hope in Premier Robert Philp's opinion that Wragge's bureau belonged in the Federal Post and Telegraph Department. "He thinks the weather prophet must be connected with the Telegraph Department to enable him to make free use of the telegraph wires for the innumerable reports which arrive from all over the country," the *Telegraph* explained.[337]

Wragge had returned to Australia in February 1901, expecting to lead the federated colonies' first national weather service. While in Britain, he prepared methodically by sending his CV to the colonial premiers promising to establish a first-class national bureau, "thoroughly scientific and practical in every respect". In London, in November 1900, he began his speech to the Imperial Institute with enthusiastic references to Federation and the new Commonwealth of Australia.[338] Writing to Drake in September 1901, he looked forward to giving Australia a meteorological service equalling, if not surpassing, that of England and the United States. He believed his bureau had already reached that standard.

I can detect his language in the meteorological section of the Queensland Post and Telegraph Department's 1899 guide that ranked his bureau among the world's foremost meteorological and signal services. This unsigned report sketched the bureau's progress from 1887 to 1898 when renowned as the hub of a weather service covering Australia and New Zealand, British New Guinea,

---

[337] *The Telegraph,* March 2, 1901, p14; the *Brisbane Courier,* March 8, 1901, p4

[338] *The Times,* London, November 28, 1990, p4; *Morning Post,* London, November 27, 1900, p4

New Caledonia and the South Pacific Islands. The writer depicted bureau staff handling a torrent of data from the state's 632 weather stations, large and small, governed by the perfectionist precepts of London's Meteorological Office and Washington's Chief Signal Office. The Brisbane bureau also received daily data from 20 Western Australian stations, 19 in South Australia and the Northern Territory, 18 in New South Wales, 13 in Victoria, 10 in Tasmania, 3 in New Zealand, 2 in New Caledonia and also observations from Manila, Singapore, Batavia and Hong Kong. By 1899, the bureau's public service included daily 4 pm forecasts for each of the Australian colonies, New Zealand, New Caledonia and the Loyalty Islands, as well as a regular Australasian weather digest giving conditions for each colony.[339]

But the guide's author overlooked the 160 weather stations established in Queensland before Wragge's arrival. The first three outside Brisbane — at Cape Moreton and the Darling Downs towns of Toowoomba and Warwick — were set up in 1869 by then-Government Meteorological Observer Edmund MacDonnell. He opened many more during his 18-year career, 30 of them in 1885, a year before his retirement.[340]

The entwinement of telegraphy and meteorology that upset Wragge's grand plans in 1901 began in MacDonnell's era. In 1871, the colony's Superintendent of Electric Telegraphs, William John Cracknell, reported having issued rain gauges to every telegraph station in the colony. Remote officers were meant to wire any

---

[339] *Queensland Post and Telegraph Guide,* January 1899, pp 214–15

[340] *The Brisbane Courier,* November 23, 1869, p3; 'Report of the Government Meteorological Observer, for the year 1885', pp 1–2, *Queensland Legislative Assembly Votes and Proceedings, 1886–87*

rainfall readings to Brisbane at 9 am. Staff at Cape Moreton and the Darling Downs stations sent wind and weather reports at 9 am and 3 pm daily for public display outside the GPO. Cracknell complained in 1871 that free shipping and meteorological messages were costing the Post and Telegraph Department £9000 annually — proportionately more than in other colonies. Bad weather had possibly skewed the costs.[341] In 1876, Postmaster-General Charles Stuart Mein pledged to restore the apparently lagging distribution of rain gauges and thermometers. He saw no excuse for lax meteorological record-keeping by telegraph officers, "Their ordinary duties being of a semi-scientific character."[342] Wragge found many observers wanting on his 1885 reconnaissance of Queensland weather stations. During the 1890s, he regularly trekked the colony as Government Meteorologist making spot checks on errant officers.

* * *

Drake was no help in his new position. Wragge's plea in September 1900, for his backing to lead a new national weather bureau was a hypothetical application for an imaginary job.[343] On March 16,

---

[341] Whitmore, R L (ed), *Eminent Queensland Engineers*, Brisbane, 1984, pp 24–25, https://www.engineersaustralia.org.au/sites/default/files/resource-files/2017-01/eqe_vol_i.compressed.pdf, accessed October 13, 2022; *The Brisbane Courier*, April 27, 1871, p3, 'Report from the Superintendent of Electric Telegraphs, 1870'

[342] Queensland Parliamentary Debates. Legislative Council (1876), Parliamentary Debates (Hansard), p1367 Brisbane, Australia

[343] Clement Wragge to Hon G Drake, MLC, Postmaster-General, Brisbane, September 28, 1900, under letterhead of Free Library, Wragge Museum, Borough Hall, Stafford

1901, *The Brisbane Courier* reported Drake had told Queensland's Acting Premier Arthur Rutledge that the Commonwealth was not obliged to take over the state's bureau. The merger of colonial post-and-telegraph departments covered the transfer to federal control of posts, telegraphs and telephones, not meteorology. Then campaigning for election to the senate, Drake avoided — in public at least — any view on Australian meteorology or Wragge's credentials to lead a federal bureau. He was possibly cautious because of accusations in the Melbourne press of partisanship in his promotion of Queenslander Robert Townley Scott (1841–1922) to the new position of Principal Deputy Postmaster-General. Drake chose Scott, previously permanent head of Queensland's Post and Telegraph Department, ahead of Victoria's experienced Deputy Postmaster-General, Frank Leon Outtrim (1847–1917).

Rutledge — a New South Wales-born barrister and ordained Wesleyan minister — promptly re-established the Chief Bureau under temporary management of the Queensland Chief Secretary's Department. He invited Wragge onto a razor gang which slashed projected telegram traffic by 70 per cent, beginning by discontinuing despatches of daily observations from Queensland's hundreds of weather stations to head telegraph offices in other states. In mid-March, *The Brisbane Courier* reported Rutledge's view that the government could not afford maintaining the Chief Weather Bureau "up to the old standard".[344] Nevertheless, a week later, Wragge told southern newspapers he would continue sending a daily weather synopsis and data from Queensland's 24 first-order and 48 second-order stations to

---

344   *The Brisbane Courier*, March 2, 1901, p6 and March 13, p4; *The Telegraph*, Brisbane, March 21, 1901, p2

government astronomers in other states. Other data, such as the wind and weather reports previously sent free interstate to head telegraph offices, would only be supplied if reply telegrams were prepaid. *The Argus* in Melbourne, the *Evening News* in Sydney, and the *South Australian Chronicle* and *Register* in Adelaide continued publishing the Brisbane office's forecasts.

In August 1901, Prime Minister Edmund Barton conceded laws establishing a federal meteorological agency were unlikely before parliament's second session. Barton told concerned New South Wales MPs that the government had greater priorities. Bruce Smith and George Cruickshank, members for Parkes and Gwydir, respectively, had complained the Postmaster-General's Department's squeeze on weather telegrams was depriving farmers, graziers and master mariners of crucial data. Barton regretted the Commonwealth could not take over any of the state weather bureaus until making necessary laws, which was "not of the first order of importance".[345] He was at odds with Premier Robert Philp, back from three months of leave in South Africa where his son Lieutenant Colin Philp and the Sixth Queensland Imperial Bushmen were fighting the Boers. Philp argued the federal government was obliged to pay for Mr Wragge's services since he helped the other states as much as his own. Queensland faced an unexpected £3000 increase in weather bureau costs in 1901–02 for essential telegrams. The estimated £4739 needed to sustain the bureau until the end of June 1902, included £1639 for salaries, instruments and repairs. Wragge's salary was £500,

---

345 *House of Representatives Official Hansard,* August 29, 1901, First Parliament, First Session, pp 4300–08; *The Brisbane Courier,* August 30, 1901, p5, report of grievance debate

his four assistants split £600 and the bureau's messenger took home £89. Speaking during the estimates debate, Philp accused the Commonwealth of treating Queensland very shabbily.[346] Incidentally, the Queensland Chief Secretary Department's budget was helped in 1901 by the transfer of military and marine operations to federal control.[347]

Wragge apologised in his daily weather reports for limitations caused by the bureau's straitened circumstances. In May 1901, he wrote for the *Pastoralists' Review of Australasia* a blueprint for a new, first-class Australian weather service. This 3000-word statement was also published in Hobart's *Mercury* newspaper and regional papers. He estimated annual running costs at £10,000, in line with the budgets of British and Canadian weather offices — £15,300 and £15,400 per annum, respectively. Logically, the national office would be a branch of the Federal Post and Telegraph Department, since practical meteorology was inseparable from telegraphy. He argued the central office would be best placed in Brisbane or Sydney to take advantage of the latest telegraphic technology. If asked to become Federal Meteorologist — which he conceded might not happen — he would give it his best shot.[348]

* * *

---

346 *Morning Post*, Cairns, September 27, 1901, p3, 'Report of Queensland Legislative Council estimates debate, September 24, 1901'
347 *The Brisbane Courier*, September 25, 1901, p6
348 *Mercury*, Hobart, June 12, 1901, p6, 'Federal Weather Bureau, Suggestions by Clement L Wragge', first published in *Pastoralists' Review of Australasia*, May, 1901

Graziers gradually realised the Queensland bureau's plight. Early in May 1901, Australian Estates and Mortgage Co manager William A Smith wrote to *The Age*, in Melbourne, seeking reinstatement of daily reports from Queensland country weather stations. He argued that to Victorian investors with Queensland pastoral interests, rainfall observations were as important as weather forecasts. In his opinion, Queensland's fanciful £2000 fee for resumption of the service was a poor outcome of Federation. He backed, as more realistic, the counter offer of £200 by Victoria's Deputy Postmaster-General Frank Outtrim. "Surely under Federation, the residents of Victoria are more entitled than ever to the gratuitous advice of such universally important data," Smith wrote.[349] Wragge's legacy in western Queensland included a network of citizen weather watchers on sheep and cattle stations. In 1890, he tried to contact every grazing property in the colony for that year's monthly rainfall figures and recent annual figures. He sent instructions for rain gauge set-up and readings, explaining he needed accurate records for flood charts and reports.[350] By 1897, most climatological observations from Queensland's tropical zones came from private stations.

In November 1901, Rockhampton's *Morning Bulletin* newspaper asked why post office staff had been banned from displaying on a public noticeboard rainfall figures from out west. On November 9, a crowd seeking news of a rumoured cloudburst in the Barcaldine district found reports solely from Rockhampton and the townships of Emerald and St Lawrence. Post office staff apologised that free notification of data from smaller stations had

---

[349] *The Age*, April 9, 1901, p6; *The Argus*, April 3, 1901, p4
[350] *The Brisbane Courier*, April 25, 1890, p4

been officially banned since March. The newspaper reported Rockhampton's veteran meteorological observer E L Hanna had until then made public any wires he received. The paper argued that Rockhampton's Chamber of Commerce must lobby for proper funding of the meteorological bureau.

Merchant seamen with trust in Wragge's forecasts protested too. In early November 1901, Howard Smith shipping line skipper Captain W Hurford condemned the Central Bureau's decision to stop posting daily coastal weather forecasts at Queensland telegraph offices. "When you go to get the only little bit of information that a practical seaman wants, you find a blank sheet staring you in the face," Hurford wrote to the *Brisbane Courier*. He was dismayed that usually courteous telegraph-office staff in the port of Maryborough could not give him a 9 am weather report on the Wide Bay Bar. Information needed for safe navigation, still widely available in other states, seemed rationed in Queensland by a user-pays arrangement — scandalous considering the money wasted on scientific cloud-chasing and meteorology, generally.

Wragge replied indignantly, denying knowledge of privately commissioned forecasts and defending his bureau's budget. "For my own part, I would most willingly and gladly supply all the information possible at every coast station ... did means and the condition of the weather service admit of this line of action," he wrote. Since being severed from the Post and Telegraph Department, the bureau had to pay for every telegram, and the flow of data had been cut by more than 70 per cent.[351]

* * *

---
351   *The Brisbane Courier*, November 4, 1901, p9; November 12, 1901, p3

Wragge muddled through 1901 with fading hopes of what Federation would bring. He shared his precariousness in public lectures through south-east, central and north Queensland, beginning in late April with a visit to Biggenden, 200 miles north-west of Brisbane, as guest of the farming town's literary and debating society. Members believed themselves instigators of his forthcoming rainmaking experiments, having in 1900 used the forum of *Queensland Agricultural Journal* to canvass the explosive options later advocated by Mackay district farmers.

Wragge's talk lasted three hours. Beyond the advertised topic, "Cloud-Shooting as Prevention Against Hail and Means of Causing Precipitation of Rain", he explored mountain-top meteorology, God the Infinite Dynamo, and merits of the Union Jack. The *Maryborough Chronicle*'s Biggenden scribe reported their famous guest's ire at the government's lukewarm attitude to his proposed rainmaking experiments and his hope that farmers, squatters, shipping companies, chambers of commerce and MPs would press for stronger official support. After all, he reminded the full house at Biggenden's Exhibition Hall, the Chief Bureau was still Australia's premier meteorological department. He wound up with some sea shanties, self-accompanied on autoharp, and hopped on a railway trike driven by a volunteer ambulanceman for a 40-mile dash down the branch line to catch the Brisbane train at Mungarr Junction. "Mr Wragge is said to have greatly enjoyed the trip and his stay among us," the paper reported.[352]

In June 1901, Wragge returned from sick leave with a 71-page hail-busting and rainmaking briefing for Queensland's Chief Secretary. It was as dull and circumspect as its title: "Report on the

---

352 *Maryborough Chronicle*, May 3, 1901, p3 and May 7, 1901, p3

Steiger Vortex or Methods of Preventing Damage to Vineyards and Crops by Hail Storms and Probably for the Production of Rain, commonly called *Das Wetterschiessen*." In essence, he planned to test *Das Wetterschiessen* — literally, weather-shooting — on the dry clouds of western Queensland. He asked the government to import 10 cannons for the experiment, built to the specifications of Austrian *Wetterschiessen* pioneer Albert Steiger. A wine merchant and mayor of Windisch-Feistritz, Steiger was unsure why his hail cannons sometimes produced gentle drizzle. Possible causes included the action of sound waves, smoke, air currents or the vortex — the gaseous ring generated on discharge of the gun.

Wragge foreshadowed also testing a then-popular theory promoted by the P&O shipping-line skipper W D G Worcester. A frequent visitor to Australia, Captain William Davey Gray Worcester detested droughts and advocated firing 100-pound bombs of liquid air into what he called non-precipitating nimbus clouds. He contended that the burst of intensely cold liquid air would instantly produce condensation needed for rain. Wragge felt Worcester's projectiles were more likely to produce good rain than Steiger's vortex. Nevertheless, he concluded that *Das Wetterschiessen* was thoroughly worthy of trial in Australia on the lines he had suggested, adding:

> It must be understood that this method is employed mainly for the protection of vineyards against damage by hail, and with respect to "controlling rainfall" it is limited to the production of a beneficent precipitation of rain.

Desiccated coastal districts reactivated rainmaking schemes as the Big Dry spread from western Queensland in 1900.

Henry Alexis Tardent (1853–1929), the Swiss-born manager of Biggenden's progressive State Farm, endorsed the Steiger Vortex Gun trial in a letter Wragge attached to his report. Tardent argued cloud-shooting must be tested in local conditions, given what he called the growing notion of its rainmaking potential.[353]

Coincidentally, Captain Oliver Haldane Stokes, a member of Wragge's Kosciuszko team, was also a true believer in pluviculture. In March 1876, he wrote to the *Sydney Morning Herald* recalling that cannon firings during artillery reviews in London were invariably followed by heavy rain, at least on overcast days. He advocated using field guns in a weather-shooting trial to break the drought then afflicting New South Wales. Nothing came of this suggestion.

In 1871, former US Civil War General Edward Powers observed in his book, *War and the Weather*, that heavy cannon fire was often followed by copious rain. In his opinion, explosive concussion caused precipitation in the upper atmosphere, where moisture was carried eastward from the Pacific Ocean. In 1890, US Senator Charles Benjamin Farwell won government funding for a series of ultimately inconclusive rain-concussion trials using bomb-laden balloons above a ranch in north-western Texas.

A radically different pluviculture theory by US meteorologist James Pollard Espy (1785–1860) was never tested. Nicknamed the "Old Storm King", Espy believed the first people of North America burnt their prairie lands to make rain. He put this theory in his *Philosophy of Storms*, published in 1843, but could not convince

---

[353] Queensland State Archives, 'Report to Chief Secretary, Brisbane, on Steiger Vortex, June 1901', Appendix A from Henry A Tardent, Biggenden, May 1, 1901

Congress to finance an experiment.[354]

Wragge's intended trials were stymied from the outset. The government could not afford 10 Austrian-made cannons and instead commissioned Brisbane ironworkers Sargeant and Co to build a £10, single prototype mortar and inverted cone-shaped barrel, unveiled in the city's Botanic Gardens on September 28, 1901. Wragge supervised eight detonations of blasting powder, none producing rain. One slightly fractured the cannon's 17-feet-tall muzzle made of 12-gauge iron plate. Two elicited a long shrill sound that Wragge was reported to have excitedly identified as "vortex whistling". In further trials on October 2 and 3, he lowered the barrel and hit a target 190 yards away with a volley of shot from the mortar. Then he raised it and with three shots dispersed what Sydney's *Evening News* described as a small, loose cumulus cloud. He reportedly told onlookers he would have needed two guns to bring about any practical results.[355]

During October, Wragge took his rainmaking sideshow to north and central Queensland, leaving the bureau in charge of a theoretically refreshed Archie Anderson. Wragge owed Mackay a visit because of the Pioneer River Farmers' Association's lobbying

---

354 *The Guardian*, Australian Edition, June 12, 2015, Scurr, Ruth, 'Rain: A Natural and Cultural History', https://www.theguardian.com/books/2015/jun/11/rain-natural-cultural-history-weather-experiment-review, accessed October 13, 2022; Smithsonian.com, September 4, 2018, Nodjimbadem, Katie, 'When the US Government Tried to Make it Rain by Exploding Dynamite in the Sky', https://www.smithsonianmag.com/history/when-us-government-tried-make-rain-exploding-dynamite-sky-180970193, accessed October 13, 2022; *The Scientific Monthly*, Vol 14 No 2, February 1922, D W Hering, 'Weather Control', pp 180–81, https://www.jstor.org/stable/6437?seq=4#metadata_info_tab_contents, accessed October 13, 2022; *The Sydney Morning Herald*, March 14, 1876, p5

355 *The Week*, Brisbane, October 4, p32, *Evening News*, Sydney, October 3, p3

for a rainmaking trial. Their 45 inches in 1900 was 26 inches below average, based on two decades of official measurements.

In May 1901, sugar planter Patrick McKenney, of Plane Creek, had asked the government to match £1 for £1 public donations to build Steiger-style weather-shooting guns in place of four crude mortars made at a local foundry in 1900. A month later, McKenney launched a statewide shilling fund, hoping to finance the importation of six Steiger Vortex cannons from Austria or Italy.

In early October, Wragge sailed north to Mackay where he tested his dented prototype, inspected weather stations and gave a couple of lantern-slide lectures. His next engagements, in Rockhampton and Mount Morgan, were disrupted at first by the absence of the cannon, accidentally left on the wharf in Mackay, and later by its partial disintegration during trials in Rockhampton. Ironworkers from the Mount Morgan Goldmining Company reassembled the riveted barrel for an anti-climactic finale on October 25, when clouds shrouding the mountain-top mine dissipated before the first shot was fired.

The North Queensland Register, never in love with Wragge, reported the crowd was unmoved by his consolation of hearing the whistling vortex. "The general idea of the crowd was that Mr Wragge was endeavouring to provide amusement, not instruction," the *Register* remarked.[356]

A Rockhampton *Morning Bulletin* reader warned the Weather Prophet was risking his scientific reputation by waging war on the clouds. The reader, a septuagenarian ex-East India Company infantry sergeant, doubted the Steiger gun was capable of

---

356 *Mackay Mercury*, October 8, 1901, p2; Queensland State Archives, 'P McKenney, Mackay, to Under-Secretary, Colonial Treasury', Brisbane, May 6, 1901; *The North Queensland Register*, October 28, 1901, pp 5–6

compressing watery vapour into copious clouds. In November, replying to a sceptical story in *The Sydney Morning Herald*, Wragge calculated he would have needed at least 10 guns to produce satisfactory results. The Queensland Government's £10 prototype was destined to provide no result whatsoever.[357]

* * *

In December 1901, Wragge's life was knotted in unknowns. The fate of the Chief Bureau was uncertain and his Kosciuszko project mired in deepening debt. His own finances were so shaky that, a few months earlier, he had offered to sell his home, Capemba, for £2500. As for his rainmaking trials, he seemed likely to keep on tantalising Queensland's parched pastoralists for as long as the drought continued. This was evident in early December, on an inspection of weather stations along the western railway line from Toowoomba to Cunnamulla.

The people of Charleville, 460 miles inland from Brisbane on the dying Warrego River, were unconcerned by the rumoured problems with his cannon. The 80 who paid a shilling each to hear Wragge speak at the School of Arts fidgeted through lantern slides from his Paris trip, but cheered up when he showed them pictures of the famous, mysterious Steiger Vortex gun. Their district had been in drought for four years, with annual rainfall totals down to one-third of the 24 inches once assumed to be an average annual fall. Wragge stressed weather-shooting existed chiefly in Italian

---

[357] *Morning Bulletin*, Rockhampton, October 29, 1901, p4, *The Capricornian*, July 24, 1909, p41; for biographical details of prolific correspondent J W Head: *The Sydney Morning Herald*, November 5, 1901, p4

and Austrian vineyards to tame hailstorms, and that making rain by separating cloud masses was a secondary function.[358] But he also hinted his so-far disappointing trials had foundered for want of a battery of guns.

On January 8, 1902, Charleville Municipal Council obliged and sent the government £10 towards six new guns to enable him to carry out trials in their town. "We are quite at one with Mr Wragge in his belief that the experiment will yield rain in western Queensland if properly initiated by an organised battery as proposed," the council told the Chief Secretary's department.[359]

Wragge wrote little about 1901 in his memoir notes. It was simply the year he, Edris and Kismet returned to Queensland from England. The title deeds to Capemba show that in November 1899, he transferred ownership of the house and grounds to Edris. In early 1900, clarifying their relationship in a letter to *Queensland Figaro*, he called Edris his housekeeper and literary and scientific assistant, and defended her character as cultured and refined. He and Nora formally separated on May 31, 1899. In July 1900, a social columnist in *The Queenslander* reported Mrs Clement L Wragge and her family had moved to Cannon Hill, Morningside, "during her husband's absence in Europe".[360]

In August 1901, Wragge gave the Queensland Government first option on buying Capemba and its surrounding gardens for use as a museum and pleasure grounds. Citing an urgent need for funds, he told Home Secretary Colonel Justin Foxton the house, observatory and three-acre grounds were on the market for £2500.

---

358   *The Charleville Times,* December 14, 1901, p2
359   *The Brisbane Courier,* January 10, 1902, p5
360   *The Queenslander,* July 7, 1900, p46

He offered the lot to the government on generous terms, throwing in 1000 palms and other trees valued at £900, and much of his scientific library and collections in conchology, ethnography, geology and natural history. "I think you will agree my entire offer is most liberal and that no bigger bargain was ever on the market," he wrote. The government declined. Neither did anything come of auctioneers' advertisements for a large, modern twelve-roomed villa within a three-acre Garden of Eden.

A fortnight later, Edris as mortgagor borrowed £175 from their mortgagee, the ageing explorer Augustus Charles Gregory, which seems to have solved the crisis. But money remained short. In January 1902, Wragge was relieved when the government agreed to meet the £14 cost of repairs to the storm-damaged observatory roof. And in May 1902, he advertised to sell his astronomical telescope and observatory.

In November 1901, the New South Wales Government, the major supporter of the Kosciuszko project, warned only £100 remained of the £400 allocated for maintenance of the high-level and sea-level stations in the 1901–02 financial year. This grant was intended to cover provisioning of the Kosciuszko observatory and also the running costs of a check station in Sydney, but costly teamsters' charges to the mountain were already threatening the budget.

That month, the then officer in charge of the Kosciuszko observatory, Philip Sydney Whelan, reported having killed 107 rats in October, and that he was seriously considering cashing in their scalps at threepence each to replenish damaged stores.[361]

In June 1901, Wragge flattered his staunch supporters, ex-New South Wales Premiers George Reid and William Lyne, by writing that their names should be "carved indelibly" in Snowy Mountains granite. Issuing a dire High Country weather forecast, he was sure his Kosciuszko observatory crew would survive the likely terrible gales and driving snow in the generous shelter financed by Reid and Lyne. That June, wild horses were said to have died of starvation in snowdrifts 60 feet deep around Kosciuszko.[362]

In the warm months of 1901 and 1902, drovers shifted many thousands of starving sheep from the parched Riverina plains to snow leases around Kiandra and Bombala. Grazier Samuel McCaughey, known as the "Wool King", reportedly sent 150,000 head to the Kiandra district. In December 1902, a drover estimated having seen more than 20,000 carcasses of dead sheep between Tumut, Kiandra and Broken Cart Flat.[363]

Wragge's newspaper presence remained irrepressible in 1901. He showered Brisbane's *Courier* and *Telegraph* with daily weather homilies and supplied basic forecasts to the *Evening News* in Sydney, the *South Australian Chronicle* in Adelaide, and *The*

---

361 *The Sydney Morning Herald,* January 16, 1903, p8 and *The Brisbane Courier,* January 17, 1903, p14, outlining worsening finances of the Kosciuszko project in 1901 and '02; *The Telegraph,* Brisbane, November 19, 1901, p2, quotes Whelan saying he is thinking seriously of claiming threepence a head on rat scalps to recoup damage to stores.

362 *The Brisbane Courier,* June 19, 1901, p5; *Catholic Press,* Sydney, June 29, 1901, p25, references to Kosciuszko snowfall and death of horses in letter from 14-year-old Lucy Violet O'Donnell, Merrettville, via Dalgety.

363 *Gundagai Times and Tumut, Adelong and Murrumbidgee District Advertiser,* MY 9, 1902, p2; *Gundagai Independent* and *Pastoral, Agricultural and Mining Advocate,* December 20, 1902, p3

Argus in Melbourne. The latter also ran predictions from Todd, Russell and Baracchi.

In 1902, a correspondent to the biweekly *Wellington Times* in central New South Wales judged that country people had more faith in Wragge than their state's astronomer Russell: "If Mr Russell ... ventures a prediction that rain will fall, people have got into the habit of saying, 'Then we may expect fine weather, but what does Wragge predict?'"[364] Russell's probabilities were said to be always deliberately contrary.[365]

Wragge's peers remained disdainful of naming cyclones. Privately, Western Australian Government astronomer Bill Cooke thought the names a gimmicky grab for public acclaim.

* * *

The historic submarine cable link between New Caledonia and Queensland lasted from 1893 to 1923. When opened at Mon Repos, near Bundaberg, in October 1893, the 721-mile wire was promoted by hopeful politicians as the beginning of a telegraph link with North America. But others were dismayed that Queensland and New South Wales had each agreed to pay an annual subsidy of £2000 for 30 years to their French partner, the Société Française des Télégraphes Sous-Marins.[366] In 1901, Britain condemned the liaison as a security risk and financed a 3862-nautical mile undersea cable to Canada via the British outposts of Norfolk

---

364   *Wellington Times,* May 15, 1902, p3
365   *Press,* Christchurch, August 5, 1898, p5, citing an undated letter to the *Hobart Mercury*
366   *Daily Telegraph,* Sydney, October 17, 1893, p4

Island and Fiji, opened in December 1902 at a cost of £2 million. The New Caledonia cable failed in 1923, could not be repaired, and was superseded in 1925 by a radio service to Sydney. Today, if you know what you're looking for, traces of Wragge's treasured cable station can be found at low tide on the beach that a French settler named Mon Repos (My Rest).[367]

---

[367] Rea, Malcolm M, 'Communications Across the Generations: an Australian Post Office History of Queensland', paper read at a meeting of the Post Office Historical Society of Queensland, November 22, 1971, pp 218–20

CHAPTER 16

FAREWELLS AND A RETREAT, BRISBANE AND MOUNT KOSCIUSZKO, 1902

Anger not the Prophet Wragge
By scoffing at his tales,
He may untie his weather bag
And loose the storms and gales.
The ill effects of drenching rain
We only can endure
By driving out the cold, 'tis plain
With WOOD'S GREAT PEPPERMINT CURE.
— Advertisement in the *Darling Downs Gazette*, Queensland, October 2, 1903, page 2.

A curious crowd packed Brisbane's Centennial Hall on the night of Saturday 17 May 1902, for an event advertised as "Mr Wragge's Farewell to a Brisbane Audience". The *Telegraph* judged the auditorium nearly full, all standing space taken. State Governor Sir Herbert Chermside, a Boer War Major-General, was there too. A month earlier he had insisted on a 15 per cent pay cut in solidarity with senior public servants being sacked to steady the government's drought-bitten budget.[368]

Wragge and his long-serving bureau officers — Anderson, Fowles and Bond — were officially told of their retrenchments on May 10, 1902. By then, he had booked the auditorium, situated about a block from their headquarters in Brisbane's General Post Office. This privately owned dance hall with a detachable skylight for steamy nights was opened in 1888, a century on from the foundation of the New South Wales convict settlement. Wragge had previously used the 700-seat hall for lectures and lantern-slide shows. In late April, he began promoting his latest presentation as a "Grand Illustrated Lecture Entertainment", but retitled it as his farewell when news broke the Chief Bureau was to close by June 30. Admission charges remained "popular prices at door", the *Courier* advised.

The *Telegraph* depicted his performance on May 17 as distraught: "Having mind and heart steeped in his particular science ... [he] could hardly help being forceful and unbosomed himself without reserve." There had been frequent cheers as he praised Queensland's leadership of meteorological work in Australia and New Zealand and condemned the federal government for not making the Chief Bureau a Commonwealth

---

368 *Telegraph*, Brisbane, July 28, 1902, p5. In April 1902, after hearing of the government's retrenchment scheme, Sir Herbert Chermside reportedly asked to forgo the fullest amount of salary provided for in the Special Retrenchment Bill. This equated to a pay cut of £750, which took effect in July 1902; Paul D Wilson, 'Chermside, Sir Herbert Charles (1850–1929)', *Australian Dictionary of Biography*, National Centre of Biography, Australian National University, http://adb.anu.edu.au/biography/chermside-sir-herbert-charles-5575/text9511, accessed October 13, 2022, published first in hardcopy in *Australian Dictionary of Biography*, Volume 7 (MUP), 1979.

institution. The *Courier* stressed his hope the bureau could be saved and his insistence that federal, not Queensland, politicians had caused his predicament.

The *Courier*'s reporter added an allegation, absent from the *Telegraph*, that a ring had been formed somewhere in the southern states to "quash" the Chief Bureau. After relating this story, Wragge exclaimed he was not a man easily quashed and, while still in Australia, whether or not employed by the state government, he would always do his best for Queenslanders. He pledged amidst loud applause to save the bureau, trusting in public sympathy for his cause.[369]

His lantern slides included coloured photos of Capemba, which was up for auction. The property had been for sale since the previous August, asking price £2500. In April 1902, the first advertisements for the auction coincided with rumours Philp had decided to abandon the Chief Bureau in hope of forcing assistance from the federal government. Capemba was passed in at auction on May 24, 1902, and advertised to let, part furnished. In January 1903, further newspaper notices appeared, offering the house to let for £2 per week, with the enticement of a "most lovely garden, gardener provided". An inventory compiled in August 1901 by nurseryman S H Eaves showed the latter included a Kismet Bed, Kismet Bank and Edris Terrace.

In 1899, Nora and the family had moved from Taringa to Morningside — a settlement about 10 miles east, on the south side of the river. Enrolment records for nearby Norman Park State School, officially opened in July 1900, show six of the Wragge children were listed as students. Bert, who turned 18 in August

---

369 *Telegraph*, Brisbane, May 19, 1902, p7; *Brisbane Courier*, May 19, 1902, p4

1900, had missed about four months of school in 1899, during adventures on Mount Kosciuszko. In November 1903, Wragge was registered on the federal electoral roll as a resident of Swann's Road, Taringa, but his notes suggest that, by then, he, Edris and Kismet had moved to Sydney.

* * *

On June 25, 1902, Philp announced a tentative reprieve for the Brisbane bureau. It was not the federal lifeline he believed was Queensland's due. Rather, it was Wragge's own scheme for a pared-down, semi-privatised weather service with barely viable working capital of £1100 and a federal government pledge to waive £3500 in meteorological telegram fees in the coming 12 months. Prime Minister Barton and Attorney-General Alfred Deakin were at first adamant any special treatment would be unfair to the other states. But on May 19, Deakin — deputising for Barton — wrote to Philp conceding the excellence of the Queensland bureau and the Australian scope of its "researches" and asking for suggestions on an equitable way of saving it. A few days earlier, state premiers meeting in Sydney had supported a motion warning the Commonwealth that maritime and industrial interests would suffer if Queensland's storm alerts were discontinued.

The closure was debated in federal parliament in Melbourne on May 20, 1902, when Queensland Labor Senator William Guy Higgs raised Wragge's retrenchment as a matter of public importance. Higgs, a former editor of Queensland's *The Worker*, praised him as Australia's de facto chief meteorologist. He found, when talking about the weather, most people asked, "What does

Mr Wragge say?" His forecasts in the daily papers were widely read. Senator Simon Fraser, a Victorian grazier, concurred that Wragge's forecasts were accepted everywhere. Not one of the state astronomers was in his league. Fraser, a 70-year-old Protectionist, argued for prompt establishment of a federal bureau:

> Why should we have six weather departments and six astronomers? Why should we not have one first-class weather bureau, managed by a first-class man? We can then afford to do the thing properly to the great advantage of the people of Australia. Great economy would result from such a re-arrangement ...

New South Wales Senator Richard Edward O'Connor, another Protectionist and Vice-President of the government's Executive Council, said Deakin had drafted letters to the state premiers asking their views on the Commonwealth control of the meteorological sections of their astronomy departments, and that he intended to gauge the cost of a national bureau. Deakin had already put to Queensland a proposal for free transmission of weather data, subject to passing of the yet-to-be-debated Posts and Telegraphs Rates Bill. Senator O'Connor said the government's action had been and would continue to be, prompted by a sincere desire to preserve, if possible, the work of the Queensland Weather Bureau for the benefit of Australia.

Wragge's Central Weather Bureau, as his relaunched Brisbane office was officially known, relied on government subsidies — £700 from Queensland, £300 from New South Wales and a maximum of £3500 worth of free meteorological telegrams from the federal government. He soon realised he had over-estimated wider support in drought and Federation-straitened 1902. Hoped-for grants from Tasmania, New Caledonia, Fiji and New Zealand that he had factored into his scheme proved to be pipe dreams, and his trickle of contributions from chambers of commerce, business houses, pastoralists and other benefactors failed to cover costs. The annual subscriptions that interstate daily newspapers paid for his forecasts and meteorological synopses left him well short of the revised figure of £1760 per annum — excluding more than £4000 in telegrams — which in August, 1902, he estimated was needed to sustain his ambitions. Reflecting on this dilemma, he conceded the US weather bureau operated on an annual budget of £180,000, the London Meteorological Office on £15,300, and Canada's weather bureau on £15,400.[370]

Wragge's plight was widely debated. In May 1902, *The Bulletin* scoffed at the tone of his pitch for help, such as Wragge's declaration: "We mean to save the Weather Bureau as a great private concern to the Australian nation." This struck the so-called Bushman's Bible as a peevish royal proclamation demanding tribute.[371] In Port Elliot, South Australia, the *Southern Argus* branded him a charlatan and took issue with his naming a recent Antarctic storm after Postmaster-General Drake. In Perth, the *Western Mail* damned him as a hand-wringing professional quack whose "whimsically grandiloquent" language had blinded the public to his poor forecasting record.

On the other hand, the *Evening Star*, in Boulder, Western Australia, warned his weather reports would be missed all over the continent, while the *Bairnsdale Advertiser*, in Victoria,

---

370 *Wragge*, August 28, 1902, p52
371 *The Bulletin*, May 31, 1902, p13

acknowledged his national standing and argued the urgent need for an up-to-date federal bureau.[372] His New South Wales supporters included members of the Royal Exchange of Sydney, the Stockowners' Association, officers of the State Agriculture Department, and petitioners from the Port of Newcastle. New South Wales Premier Sir John See, an ex-coastal shipping company owner, told *The Sydney Morning Herald* he regarded Wragge's service as indispensable. His state's contribution of £300 in two instalments covered the wages of Anderson, Fowles and Wragge's eldest son, Eggie, who joined after Anderson's retirement in September. Bond and O'Mahoney both transferred to other public service positions. Eggie, then 22 years old, had recently returned to Brisbane from a fraught assignment on Mount Kosciuszko.

Andrew Barton (Banjo) Paterson made passing reference to "disciples of Wragge" in a story published in the *Sydney Morning Herald* on July 25.[373] Coincidentally, a few days later in Melbourne, a belligerent New South Wales federal MP, Alfred Hugh Beresford Conroy, likened Wragge's followers to naïve Hottentots in thrall of a self-appointed rain god. Speaking in the adjournment debate on July 30, Conroy railed against his subsidised telegrams, alleging his forecasts were almost invariably wrong. He continued:

> I believe ... he spent a portion of his time amongst the blacks under the impression that they possessed a knowledge of storms

---

372 *Evening Star*, Boulder, Western Australia, June 25, 1902, p2; *Bairnsdale Advertiser*, Victoria, May 21, 1902, p2; *Southern Argus*, Port Elliot, South Australia, June 12, 1902, p3; *Western Mail*, Perth, May 24, 1902, p50

373 *The Sydney Morning Herald*, July 25, 1902, p7, Paterson's aside in his account of a trip to the New Hebrides: "There are no disciples of Wragge among the heathen, so there are no official [rainfall] figures."

which the white man did not. Why do the Government not select one of the rain-gods in South Africa, and send telegrams to him? Just as much reliance is to be placed in one as in the other.[374]

Wragge retaliated by naming a "black and suspicious" Antarctic disturbance after Conroy. On August 20, speaking in Melbourne during the second reading of the Post and Telegraphs Bill, Conroy conceded he had never met Mr Wragge, but had studied his forecasts and found them wanting. While supporting free weather telegrams to and from state meteorological departments, he opposed an amendment by Darling Downs MP Littleton Ernest Groom to extend the privilege to state-subsidised, semi-private concerns such as Wragge's. Surely, the government seal should be reserved solely for sound and certain science:

> If Mr Wragge confined himself strictly to the statement of what he knew — and I presume he ought to know something — I should have no objection, but he goes entirely beyond the range of recognised scientific limits in his weather prophecies … We never find Mr Baracchi, of Victoria, Sir Charles Todd, of South Australia, or Mr Russell, of New South Wales, putting forward their predictions in the same ridiculous way.[375]

Conroy, a surveyor turned lawyer, represented the Southern Tablelands seat of Werriwa from 1901 to 1906 and again from 1913 to 1914. Friends remembered him after his death in 1920, aged 58, as "a sort of Ishmael in politics", an outsider to the party line with

---

[374] House of Representatives (Hansard), July 30, 1902, 'Adjournment, Order of Business, Weather Telegrams, Alfred Conroy'

[375] House of Representatives (Hansard), August 20, 1902, 'Post and Telegraphs Bill, Second Reading'

a "disabling but valiant habit of fighting for hopeless causes".[376] Decades later, two more tough nuts, Labor Party leaders Gough Whitlam and Mark Latham, represented redrawn, pared-down versions of the Werriwa electorate — Whitlam from 1952 to 1978 and Latham 1994 to 2005.

Conroy let Groom's amendment stand when he realised none of his colleagues shared his Wragge aversion. Opposition frontbencher Joseph Cook observed: "The honourable and learned member stands alone; it is a case of Conroy *contra mundum*." In fact, Cook continued, nearly everyone along Australia's coastline believed intensely in Wragge and followed his directions. If his predictions were not on the whole accurate, they would not be acted on, because it sometimes cost hundreds of pounds to follow his advice.

* * *

On August 21, 1902, the day after Conroy's vexation, Wragge made public two letters from Eggie that described his forced abandonment of the snowbound Kosciuszko observatory. In the letters written on July 15 and 24 from Kosciuszko and Jindabyne, respectively, Eggie told how he shifted three hundredweight (340 pounds) of meteorological instruments, field books and personal gear off the summit in backpacks and a broken sled. Wragge published the letters in *Wragge*, his Brisbane bureau's recently launched weekly gazette.

In May, the New South Wales government had discontinued its £400 per annum funding of the Kosciuszko project, arguing

---

376 *Land,* Sydney, December 10, 1920, p12

disbandment of the Queensland bureau voided the states' partnership that had been in place since 1897. Wragge responded by launching an appeal in metropolitan and regional newspapers for £100 to run the Kosciuszko and Merimbula observatories until the end of 1902. Eight months of provisions were already on hand and he hoped educated people would help his valuable work to continue and ensure publication of unique records, then in manuscript form. Surely, such help would be as patriotic in the cause of peaceful British science as that accorded the Empire in time of war?[377]

Meanwhile, the New South Wales government insisted both observatories must be closed at once, and that surplus stores should be taken to Jindabyne — the closest settlement — and sold to help clear debts incurred in 1901–02. Wragge failed to win a postponement until summer, and Eggie began moving out in heavy snow on July 15, assisted by his second-in-charge, Brisbane adventurer Frank Davies, who was an experienced mountaineer.

Wragge serialised and annotated his son's Kosciuszko letters in three consecutive issues of the gazette. Those published in the August 14, 1902, issue of *Wragge*, were written both well before and immediately after July 4, 1902, when the order to quit the observatory reached Eggie. These despatches set the scene for subsequent letters — published on August 21 and 28 — on the nine days it took the Kosciuszko crew to cart the most precious observatory gear to relative safety.

Wragge's threepenny scientific journal, dedicated to *God, King, Empire and People* was the sole Brisbane publication with any news of their rugged retreat from Kosciuszko. Their story was

---

[377] *The Brisbane Courier*, June 13, 1902, p5

overshadowed in Queensland by controversy over the closure of the weather bureau. There was greater interest in New South Wales because of the state's investment of more than £2000 in the Kosciuszko project since 1897. Wragge's plea in mid-June for the powers-that-be to consider the risks of a mid-winter shutdown was backed by *The Sydney Morning Herald*, which endorsed his public appeal for £100 in shilling donations to meet the honorarium due to his observers in lieu of salaries and to provide living expenses. Wragge hoped this would forestall the likely abandonment of the Kosciuszko station's library, observers' belongings and provisions. As for the value of the observations, questioned from the outset by H C Russell, the *Herald* accepted Wragge's assurance the project had been endorsed by the International Meteorological Conference in Paris in 1900.

On August 6, the *Herald* ran an edited version of Eggie's account of his retreat from the mountain. He wrote the story, published two days earlier in the Cooma-based *Manaro Mercury*, while marooned in Jindabyne, waiting for an advance from the state government to pay Harris, get Davis back to Brisbane, and square with the hotel where they had been stuck for the previous fortnight. Eggie finished by protesting he felt abandoned: "My very best efforts have been put forward to ensure the safety of government property taken from the observatory and now I find myself in this fix."

The New South Wales Treasury grudgingly sent £30, enabling Eggie to pay the outstanding accounts and return to Brisbane in early September 1902. Later that month, writing in his magazine, Wragge attacked the New South Wales government's disrespect for true science. He feared the worst for the valuable books, equipment and stores left behind at the mercy of the wind, weather and vandals. How could a relatively trifling sum not have been available to at least enable observations to continue until the end of 1902? He was embarrassed at the sudden curtailment of meteorological data from Kosciuszko that he had promised to organisers of British and German Antarctic expeditions then under way.

In February 1903, the *Sydney Mail* reported the observatory had been trashed by a procession of scavengers from the Murray River equipped with packhorses. A roving journalist found the observers' hut occupied by two tramps who were sharing with a multitude of rats a diminishing stockpile of biscuits and preserved meat and milk. The library books and a New South Wales government theodolite were untouched at that stage. In April 1904, police found 40 cans of stolen kerosene stashed in a rocky outcrop near the summit.[378]

---

378 *Sydney Mail*, February 4, 1903, p279; *Grey River Argus*, Greymouth, New Zealand, April 12, 1904, p2

CHAPTER 17

CHARLEVILLE DEBACLE, EGESON REVISITED, THE FEDERATION DROUGHT ENDS, 1902–03

You look at me. You can't see me. You can only see the suit of clothes — and God knows that's not much. You can't see the man, the brain, the mind. You can only see my shell. — Clement Wragge, from a lecture reported by the *Windsor and Richmond Gazette*, December 21, 1912.

In mid-August 1902, a freight train trundled through Queensland's parched outback, carrying four kegs of blasting powder, four cases of gunpowder and six custom-made cannons. The people of Charleville at the end of the line had ordered the guns eight months earlier to shoot at clouds that might break the drought. Harvey and Son's Globe Ironworks in Brisbane built them to specifications from Mayor Albert Steiger of Windisch-Feistritz, Austria.

When erected, the Steiger Vortex guns' sheet-metal funnels looked like a brace of giant uptilted megaphones, each anchored to a mortar box. Writing in his magazine, Wragge dubbed them the "Billington Battery", acknowledging Charleville Mayor William Billington's enthusiasm for Steiger's cloud-shooting theories. He also christened each gun individually, naming one after himself and the others in honour of Steiger, Austrian engineer Herr Suschnig, Brisbane engineer Enos James Harvey, Premier Robert Philp, and Charleville's MP John Leahy. Charleville Municipal Council's appeal raised £100, covering nearly all construction costs, and the Queensland Government gave another £16 — the cost of a single cannon — donated the explosives and waived freight charges for the 500-mile rail journey.[379]

Life was spreading Wragge thin that August. His bureau was understaffed because of Archie Anderson's illness and the Kosciuszko debts were mounting during Eggie's forced retreat. Consequently, he asked Billington to set up the cannons, assuring him he would take charge when the weather was favourable.[380] Described by Brisbane's *Truth* newspaper as Charleville's big gun, Billington had been elected mayor unopposed in January 1902, and soon converted Brisbane business contacts to the rainmaking cause. But when Wragge arrived on September 12, he found Billington in a cranky mood and unwilling to help erect the guns, then scattered around the town. "He would do nothing, be responsible for nothing," Wragge recalled. Billington appears to have been spooked by mishaps in trial firings and public backlash on the council's investment.[381] Nevertheless, Wragge had better luck with council staff and the local telegraph manager, who helped erect the battery as specified by Steiger — three abreast in two lines, each three-quarters of a mile apart, all mounted on wooden stands.

---

379 *Wragge*, August 7, 1902, p27
380 *Ibid*, p43
381 *Truth*, Brisbane, October 5, 1902, p7; obituary of W G Billington, *Brisbane Courier*, February 3, 1926, p19; obituary of G Barber, *Brisbane Courier*, April 11, 1932, p13; *Wragge*, October 2, 1902, pp. 89-90

They finished preparations on the morning of Sunday 14 September, with promising clouds building in the north-west. But then an authority unnamed by Wragge stepped in and demanded a postponement in deference to the day of rest. "The church bells were ringing, the Salvation drum beating, and the clouds might gather yet more and more, and nobody cared," he grizzled later.[382] In a different version of his visit, published in Brisbane's *Telegraph*, the experiment was blighted by a strong westerly wind that covered the town in great clouds of dust. The *Telegraph*'s Charleville correspondent also blamed the weather for the flop of two lectures Wragge had planned for the town's Albert Hall — on rainmaking and astronomy.[383]

Regardless, Wragge tested one of the guns on Sunday afternoon and persuaded Billington to fire another trial shot the next morning. Wragge and an assistant then returned to Brisbane, leaving instructions on what to do when suitable conditions next arrived. The mortars required exactly seven-and-a-quarter ounces of powder each and were to be fired at two shots per minute. On September 26, when clouds again massed over Charleville, a revitalised Billington prepared for a full-scale barrage, beginning at noon. He had 10 shots rapid-fired from each of the six cannons, yielding a few raindrops and, at 2 pm, a slight shower. In a reprise at 4.30 pm, two of the cannons imploded as the clouds resisted Billington's bullying.[384] The October 9 issue of *Wragge* gave this peeved summary of the mayor's experiment:

Since our return to Brisbane, news arrived that the Mayor

---

382 *Wragge*, October 2, 1902, p90
383 *Telegraph*, Brisbane, September 16, 1902, p2
384 *Wragge*, October 2, 1902, pp 89–90

had initiated experiments on monsoonal clouds, and that two guns burst. Ten shots only were fired. We at once telegraphed our friend the telegraph manager, and found that despite the foregoing explicit instructions, the mortars had been overcharged. *What more can one do?*

A columnist in *The Queenslander* congratulated Billington on his courage, but asserted the experiment had proved nothing except that the guns were dangerous to the general public. That no-one was killed was probably due to good luck and the native caution of Charlevilleans.[385] In coming decades, witnesses embroidered Charleville's drought-defying renown. In 1920, *Freeman's Journal* recalled the town had turned out under umbrellas during the firing of four of the cannons, placed in rectangular formation around its perimeter:

> The four monsters were fired simultaneously, the blast being heard 30 miles away. Some of the spectators were knocked over by the concussion, windows were shattered in all directions. Two of the guns burst, and the promising clouds scattered and disappeared. The oldest inhabitant is laughing yet.[386]

In 1934, Brisbane's *Telegraph* published a letter from one of the gunners, who recalled Charleville had been encircled by six or eight cannons, spaced a quarter of a mile apart under Mr Wragge's supervision. Several days of firing had yielded absolutely no rain.[387]

In 2019, two of the cannons were still in Charleville, mounted

---

385 *The Queenslander*, October 4, p737
386 *Freeman's Journal*, January 1, 1920, p11
387 *Telegraph*, Brisbane, July 12, 1934, p6

at the entrance to the town's Bicentennial Park. Murweh Shire Council listed their installation as a Local Heritage Place, signifying the guns' contribution to Charleville's sense of identity and the South-West Queensland shire's unique history.

* * *

Wragge complained to Brisbane journalists about the Charleville debacle and grumbled in his weekly magazine over his cold-shouldering by the mayor. In the October 2 edition, he joked Charleville was under the spell of a cyclical solar quirk that he now blamed for prolonging the drought. The sun's abnormally long quiescence — in other words, absence of sunspots — had not only dried up the rain but also becalmed the town. The mayor's waning interest in the Steiger cannons was surely evidence of his sunspot-minimum-induced torpor. The *Brisbane Courier* ridiculed this theory, calling it Wragge's latest craze, concluding: "If the Charleville experimenters erred, they maintain it was through ignorance, not the want of interest."[388]

From early March 1902, Wragge devoted many of his daily weather columns to puzzling over the drought. Writing in the *Courier* on March 5, he noted that during the dry summer which had just ended, the flow of anticyclonic systems that typically brought rain to eastern Australia had dipped south of the continent. He continued, "We believe it to be connected with distribution of ice around Antarctica ... but primarily referable to periodic changes

---

388 *Wragge*, October 2, 1902, p43; *Brisbane Courier*, September 24, 1902, p8, and September 29, 1902, p8; *The Queenslander*, October 11, 1902, p793, and October 25, 1902, p905; *Truth*, Brisbane, August 17, 1902, p6

in progress in the sun's atmosphere." He revisited research by British father and son astronomers Sir Norman and Major Jim Lockyer on the pattern of recurring floods and famines in India and Mauritius. In 1900, the Lockyers completed a study correlating maximum and minimum solar disturbances with fluctuating rainfall in both countries between 1877 and 1886. They suggested the sun was hotter at sunspot maximum and cooler at minimum. But they cautioned they had not yet investigated the impact of sunspot minima and maxima on rainfall in Australia and the Cape Colony.

In 1902, Wragge stepped up his denouncements of ringbarking and pleas for conservation of forests and urgently needed tree-planting schemes. He warned Australia's "suicidal" land-clearing would bring devastation similar to Spain's hideously eroded Plains of Castille, which he had seen first-hand and described as "weary miles without a glimpse of a once grand forest". He concluded:

> Depend upon it, interference with Nature on an extravagant basis, and without bringing to bear counteracting agencies in accordance with science which is wedded to Nature, is a profound mistake, and must bring in the long run a punishment that shall "fit the crime" and from this there can be no escape.[389]

* * *

On May 8, 1902, the eruption of the Mount Pelee volcano in the West Indies, killing up to 40,000 people, alerted him to the Lockyers' theory of a yet-to-be-defined link between solar

---

[389] *Newcastle Morning Herald*, April 12, 1902, p5; *Brisbane Courier*, January 10, 1902, p6

and seismic activity. Sir Norman Lockyer told *The Times* the disaster on the French island of Martinique had coincided with a pervasive, well-defined sunspot minimum. His study of seismic events since the 1830s showed — conclusively, in his opinion — that the worst earthquakes and volcanic eruptions had occurred around the dates of sunspot maxima and minima. Intending to further refine his theory, Lockyer appealed to observers on islands around Martinique for their barometric readings for the preceding two months.

Wragge commented in special weather notes published in Brisbane a few days later, on the conjunction of the drought, the Martinique catastrophe and unusual conditions in Antarctica:

It is as though our globe were rising in protest, and simply as we deplore the West Indies catastrophe, we have reason to be thankful, for the eruptions prove the earth as a planet still has sufficient vital force left for support of the human race for ages yet to come, ere she becomes a lifeless blank, as our grotesque satellite, the moon.[390]

Through 1901 and 1902, Southern Ocean and South Pacific sailors reported seeing masses of giant icebergs. In August 1901, the New Zealand Shipping Company's London-bound steamer *Rimutaka* threaded through an estimated 700 along the 50th parallel, between the 148th and 123rd longitudes, on the first leg of their voyage to Cape Horn. In Sydney in December 1901, H C Russell recalled past reports of innumerable icebergs had occurred in the remarkably hot years of 1895, 1896 and 1897. He wondered

---

390 *The Telegraph*, Brisbane, May 12, 1902, p4

if the latest flotilla had been caused by a volcano or earthquake shattering the edge of the Antarctic ice cap.[391]

* * *

In May 1901, Lockyer's son William, proposed a 35-year sunspot and weather cycle, quoting Eduard Bruckner and the almost-forgotten Charles Egeson in support of his theory. In a paper to the Royal Society, Lockyer examined variations in the occurrence of maximum and minimum solar activity in the 67 years between 1833 and 1900. Using sunspot and solar-magnetic records for the six 11.1-year solar periods involved, he found recurrent patterns when comparing the first bracket of three solar periods with the second. He observed an alternate increase and decrease in the length of the sunspot period, reckoning from minima to minima.

The intervals between the sunspots' occurrence also advanced and contracted. Lockyer — then-assistant director of the British Solar Physics Observatory at Kensington — praised Egeson's *Weather System of Sunspot Causality*, noting Egeson had published it a few months before Bruckner's *Klima-Schwankungen seit 1700 (Climate Fluctuations since 1700)*. Bruckner had used weather records from nearly every part of the civilised world in deducing a mean periodical variation of 34.8 years (plus or minus 0.7 years) in climates in all parts of the globe. But he had been unable to harmonise it with Rudolph Wolf's contentious hypothesis of a 55-year cycle.

On the other hand, Egeson had correlated occurrences of sunspot maxima with historic instances of wild and windy

---

[391] *Evening News,* Sydney, December 14, 1901, p9

weather in Sydney, such as in April 1837 and 1870. Lockyer wondered if the approaching 35-year sunspot maximum would be marked by solar, meteorological and magnetic phenomena like those reported roughly 35 years and 70 years earlier. He did not doubt that during the seven decades his study spanned, aurorae, magnetic storms and meteorological events harmonised with secular variation of sunspots.[392]

In his weather notes on May 20, 1902, Wragge told of having begun to watch the sun closely from his home observatory for any hint of respite from the quiescence he now believed was prolonging the drought. In retrospect, he conceded that in 1889 he had slighted Charles Egeson's science, when, in fact, the young meteorologist's only fault had been expressing his views somewhat rashly. He now accepted Egeson's research had a rational foundation and that his conscientious labours had deserved better treatment from his superiors.

* * *

The folklore of Egeson's sunspot theories flourished as the drought continued. In December 1901, the *Pastoralists' Review* in Melbourne recalled his warning that the coming sunspot epoch would bring famine as dire as any in the history of New South Wales or the dried-up river days of Aboriginal legends. Wragge's nod to Egeson's prescience was widely reported. In July

---

[392] Lockyer, William J S, *The Solar Activity, 1833–1900*, communicated by Sir Norman Lockyer, received April 29, read May 23, 1901, Proceedings of the Royal Society of London, published September 1, 1901, https://royalsocietypublishing.org/doi/pdf/10.1098/rspl.1901.0047, accessed October 14, 2022

1902, Sydney's *Daily Telegraph* rehashed the story of Egeson's dismissal and concluded that regardless of H C Russell's scorn of the 35-year cycle, his whipping boy's seasonal forecasts had been at least partially fulfilled.[393]

In October 1902, the *Newcastle Morning Herald's* Adelaide correspondent related "a sort of Egerson-worship [sic]" based on a belief he had predicted that year's culminating horror of the Federation Drought. Egeson had — possibly posthumously — dethroned Clement Wragge as the greatest of Australia's weather prophets and would, should he manifest himself in Adelaide, be surely a certainty to replace Todd by popular acclaim as Chief Meteorologist.[394]

None of this helped Egeson. In 2010, a reappraisal of his work found circumstantial evidence that sometime after his dismissal he had been confined in a lunatic asylum and died young. US chronobiology scientist Franz Halberg, lead author of *Egeson's (George's) Transtridecadal Weather Cycling and Sunspots*, sifted traces of the young meteorologist's life and concluded he had bucked his era's conventional scientific wisdom with tragic results. Halberg found the nub of Egeson's griefs scrawled on the back of an 1887 photograph of seven reclining Sydney Observatory staff flanked by the bald and portly H C Russell, hands in pockets, and the tall and vaguely haughty Egeson, with crossed arms and polished shoes:

> Mr Charles Egeson was the acting meteorologist for 12 months while Mr Russel [sic] was in Europe. Mr Egeson was the originator of daily weather forecasts in 1887. Mr Russel tried to abolish them. The Evening News agreed with Mr Egeson. Fought together in 1890 & won.

---

393 *The Pastoralists' Review*, December 12, 1901, p703; *The Sydney Morning Herald*, March 11, 1902, p6; *Daily Telegraph*, Sydney, July 4, 1902, p5

394 *Newcastle Morning Herald*, October 8, 1902, p7

Halberg gleaned Egeson's story from archived newspapers, including rumours in 1902 and 1903 of his confinement and likely death in a lunatic asylum. In February 1903, Dudley Eglinton, an amateur astronomer and founder of Queensland's technical education system, paid tribute to Egeson as a man of genius who had died in a lunatic asylum. Writing in the *Brisbane Courier*, Eglinton endorsed Egeson's weather system as a thoroughly scientific work based on most elaborate data.[395]

Wragge and Egeson were surely kindred spirits. In 1887, they upset the denizens of Australia's old-school meteorology with their pioneering forecasting for daily newspapers in Brisbane and Sydney, respectively. They were outsiders, both self-taught and neither particularly afraid of failure. Forerunners of 21st century celebrity forecasters, they inevitably risked notoriety, given the public's thirst for weather news. They were also alike in their apprehension of God in nature. Egeson was, according to the *Evening News*, "As orthodox as could possibly be desired, ascribing everything as appertaining to God."[396] Religion was science and science religion, Egeson declared at a public lecture in Sydney in June 1890.

Wragge repeated in many of his lectures that atheism was not an option for true scientists. Egeson's pursuit of long-range forecasting in 1888 and '89, which Wragge picked up in 1902, was grounded in faith in the God of nature. He assumed the faultless machinery of weather and climate could be seen and understood in recurring patterns of nature, chronicled through the ages. In 1889, in the conclusion of his sunspot pamphlet, Egeson wrote:

Climate, in its broadest sense, includes the changes of everything

---

395 *Brisbane Courier*, February 28, 1903, p6
396 *Evening News*, June 17, 1890, p4

on the earth, and in space around us. Climate is the language of God, a language, once spoken, reverberating for all time ... Science is our alphabet — imperfect, it is true, but still enabling us to decipher the sublime records, and child-like we rejoice to read even a single word.[397]

Wragge was also convinced of hidden truths in nature, to be discovered through disciplined physics and chemistry. That was his Theosophist yen. He wrote in Queensland's *The Worker* newspaper in 1898:

> All Nature speaks to us of the Infinite Dynamo of whom you and ourselves form a part — every zephyr, every sand ripple, every list in the trees ... has its own tale to tell — some sublime law on which its evolutions depend.[398]

In November 1890, Egeson wrote he considered Wragge a superior meteorologist to H C Russell. Had Egeson wished to better his grasp of meteorology, he would have joined Wragge in Brisbane, not worked at the Sydney Observatory, where he complained he had learnt nothing.[399] Wragge, in his apology to Egeson 12 years later, described him as a brother scientist.

Egeson's last hurrah, at least the last recorded in the National Library of Australia's newspaper archive, came later in December 1890 when he published a weather almanac replete with parting shots at Russell and Public Instruction Minister Joseph Carruthers. Egeson left almost no record of his life before joining the Sydney

---

397 Egeson, pamphlet on sunspots, 1889, p63
398 *The Worker*, Brisbane, May 28, 1898, p11
399 *Freeman's Journal*, Sydney, November 8, 1890, p20

Observatory as a clerk in 1884. In October 1889, in a letter to Sydney's *Evening News*, he identified himself simply as a young man with four or five years of experience in meteorology. Shipping notices show a Mr and Mrs Egeson and six children arrived in Sydney on April 24, 1884, from Colombo, Ceylon.

Egeson's sunspot book is the work of a thoughtful, well-read individual. Halberg applauded the scholarship evident in his calculation of transtridecadal cycle lengths and his resourceful collection of historical data.[400] Egeson himself believed his study of sunspot causality was more extensive than any previously made. It's telling his acknowledgements included thanks to attendants at Sydney's Free Public Library for their help in his search for old weather records. He strove in Sydney to better himself, support his largish family and make meteorology more truly useful to humanity. To the latter end, Egeson quoted in his book the views of Herbert Spencer and Sir William Herschel on science and weather forecasting, respectively. Spencer believed the object of all science was to enable the prediction of natural phenomena. Herschel said of meteorology: "A probability of a hot summer or its contrary would always be of greater consequence than the expectation of a few rainy days."[401]

* * *

---

[400] *History of Geo and Space Sciences*, 1, 49–61, 2010, 'Egeson's (George's) transtridecadal weather cycling and sunspots', F Halberg, G Cornelissen, K-H Bernhardt, M Sampson, O Schwartzkopff, and D Sontag, 2010, https://hgss.copernicus.org/articles/1/49/2010/ accessed October 17, 2022

[401] Egeson, C, *Egeson's Weather System of Sunspot Causality: Being Original Researches in Solar and Terrestrial Meteorology*, Sydney, 1889, p31, National Library of Australia digital copy, https://nla.gov.au/nla.obj-2558311020/view?partId=nla.obj-2558321668#page/n8/mode/1up, accessed October 24, 2022

In September 1901, Russell elaborated for the Royal Society of New South Wales on his long-held theory that Australia's dry spells followed a 19-year lunar cycle. In effect, he argued rainfall ebbed and flowed with the moon's gravitational pull. His correlations of the moon's motion and declination with rainfall suggested that when the moon's course was southward in the Southern Hemisphere, more rain fell than when the moon moved to the north.[402] In Victoria, Pietro Baracchi conceded the 11-year sunspot cycle might influence rainfall, but could not spare time for a thorough study. Todd remained sceptical of sunspot theory, as did many other mainstream meteorologists. In India, retiring chief meteorologist John Eliot had begun collecting data from wide expanses of the Pacific and Indian Oceans that his successor, Gilbert Walker (1868–1958), used to identify strong alternating patterns in global weather. Walker recalculated the Lockyer's sunspot-rainfall correlations, found them unreliable and discounted any scientific basis for seasonal forecasts based on solar cycles. But he used Jim Lockyer's observation of a "barometric see-saw" between India and South America in his identification in 1924 of a periodic Southern Oscillation, bringing what today are identified as recurrent La Niña floods and El Niño droughts to Southern Hemisphere continents.

* * *

In November and December 1902, fringes of a monsoonal build-up drenched western Queensland in more than four inches

---

402 Miller, Julia, 'What's Happening to the Weather? Australian Climate, H C Russell and the Theory of a Nineteen-Year Cycle', *Historical Records of Australian Science* 25, 2014, p24, http://dx.doi.org/10.1071/HR14006, accessed October 17, 2022

of rain. The *Western Champion* in Barcaldine — 100 miles north of Charleville — hoped the worst was over after six years of drought: "It may truthfully be alleged the drought, if not altogether broken, is checkmated at least." But the southern states missed out. In October 1902, Melbourne's Lord Mayor opened an appeal for drought relief that raised £19,000 and helped 1670 families. Three months later, the Lord Mayor of Sydney began a relief fund that yielded £23,000 distributed through 1903.[403] Sailors arriving in Sydney from New Zealand in early 1903 saw dust clouds 150 miles out to sea.

In December 1902, water authorities from New South Wales, Victoria and South Australia issued an inconclusive report on sharing the Murray River's droughty trickle. They took on fraught issues of supply and demand — still intractable in 2022 — in an ambitious interstate Royal Commission on the conservation and distribution of water from the Murray and its tributaries. Commissioners Joseph Davis of New South Wales, Stuart Murray of Victoria, and Frederick Newman Burchell of South Australia examined 294 witnesses at 63 meetings held in all three states. The most tangible result, some years later, was the development of the Murrumbidgee Irrigation Scheme, harnessing water from the Barren Jack Dam (later known as Burrinjuck) 250 miles upstream from the Murrumbidgee River's junction with the Murray. The Upper Murrumbidgee and Upper Murray both drained a pristine, dependable annual torrent from the Snowy Mountains' summer thaw.

* * *

---

403   *Western Champion,* December 7, 1902, p4

A recent reassessment of the Federation Drought put its length Australia-wide at seven to eight years, from 1895–1903, and argued it had been caused by three El Niño events that closely followed each other. The first, from 1895 to '98 brought heatwaves, bushfires and dust-storms to New South Wales, Victoria, Tasmania and South Australia. A weak La Niña then brought some relief, before another El Niño event in 1899–1900, followed by another short break and a profound El Niño event beginning in 1901 and ending during 1903.[404]

On March 9, 1903, in Townsville, north Queensland, the break-up was signified by a cyclone that hit without warning, killing ten people, seven of them inmates of a collapsed hospital ward. Wragge was on a lecture tour of Victoria at the time, still raising funds in hope of saving the Brisbane weather office and paying debts from the ill-fated Kosciuszko project. His long-time assistant Edgar Fowles later apologised for having lost track on February 27 of the storm he named Leonta. He reported that the neglect of Queensland's cyclone-warning network since Federation had hampered monitoring of dangerous weather systems. Leonta had been first located over the Coral Sea north-west of New Caledonia on February 19. He complained that only four of the thirteen observation posts down the coast from Cooktown still sent barometric readings to the Brisbane bureau — daily at 9 am, bereft of wind speed data, never by urgent telegraph

---

[404] Garden D, 2010, 'The Federation Drought of 1895–1903, El Niño and Society in Australia', *Common Ground: Integrating the Social and Environmental in History*, Massard-Guilbaud G, Mosley S (eds), Cambridge Scholars Publishing, United Kingdom, 270–92, https://climatehistorydotcomdotau.files.wordpress.com/2020/02/0a96b-garden_book_chapter_2010.pdf, accessed 25 October 2022

and never on Sundays. Leonta was then the most destructive cyclone experienced in any of Queensland's coastal cities and towns, albeit less damaging than Mahina, which, in March 1899, flayed the pearling fleet in Princess Charlotte Bay with a death toll of 400.

Wragge wrote the leading article for the March 12, 1903, edition of his eponymous magazine knowing only that the Townsville weather station had been off air for some days. He attacked the Queensland Post and Telegraph Department's neglect of his bureau's weather data and the federal government's virtual sanctioning of slipshod meteorology. Couldn't Australia's commercial interests see how much they lost through the shambles of disorganised and under-staffed state-run weather bureaus? Australia needed a central office in daily contact with all parts of the Commonwealth and its outposts beyond. He continued, "Why cannot we do likewise? We must move forward with the times or be left behind. The cost of such a department would be but a premium on life and property." In the United States and Europe, meteorological records were obtained and stations extended, regardless of cost. There could not be too many well-placed recording stations. Undoubtedly, the loss to Australia and to science by the closure of his Kosciuszko observatory was incalculable.

* * *

On December 1, 1902, Wragge began his lecture tour of Melbourne and country Victoria in growing anxiety. He was puzzled by the New South Wales government's quibbling over £384 in outstanding debts on the Kosciuszko project. Despite having justified — he believed — the extra costs entailed in abandoning the observatory in mid-winter, he had received yet another please explain from the New South Wales Treasury. The government led by Sir John See had sent all the unpaid Kosciuszko accounts to Queensland Premier Philp, anticipating Wragge had set aside funds to foot the bill. In fact, there was only £16 left in the Kosciuszko kitty. Liabilities included accounts for supplies bought shortly before the observatory's forced closure, in expectation of business as usual through 1902–03. Beyond these frustrations, Wragge was disappointed that John See had declined to fund publication of data from the high-level and low-level observatories.

Wragge's tipping point came in November, when a Jindabyne storekeeper demanded he at once honour the £20 promissory note he had meant to cover payment of Frank Harris, Eggie's hard-working guide. Wragge left by train for Melbourne in a defiant mood, armed with a cache of his latest lantern slides, leaving Fowles in charge of the Brisbane bureau. While in Sydney, waiting for the train south, Wragge tried without luck to confront the New South Wales Premier. "I called at the Treasury offices in Sydney, I haunted the Premier's office, but I could get no interview with Sir John," he later told a Melbourne journalist.

His subsequent tour of 40 or so Victorian country towns was a wing-and-prayer crusade to keep the Brisbane bureau afloat, pay the Kosciuszko liabilities and personal debts too. He explained, "I hoped thereby to raise the wind and keep things going generally — myself and those dependent on me included."[405] The dependants

---

405 *Wragge*, February 5, 1903, Supplement, 'The Kosciuszko Observatory — To the People of the Australian Commonwealth, and Colleagues in all Parts of the World', p3

included Nora and most of their eight children, the youngest of whom, Lindley, was 11 years old. Allowing for touring costs, Wragge had faint hope of breaking even from one-night stands in Mallee and Gippsland towns — tickets two shillings or one shilling a head, special rates for school children. He was unflagging, though, banking on sympathy from Melbourne audiences at least. Advertisements for his four lectures at the 800-seat Athenaeum Hall promoted him as "*facile princeps* amongst Australian platform orators" with subject matter of great public importance. He had hired Melbourne theatrical manager Joseph Gibbs to promote the tour.

The first two nights, in early December, were well attended but neither pleased *The Age*, which expected much more from a self-proclaimed expert meteorologist. His facile arguments for water conservation and irrigation were most disappointing, given the Victorian government's waste of £4 million on irrigation schemes had been inspired by similar impractical gush. *The Age* was incredulous at his squandering of the chance to seriously discuss options for intelligent mastery of the weather. Beyond that, his staccato-rhythmed speech was vexing to follow — now loud and rapid as a small tornado, next soft as a whisper in a cave of winds. And had he really needed those slides of semi-naked South Sea Island women? The newspaper scolded him for lacing his lectures with what it termed his irresistible fascination of the eternal feminine, for example a bizarre aside on the curse of corsets, so impractical in the Antipodes.[406]

Six weeks down the track on the country leg of his campaign, a stringer for *The Bulletin* in Sydney described the lectures this way:

---

406 *The Age*, Melbourne, December 3, 1902, p6

[He] starts off with a spout of loyalty, in which he mainly avers a scientist must necessarily be loyal to the Empire, wanders through a universe of chaos in which rant, scriptural quotation, French phrases (always translated for the audience), scraps of Latin, repetition of adjectives and gibes at John See form the staple.

* * *

Sir John See had a sole, significant soft spot for Wragge. He trusted his storm warnings. A coastal shipping company owner, See was ever-ready to vouch for his life-saving forecasts. Hence in June 1902, he helped re-establish the Brisbane bureau as a semi-private entity by pledging £300 to top up the Queensland Government's offered £700 subsidy. This was a pragmatic decision, anticipating a federal bureau of meteorology was likely a long way off. See and Wragge shared a couple of prosaic similarities in being English-born, middle-aged and having many offspring — ten and nine, respectively. But they were as different in social background, politics, character as they were in physique. See was a Progressive Protectionist and persistent critic of Wragge's benefactor, Free-Trader Sir George Reid. When in Opposition in New South Wales, See frequently ribbed Reid for alleged fiscal failings. In the Federation reshuffle, Reid became Leader of the Opposition in the new Commonwealth Parliament and See succeeded his Melbourne-bound Protectionist colleague Sir Joseph Lyne as New South Wales Premier. See was also a founder and part-owner of the Protectionist-sympathising *Australian Star* newspaper. His hard line on the Kosciuszko observatory was probably grounded in disapproval of Reid's misguided largesse. Between 1897 and 1902, New South Wales paid £2231 for the establishment and maintenance of the Kosciuszko station and the low-level observations in Merimbula and Sydney.

Writing in his magazine, Wragge fretted over See's apparent enmity. Was he part of a conspiracy by other, unnamed, Kosciuszko critics — that army arrayed against genuine scientific research, "Those to whom my name is as a red rag to an enraged bull?" He concluded, "Well may we pity science in Australia."[407] On December 24, 1902, Wragge was in Camperdown, 120 miles south-west of Melbourne, preparing to lecture on Noble Astronomy and Meteorology, when a solicitor's letter arrived demanding immediate payment of £20 plus interest owed to the Jindabyne storekeeper, or else face legal proceedings. The letter from a Sydney solicitor had taken some time to reach Camperdown, having been first sent to Brisbane. The solicitor also asserted the New South Wales government had repudiated all of the Kosciuszko debts.

Wragge immediately wrote letters of protest to the Melbourne, Sydney and Brisbane press, goading See into an inflammatory clarification. *The Age*, in Melbourne, reported on January 8, 1903, See's contention that Wragge had established the observatory by subscription and, to his knowledge, the New South Wales government had contributed only £200, and never formally accepted responsibility for the project. See's Sydney newspaper, the *Australian Star*, weighed in with two censorious column pieces, describing Wragge on January 8 as a poetic soul too lofty for paltry financial details, then on January 10 as a bounder upset

---

407 *Wragge*, February 5, 1903, Supplement, p3

in his quest for a fat job as Commonwealth meteorologist. The latter column strongly implied Wragge was double-dipping in asking New South Wales to pay the £384 over-run in Kosciuszko accounts, having seemingly taken for granted the See government's £300 subsidising of the Brisbane weather bureau in June 1902.

On January 16, 1903, metropolitan newspapers ran a New South Wales government statement clarifying See's estimation of the state's stake in the project and confirming that since 1897 its contribution had been £2231. The state had handsomely supported the project for five years, but had warned last year it could not meet expenses in connection with the observatories beyond June 30, 1902. Wragge had been told that if his outlays exceeded the parliamentary vote of £400, only the barest of margins would be allowed. How could he have seen fit to incur an unacceptable £384 debt, knowing it could not be borne by the New South Wales Treasury?

Public opinion fluctuated. Melbourne's *Herald* was mystified why New South Wales had supported the Queensland meteorologist's folly, especially as the Kosciuszko station had been established against the advice of all the Australian astronomers.[408] On March 11, Wragge returned to the Athenaeum Hall for his fifth and final Melbourne lecture. *The Age* was kinder this time, sympathising with his struggle to keep a foothold on Australia's rooftop and his gripe with the nation's neglect of science. It was a sweltering night in the historic auditorium where, in 1895, Mark Twain gave his final Melbourne lecture, and where, in 1896, motion pictures were screened for the first time in Australia. *The Arena Sun* reported Wragge had captured his large audience with an unorthodox lecture illustrated by fine lantern views of

---

408 *Herald*, Melbourne, January 8, 1903, p4

Kosciuszko and Merimbula. In the *Sun's* view, he had a knack, like George Reid in his Free Trade discourses, of winning over scoffers:

> Those that filled the Athenaeum last Wednesday evening at least understand that meteorological science demands the courage, the heartbreak and humiliation that her votaries in Australia may have to experience. [409]

Next, Wragge took his "Save Kosciuszko, Help Science" campaign to Sydney, virtually on See's doorstep, with lectures at the Lyceum Theatre and Queen's Hall on March 26 and 27. He was helped by theatrical manager Bland Holt[410], who offered free use of the Lyceum when Sydney City Council declined to subsidise hire of Sydney's town hall. The *Daily Telegraph* reported the small audience at the first lecture had frequently applauded Wragge's aggrieved speech. He had called on the Premier, as a fellow Britisher, to follow the Golden Rule and treat him justly, as he himself would wish to be treated:

> If Sir John See had done his duty to me and to the Kosciuszko observatory, there would have been no reason for me to have stood on this platform to ask for help in liquidating the debts which are really the debts of the State. I have worked night and

---

[409] *The Age*, Melbourne, March 12, 1903, p8; *Arena-Sun*, Melbourne, March 19, 1903, p8; 'Melbourne Athenaeum', Wikipedia, https://en.wikipedia.org/wiki/Melbourne_Athenaeum, accessed October 17, 2022

[410] Joseph Thomas (Bland) Holt (1851–1942), a producer, theatre entrepreneur and actor, leased the Lyceum Theatre in Sydney and Theatre Royal in Melbourne. Shoesmith, Dennis, 'Holt, Joseph Thomas (Bland)', *Australian Dictionary of Biography*, National Centre of Biography, Australian National University, https://adb.anu.edu.au/biography/holt-joseph-thomas-bland-3785/text5985, accessed October 17, 2022.

day to serve the people of Australia and I think I deserve some consideration.[411]

The *Australian Star*'s brief account omitted Wragge's vexation with See and blamed the disappointing turn-out on wet weather.[412] Neither of the papers reported the Premier had agreed to meet Wragge and a group of his supporters in Parliament House before the second of the Sydney lectures.

---

[411] *Daily Telegraph*, Sydney, March 27, 1903, p6
[412] *The Australian Star*, March 27, 1903, p7

CHAPTER 18

THE WHOLE KOSCIUSZKO STORY, 1903

When sorrows come, they come not single spies, but in battalions. — William Shakespeare, *Hamlet,* 1602.

Eggie digested conflicting versions of Sir John See's back-down before writing to Sydney's *Daily Telegraph* in early April 1903. The first news, on March 28, the day after his father and supporters confronted Sir John, had seemed wonderful. The *Sydney Morning Herald* reported See had agreed to honour payment of the contentious Kosciuszko bills and would ask the other states to consider jointly reopening the observatory. In this version, See had scolded Wragge for the budget over-run in 1901–02 but, in appreciation of the station's storm-warning role, he was willing to help all he could — if the other states kicked in. The delegation led by John Hurley, MP, reportedly applauded Sir John's vow that if he could save one life by spending £400, he would gladly do so.[413]

But two days later, Sydney's *Daily Telegraph* revealed See had interrogated Wragge over each of the items on the provisions account. He read them out, one by one, incredulously: "Eno's Fruit Salts, 20 pounds of Signet tobacco, twelve-and-a-half

---
413 *The Sydney Morning Herald,* March 28, 1903, p5

pounds of cigarette tobacco, sundry Old Judge cigarettes, shotgun cartridges, bottles of stout and whiskey ..." Had any of this been really necessary? In addition, John See listed clothing purchases: tweed suits, gloves, overcoats, shirts, braces, a sweater and silk handkerchiefs. What wilful extravagance! He was particularly riled over a £20 bill for whiskey. Wragge replied the crew usually had a couple of bottles on hand, strictly as medical comforts. And, surely, his poor lads enduring the fearful solitude of the mountain deserved some consolation — a few smokes at least. This goaded See into a tirade on the folly of past New South Wales governments in subsidising to the tune of £2231 a project supposedly of benefit to all Australians. How dare Mr Wragge complain of shameful treatment!

In reply, Wragge pleaded his lecture tour had flopped and he could barely provide for himself and family, let alone pay off the Kosciuszko debts. Hearing this, See was apparently calmed and relented with a promise that New South Wales would meet its obligations. But See repeated his belief all the states should have shared the observatory costs.[414]

Brisbane's *Telegraph* republished this report on April 3. The same day, Eggie wrote to Sydney's *Daily Telegraph* defending provisioning during his four-and-a-half years of involvement with the project. He had been officer-in-charge from April 1902 until the abandonment of the main station in July 1902, and justified purchase of everything on the contentious list. Warm clothing had been a must at 7328 feet above sea level, where temperatures varied from freezing to scorching heat. Even Sir John might gladly wear one or two sweaters on Kosciuszko in winter. As for gloves, try recording four-hourly observations with frozen hands. The fruit salts, stout

---

414 *Daily Telegraph,* Sydney, March 30, 1903, p3

and whiskey were for medical purposes only, for example, in case of food poisoning. His father had issued strict orders in December 1897 banning habitual use of alcohol. The tobacco and cigarettes were an estimated year's supply for two observers, in line with the navy allowance of three pounds of tobacco per month.

Eggie argued the Kosciuszko crew members had accepted modest conditions compared with their counterparts in the British-German Antarctic expedition — whom they were supplying with meteorological data. The officer-in-charge was paid a small salary, an allowance of £10 per annum for clothing, plus food, tobacco and medicinal needs. His assistant received the same clothing allowance and an honorarium of £10 every six months. He concluded: "The salary is so small that, if the figures were given, New South Wales would blush carmine at the thought of the refusal to pay the accounts."[415]

* * *

Wragge conceived his Kosciuszko project with dizzy self-confidence, unfazed by a pitiful budget. His volunteers began observations on the summit on December 11, 1897, and Eggie recorded the first low-level data at midnight on December 31, in a makeshift base station in the port of Merimbula, 80 miles east. The excitement of it all fired Wragge's faith in Providence. He anticipated generous government support once analysis of the observations showed — as he expected — the true worth of vertical air-pressure gradients to accurate weather forecasting. From the outset, it was an ambitious reprise of the project that won

---

415 *Daily Telegraph, Daily Telegraph*, Sydney, April 10, 1903, p3

him fame on Ben Nevis in 1881. In a letter thanking Merimbula businessman Armstrong Lockhart Munn for making land available for the seaside site, he asserted, "The eyes of the whole civilized scientific world will be turned to the undertaking."[416]

But Wragge underestimated the costs and complications of round-the-clock weather-watching on Kosciuszko. Official responsibilities and his private life in Queensland left little time for much direct involvement. While his Ben Nevis adventures were under the watch of the Scottish Meteorological Society, none of Australia's scientific societies were able to rescue the Kosciuszko project. He lacked private means to pay bills, such as the unexpectedly steep carrying charges for cartage to and from Jindabyne — the nearest town — and probably over-estimated his unpaid observers' appetite and capacity to match the celebrated zeal of his trekking on the Ben 17 years earlier.

In November 1898, observers Bernard Ingleby and Harald Ingemann Jensen quit when Wragge revealed the 1898–99 budget was all spent and he had no hope of paying them. They packed their bags and walked to the railway at Cooma, 50 miles north-east. Jensen wrote 60 years later that Ingleby — Wragge's nephew — had, by then, been on the mountain as chief observer for nearly a year and was fed up with evasive replies on pay. Jensen supported him by co-signing a letter threatening resignation, while conceding his duties had been a labour of love. He recalled Wragge's welcome in Brisbane in January 1898: "If things go well, I can pay you £1 a week, but can guarantee nothing except your fares, your winter outfit in clothes and your food." Jensen was 18 years old, had just

---

[416] *Bega Standard,* October 1, 1897, quoting from a letter from Wragge, Chief Weather Bureau, Brisbane, September 20, 1897, to A L Munn.

matriculated from Brisbane Boys' Grammar School and narrowly missed a government scholarship to study science at Sydney University. In the couple of months before leaving Brisbane for the Snowy Mountains, he learnt basic meteorology from weather bureau staff. He reached the observatory in late March aboard a bullock-drawn dray of supplies from Jindabyne.

In his memoir, *Reminiscences of a Geologist*, written about 1957, Jensen asserted Wragge's financial problems were compounded by drinking sprees, two of them between January and March 1898 — the first lasting a week and the second a fortnight.[417] Jensen believed Wragge had bought booze with proceeds from observatory benefit lectures meant for paying off debts incurred in erecting the observatory. Jensen regretted having had to wire his father for £10 to get back to Brisbane, given Wragge's promise to cover fares, but felt he had no real grounds for complaint, since he had signed on for Kosciuszko knowing he might not be paid. Besides, Jensen liked Wragge's non-conformism. He recalled his arrival in the Brisbane office after an unexplained absence, bristling with disdain for a flat-earth society magazine found in a stack of recently lobbed periodicals:

> Read that, Jensen. It will show you how many idiots there are in the world still. Why, Copernicus and Kepler would turn in their graves if they knew such fools existed today.

Jensen remembered Wragge as a remarkable man, well versed in astronomy, geology and botany as well as meteorology, and with

---

417 Jensen, Harald Ingemann, *Reminiscences of a Geologist* [unpublished manuscript], circa 1957, State Library of Queensland, Harald Ingemann Jensen Records, 1905–65, Call No OM69-29, pp 17-31

a genius for selecting and training staff, whom he infused with his own love of science. In 1908, Sydney University awarded Jensen a Doctor of Science and university medal. He served as the Northern Territory's director of mines from 1912 to 1915 and worked as a government geologist in Queensland from 1917 to 1922.

Bernard Ingleby volunteered early in 1897, aged 19, for what was to have been a no-frills, three-month weather-watching trial on the mountain. In April 1897, Wragge wrote to potential sponsors around Australia that a permanent high-level station would sharpen weather forecasting for shipping, pastoral and agricultural interests — especially by tracking storms from Antarctica.[418]

In June 1897, the ageing Adelaide philanthropist Robert Barr Smith responded with £150 for the cause, which he abbreviated in his covering letter to "plonking someone on top of Kosciuszko" to achieve important meteorological findings.[419] Reputedly the wealthiest man in South Australia, Barr Smith loved hiking and had scaled Kosciuszko in 1892, aged 67. His gift paid for a large waterproof, hurricane-rated Arctic tent and heavy-duty tanned-sheepskin sleeping bags.

Wragge's correspondence in July and August with Queensland's Post and Telegraph Department shows he then still envisaged a three-month trial only.[420] He introduced his

---

418 National Archives of Australia: J1, Q400/1/22, Correspondence, Chief Weather Bureau, Brisbane, appeal to potential sponsors, April 30, 1897

419 National Archives of Australia: J1, Q400/1/22, Robert Barr Smith to Clement Wragge, June 12, 1897

420 National Archives of Australia: J1, Q400/1/22, Clement Wragge to Under-Secretary and Superintendent of Post and Telegraph Department, July 13, 1897; *Adelaide Observer,* September 18, 1897, p33

nephew as an able observer prepared to work without salary for that period and reported having also received various offers of help from individuals in New South Wales.

In September, the *Adelaide Observer* reported Mr Wragge's eldest son, Egerton, was likely to join the project as an assistant meteorologist. Eggie was then 15 years old and — like Harald Jensen — attending Brisbane Boys' Grammar School, where he was enrolled until the end of the first quarter in April 1897. The youngest of the first Kosciuszko volunteers, Eggie helped his father set up on the summit in early December 1897, then spent most of 1898 in charge of sea-level observations at the Merimbula station. In August 1899, Eggie and his then 17-year-old brother Bert began an eventful three months in the High Country as relieving observers. Wragge sent Bert from Brisbane and Eggie from Merimbula to help the Kosciuszko superintendent Basil de Burgh Newth who had been short-handed for several months.

Newth, 23, had, like Eggie and Bernard Ingleby, volunteered for the Kosciuszko adventure in 1897. Wragge recruited him during his reconnaissance of the South Coast hinterland, initially to synchronise chronometers for the high- and low-level stations. Remembered by Jensen as a serious-minded mathematician, Newth had excelled in 1895 high school exams for land surveying, astronomy and mathematics and won a gold medal for trigonometry. His father, the Reverend James Newth, offered land beside his rectory in the village of Candelo, about 20 miles inland from Merimbula, suggesting Basil, then studying theology, could find time for observations. Wragge declined, as Candelo was too far inland and 347 feet above sea level. He later chose a hillside site in the port of Merimbula. By then, he realised the rector's son would be of more use on the mountain, given his reputation in practical trigonometry and surveying. On December 4, 1897, Newth junior caught up with the advance team on the summit for what he believed would be a brief assignment as an assistant observer. New Zealand-born and raised on the Southern Tablelands of New South Wales, Newth was more acclimatised to the High Country than the rest of the founding observatory crew. The overseer, 52-year-old former master mariner, Captain Charles Iliff, had a heart condition and had been living in retirement in Brisbane when recruited.

* * *

On December 4, 1897, the advance party drank toasts to their guide, Bombala-born photographer Chas Kerry, and began pitching tents on the summit. Wragge confirmed their arrival in an ecstatic telegram to Brisbane a day later, sent by messenger from Jindabyne: "Kosciuszko is a magnificent & unique position for meteorological astronomical and physical observatory ... cold by night great sun scorching by day cloud effects splendid beyond description."[421]

Their setting up took a week, complicated by having to transfer gear from a dray stuck in rough country onto packhorses for the final ascent. Wragge used many of his own instruments, relics of his Ben Nevis and Adelaide days. The founding crew of Iliff, Ingleby and Newth recorded their first observations on December 11. Three days earlier, Queensland's Postmaster-General James

---

[421] National Archives of Australia: J1, Q400/1/22, 'Electric Telegraph, Queensland', Wragge to Under-Secretary and Superintendent of Post and Telegraph Department, December 5, 1897

Robert Dickson announced that regardless of Mr Wragge's zeal, the colony could not afford to pay him or his staff a penny more. Dickson set the meteorological branch budget at £1640 for 1898 and suggested the chief take private students in meteorology for extra income. Commenting on the Kosciuszko project, he said Wragge was a "splendid advertisement for Queensland", showing the colony's advance in science could help the whole community.

Meanwhile, Wragge visited Sydney before setting up the Merimbula station and persuaded George Reid to grant £100 towards scientific research at both observatories. Reid pledged more aid, pending evidence of valuable results. On December 21, his government voted £600 for improvements to the track from Jindabyne to Kosciuszko, to cater to High Country tourists. Reid and his family spent their Christmas holidays on the Monaro, visiting Cooma, Adaminaby, Kiandra and the Yarrangobilly Caves, but not Kosciuszko. In 1897, reaching the summit from the railhead at Cooma meant a coach trip to Jindabyne, followed by a two-day packhorse ride to Mount Twynam, accompanied by guides with tents.[422]

In January 1898, the Sydney press reported that 20 or so intrepid tourists who reached Wragge's station that month had included two cyclists from Victoria.[423] Wragge's crew kept a visitor book and also a subscription register for cash offerings for their noble cause. Stories spread far and wide of the volunteers' Snowy Mountains exploits, especially, perhaps, because the inland was

---

422 *The Australian Star*, Sydney, December 29, 1897, p6; *The Australasian*, Melbourne, January 1, 1898, p50

423 *Australian Town and Country Journal*, January 15, 1898, p25; *The Sydney Morning Herald*, February 22, 1898, p4; *The Sydney Mail and New South Wales Advertiser*, January 1, 1898, p131

so parched that summer. Wragge joined a debate in the popular weekly *The Queenslander* on how to pronounce Kosciuszko: *Kust-yoosh-ko*, with a longingly accentuated *yoosh*. Wragge preferred this to *Kos-y-usker*, then — and now — the most popular rendering of 18th century Polish patriot Tadeusz Kosciuszko's surname.[424] In 1906, the New South Wales government gazetted 100 square miles around Mount Kosciuszko for public recreation and preservation of game. Three years later, the state built the Hotel Kosciuszko at Diggers Creek, 16 miles from the top, by a recently completed gravel road.[425]

A correspondent for the *Cootamundra Herald* who visited the summit in mid-January 1898, with Snowy Mountains guide James Spencer, found Iliff, Ingleby and Newth enjoying life in the clouds. The writer described Ingleby as a student of several sciences who had his own meteorological observatory and ethnological museum at home in Adelaide. Captain Iliff was not a weather scientist, but his life as a master mariner had equipped him well to be commissariat of the party. They had volunteered knowing they could be snowbound for up to five months, with no telephone line to Jindabyne. The writer concluded:

> In the interests of science they go and dwell in this outlandish place, and accept all the risks. In case of sickness, the situation is peculiar. They have no horses; and without the assistance of

---

[424] *The Queenslander*, December 31, 1898, p1265. In 1840, Polish explorer Paul Edmund de Strzelecki named the unobtrusive, 2228m (7310 ft) Main Range mound after Kosciuszko.

[425] *The Sydney Morning Herald*, December 1, 2007; Scott, David, *Tourists on the Summit, 1875–1914*, for the Kosciuszko Huts Association, 2013, pp 18-20, https://www.khuts.org/images/stories/history/TouristsOnSummit_DScott_14aug2013.pdf, accessed October 17, 2022

a guide it is questionable whether they could in any case find their way down to Jindabyne.[426]

* * *

Two hours before daybreak on February 14, 1898, a storm that Wragge nicknamed Blastus shredded every tent on the summit, blew Captain Iliff off his feet and forced a temporary retreat from the observatory. They stashed their equipment and food between boulders and fled downhill to shelter at the nearest shepherds' camp, a few miles beyond the flooded Snowy River. Newth and Ingleby later walked, separately, 34 miles to Jindabyne to borrow tents and telegraph the news.[427]

A clamour ensued. Wragge pleaded with Reid for funds to extend observations through autumn, winter and spring, eight months from April to November. He argued collecting winter data was of crucial scientific importance. Beyond that, his project would surely draw even more tourists to the New South Wales Alps.[428] Reid quickly pledged £336 for a permanent building on the summit and purchase of skis for the observers. Wragge, celebrating what he saw as the observatory's now permanent footing, contracted a Cooma builder to erect a two-roomed weatherboard observers' hut, completed in mid-April.

A few days later, Iliff resigned and returned to Brisbane.

---

426 *Cootamundra Herald*, January 12, 1898, p6
427 *The Queenslander*, March 12, 1898, p502, 'A Hurricane on Kosciuszko', by Bernard Ingleby
428 National Archives of Australia: J1, Q400/1/22, Wragge to Reid, February 11, 1898

Despite the project's promises for science, shipping and agriculture, Iliff was not prepared to spend winter on the summit. Wragge promoted Ingleby to take Iliff's place and in early May gave Harald Jensen a start as an assistant, with vague promises of honorariums for both. Ingleby and Jensen lasted until early October. Jensen recalled in his 1950s memoir Ingleby's statement, "I cannot waste a lifetime here without salary and neither can you." Newth tried to quit a week or so later, but Wragge convinced him to stay as manager, promising food and equipment, "consistent with the state of the funds", and warning that economy and thrift must be practised as essential virtues in every respect.[429]

Each of the observers survived misadventures that winter with courage their mentor admired. Old hands rated the vast, deep blanket of snow covering the main range in 1898 as the biggest in 20 years. In May, Newth got lost in a storm between Jindabyne and the summit, and survived 14 hours wrapped in a blanket in a hole that he and a mountain guide had dug in the deep snow. When thick fog lifted, they backtracked to Jindabyne, in Newth's words, more dead than alive. He waited there two weeks for the delivery of skis to help reach the summit. Wragge bought skis for his crew when the New South Wales government baulked at building a telegraph line to Kosciuszko. The observers agreed to take turns once a month to ski to Jindabyne and telegraph the weather data to Brisbane.

Every four hours, from midnight onwards, they took readings of air pressure, temperature, humidity, wind speed, cloud mass,

---

429 Basil Newth, letter to 'My Dear Johnnie' [believed to be John H Ellison, 10 Bougainville Ave, Forrestfield, Western Australia], November 10, 1959, held by The Royal Australian Historical Society, Sydney; *Evening News*, Sydney, November 5, 1898, p5

precipitation and surface ozone. The pace increased between 8 am and noon, when half-hourly data were required. That all added up to 13 sets of observations and more than 200 data entries every 24 hours.[430] Jensen recalled that in a two-man operation — which was generally the case — they took turns at recording day-about: "The one who took the readings for the day also made the food and tidied up the house." From late May, when their skis arrived, they practised skiing daily and tried to train Zoraster to draw a sleigh to haul supplies from Jindabyne. The wily Newfoundland dog treated this as a joke. The observers lived on porridge, canned vegetables, soup and meat, and pots of tea and coffee. Zoraster hunted hares and rabbits and raided the pantry now and then.

Jensen recalled that between May and October, nearly all precipitation came as snow and sleet, save for a rainy day in June and another in September. On July 23, Ingleby sent Jensen and Newth to Jindabyne together for the mail, judging the weather too bleak for either to face it alone. They left on skis, mired in a fog that lasted four days and returned more than a week later, on August 5, in a state described by Ingleby as "knocked up, hungry and almost prostrate". Three Sydney Alpine Club members dragged Jensen up the final four-mile ascent from the Thredbo River after finding him and Newth sheltering in a shepherds' hut.[431]

On August 6, the morning after Jensen's ordeal, Ingleby set off for Jindabyne with the Sydney skiers, who had stayed overnight. They slid together down the icy mountainside, now

---

430 *The Monaro Mercury and Cooma and Bombala Advertiser,* January 30, 1899, p3, 'Work of the Observers', by Mr B DeBurgh Newth, Chief Observer, "Altogether we make at least 208 notes daily."

431 *The Sydney Morning Herald,* August 17, 1898, p5

called Crackenback, to Friday Flat on the Thredbo River. Then the Sydney visitors picked up horses that had been left for them on the flat and suggested Ingleby follow their trail to the shepherds' hut where Newth and Jensen had sheltered a few days earlier.

This proved tricky through unfamiliar country in snow-melt slush. By nightfall, Ingleby was lost without fire, blankets or food, but still with Zoraster for company. They curled up in the snow and at sunrise walked for another half day to the nearest sheep run, Wollondilby Station, where Ingleby had his first food in 30 hours. Refreshed, he and Zoraster walked to another property, Jindabyne West, within cooee of their destination.[432] A fortnight later, in Sydney, Ingleby told reporters that Kosciuszko's locals viewed the observers as madmen: "They have a worse opinion of the dangers and difficulties of a residence upon the mountain than even the Sydney people have."[433]

* * *

Ingleby's trip to Sydney that August was unexpected. He was summoned there by Wragge, who had used a fortnight of sick leave to visit the Merimbula observatory and had organised two fund-raising lectures in Sydney on the way home. Ingleby's arrival by rail from Cooma on August 15 was well reported because of his Kosciuszko adventures. On August 9, Wragge announced Reid had given a further £100 for general maintenance of the observatory and would cover the cost of printing the first six months of observations from the summit and allied stations. Reid's

---

432 *South Australian Register,* August 22, 1898, p5
433 *The Sydney Morning Herald,* August 17, 1898, p5

largesse bothered a columnist from *The Bulletin* in Sydney, who asked how Mr Wragge had managed to get the Premier "firmly by the wool, by the coat-tail, or somewhere", considering the doubts over the true value of high-level observatories.

The *Brisbane Courier* also reported Reid's readiness to pay for the printing in Sydney of six months of observations from Kosciuszko and sea-level stations to be distributed to various scientific stations around the world. Sadly, this never happened. The government printing offices of New South Wales and Queensland both baulked at having to deal with Wragge's fiddly columns of figures. Queensland's Post and Telegraph Department eventually published just one statistical bulletin, giving observations for January 1898 from Kosciuszko, Glebe Point (Sydney), Merimbula and the Victorian town of Sale. Setting-up costs ran well over the £50 budget and the department had many greater priorities. In December 1889, Wragge complained to Queensland Home Secretary Justin Foxton that the publication of monthly bulletins was crucial, otherwise the Kosciuszko project might as well have never been undertaken. In April 1902, the *Daily Telegraph* in Sydney deplored both governments' neglect of the data, arguing that their inaction had largely nullified the observers' years of dedicated work. The fate of the records, for many years in Wragge's safe-keeping, is unclear. They seem most likely to have been destroyed by fire in Auckland in 1928 — six years after his death — along with his lifetime collection of records and curios.[434]

* .* *

---

434 *Brisbane Courier*, August 9, 1898, p4; *Daily Telegraph*, Sydney, April 10, 1902, p3; National Archives of Australia: J1, Q400/1/22, December 4, 1899, Wragge to Post and Telegraph Department on cost of report on Mount Kosciuszko, with memo to the Home Secretary

Bernard Ingleby returned to the High Country after speaking at the two Sydney lectures "in mountain costume". The editor of *The Bulletin*'s weekly Woman's Letter overlooked him, but described Wragge as a lean, lonely, weather-beaten man with a sandy beard, a preoccupied eye and a dry, tuneless voice. He had, nevertheless, made a splendid pitch for the Australian Alps, heard by the State Governor Lord Hampden. During September, Wragge pressed on with promoting the project in Queensland. On September 12, Sir Samuel Griffith — now Chief Justice — was guest of honour at his Kosciuszko slide show, telling a fair-sized crowd at the Centennial Hall that their famous meteorologist needed no introduction anywhere in Australia. On September 23, Wragge's audience in the Darling Downs town of Warwick reportedly applauded every wintery image of the Snowy Mountains.[435]

* * *

Years later, Ingleby called his time as chief observer an epiphany. In 1915, then an advertising copywriter and well-regarded poet, he wrote in the Christmas edition of *The Bulletin* magazine:

> I was in my teens ... and convinced with all the earnestness of a budding moustache, that I was a personage of much moment in the world of High Level Meteorology ... It's all very laughable to look back upon, but with me there is no sneer in the laugh ... Our pretty little standards of failure and success ... well, they

---

435 *Telegraph*, Brisbane, September 13, 1898, p6; *Darling Downs Gazette*, September 24, 1898, p3

don't seem to matter much when you've thrilled to Kosciuszko, I tell you. [436]

On October 30, 1898, Ingleby sent his final despatch from Kosciuszko, reporting the demise of another Arctic tent. He likened the sound of its canvas shredding in a gale to the death shriek of the observers' intended summer lodging. Writing for Sydney's *Evening News*, he signed off by advising tourists to visit the summit in spring: "Then there is an immense quantity of snow, the Alpine flowers are at their best and the temperatures no more than pleasantly chilly."[437]

* * *

Newth served as chief observer until December 31, 1899, when succeeded by Murray Leith Allen, a retired schoolteacher from the Southern Tablelands cathedral city of Goulburn, 150 miles north. Allen, who was aged 57 when recruited, ran the observatory through the winter of 1900, which old-timers from the high country judged even bleaker than 1898's. In September 1900, he was evacuated back to Goulburn and eventually replaced by one of his fellow observers, the much-travelled, 27-year-old Philip Sydney Whelan. Staffing details from mid-1900 to the early months of 1901 are scrappy, because of Wragge's absence overseas.

Four decades later, Whelan recalled having joined the observatory as an assistant "about 1900", and that he had later risen to officer-in-charge, a position he believed he had held for two-and-a-half

---
436 *The Bulletin*, December 11, 1915, Christmas Edition
437 *Evening News*, Sydney, November 3, 1898, p3

years. He resigned in April 1902, leaving Eggie to perform last rites. Based on newspaper reports of Allen's illness, Whelan was in charge probably at most for one-and-a-half years. He learnt his meteorology under sail through years of crewing on full-rigged ships, beginning — like Wragge — on a collier out of Newcastle, New South Wales, but bound for Chile, not North America.

Allen was an Eton-trained scholar of modern languages, mathematics and the sciences. Elected a Fellow of the Royal Meteorological Society in 1900, he was a keen amateur observer renowned for his monthly weather reports in the *Goulburn Herald*. Between them, these three makeshift supervisors led a procession of keen youngsters in the slushy footsteps of Bernard Ingleby and Harald Jensen. In all, 15 volunteers gave heart and soul to the observatory between 1898 and 1902.

* * *

"I do not think I shall care to stay long here," Eggie Wragge wrote to his father on August 15, 1899, of his fate of having been sent back to the summit. He hoped he could be replaced soon. The night before, he and Bert had collapsed in the observers' hut after a three-day trip by horse and on foot from Jindabyne with Newth and mountain guide Frederick Collins. Kiandra photographer Donald McRae rode to join them on the second day. All five spent the third day hauling their gear on an improvised sledge five miles uphill from the Thredbo River to the observatory — climbing 3000 feet. Newth estimated the load at 280 pounds — roughly the weight of two men. It included clothing, skis, books, photographic equipment and 98 pounds of pressed beef. He reported both boys had been exhausted and that he had carried 17-year-old Bert from the Kosciuszko Saddle as night neared. Eggie, in his letter, said Collins had pushed him on through deadly snow-sleepiness and had gotten him safely to the hut.[438]

Bert travelled from Brisbane, where he was still at school, and Eggie from the Merimbula observatory. Wragge conscripted them in late July when Newth's sole assistant, 21-year-old Robert Leslie Burcher, was floored by a chest infection. Knowing Burcher was a dairy farmer's son from Candelo, near Merimbula, Wragge suggested that changing places with Eggie might further his recovery. He added Bert to the threadbare summit crew as a replacement for 18-year-old Noel Alfred Carr of Maitland, the third observer from late December 1898 to mid-March 1899.

Unfortunately for Newth, the Wragge boys did not stay long either. Eggie quit on October 16, a day after nearly freezing to death on the foggy Kosciuszko plateau. Bert and Newth found him prone on a shepherds' track near the Thredbo River, in the latter's words, "stretched out like one dead". They had begun searching seven hours earlier, before daybreak, in pouring rain and thick fog. Newth wrote Eggie had looked so haggard that Bert had at first not recognised him. A hat, coat, trousers and gumboots were all that was left of the clothes he discarded to swim the flooded river. A day earlier, Newth had feared Eggie was in strife, reckoning from heliotrope signals from Jindabyne that he was overdue on the return leg of the mail run. They dragged him three-quarters of a mile to the shepherds' hut that he had been too exhausted to reach and revived him "by comfort, strong tea and food".

---

438   *Manaro Mercury*, September 1, 1899, p2; *Queenslander*, October 7, 1899, p713; *Manaro Mercury*, October 20, 1899, p2

Bert stayed for another month before resigning too, with praise from Newth for his stamina. He returned to Brisbane in early November, his place having been taken by Ernest McLure, the 17-year-old-son of Jindabyne's postmistress. Newth, who was also keen to move on, did not blame Bert or Eggie for leaving. He wrote in Cooma's *Manaro Mercury* that for a lad of 19, unused to mountaineering, Wragge's eldest son had shown great strength in surviving his terrible experiences and not succumbing to worry, hunger and exposure.[439]

In January 1900, reporting the New South Wales government's continued funding of the Kosciuszko project, Brisbane's *The Week* newspaper noted Mr Wragge, junior, remained in charge of the low-level observatory. Eggie appears not to have returned to the summit before overseeing its dismantling in 1902. There is scarce evidence of his life in Merimbula, but the weather-watching ritual must have been tedious. He is said to have had a fling with Olive Munn, the younger of A L Munn's two daughters, who was aged 16 when they first met. The observatory was on the same hillside as Munn's grand home, known as the Tower House, and also as Courunga. Munn founded and owned the Maizena Cornflour Company — the town's major employer.[440]

In early 1901, Eggie accepted an invitation from his father's friend, the British adventurer Thomas Caradoc Kerry, to join him on a voyage from Sydney to Dunedin, New Zealand, "and thence

---

439 *Manaro Mercury*, October 20, 1899, p2
440 Email from Pat Raymond, citing an oral history account of C L E Wragge's time in Merimbula, November 29, 2020; NSW Environment Department, heritage listing of Courunga, Merimbula, https://www.environment.nsw.gov.au/heritageapp/ViewHeritageItemDetails.aspx?ID=5045262, accessed October 17, 2022

elsewhere to parts not determined on". On March 24, 1901, he was among a dozen men aboard Kerry's schooner-rigged luxury yacht *Ariadne* when she ran aground at the mouth of the Waitaki River, near Oamaru, on the east coast of New Zealand's South Island. A month later, he identified himself to a shipping casualty inquiry in Christchurch as an ex-meteorological station superintendent who had crewed on the voyage as the owner's guest.

In January 1902, Eggie returned to Christchurch as a minor witness in the trial of Kerry and his skipper, George Mumford, who were alleged to have conspired to defraud insurance underwriter Lloyd's. Kerry was acquitted after the skipper admitted to having deliberately grounded the yacht and conning a Lloyd's investigator into paying him £400 for a bogus confession implicating Kerry in the scam. Mumford was jailed for four years and, in October 1903, the British High Court of Justice in London upheld Kerry's claim for his £10,000 insurance entitlement.[441]

---

441 *An Encyclopaedia of New Zealand*, 'Wreck of the *Ariadne*, 1901', Wellington, 1966, https://teara.govt.nz/en/1966/trials-notable/page-9, accessed October 17, 2022; *Otago Witness*, January 29, 1902, p28; *Daily Telegraph*, Sydney, December 12, 1903, p14.

CHAPTER 19

GOOD RIDDANCE, AUSTRALIA, 1903-04

And now, I leave Australia a saddened and broken-hearted man. I shall never come back. — Clement Wragge, *Auckland Star*, August 10, 1904, interviewed before leaving Auckland for Tahiti.

In April 1903, when Eggie vouched for the thrift and sobriety of the Kosciuszko observers, he and Edgar Fowles were beginning their fifth month of running the Brisbane weather bureau. Wragge's lecture tour that began in Melbourne the previous December continued from April to June 1903 through country New South Wales. By June, he had visited close to 80 towns. That month, answering critics of his prolonged absence from Queensland, he pleaded his six months of incessant travel had been necessary to promote the bureau's work and needs. He wrote to the *Brisbane Courier* on June 5 that few people truly knew the predicament facing his theoretically private agency. Subsidies totalling £1000 pledged by the Queensland and New South Wales governments had been inadequate and, if not for some private donations and income from the forecasts he sold to the press, the bureau would surely have had to close.

Days earlier, Fowles complained he had been unable to cash

his May pay cheque because New South Wales had not deposited an expected £150 in the bureau's account. Wragge wired the Queensland government from Narrabri, warning he could not pay the June salaries and would close the office at once unless New South Wales honoured its pledge. This was too much for Sir John See who resented what looked like grandstanding over an administrative glitch. He replied there was not one atom of truth in the inference that his government had reneged on supporting the bureau. In fact, it had done everything asked of it, and more. A week later, Wragge apologised, telling the *Courier* he was deeply grateful to See for sustaining the bureau and paying the Kosciuszko debts.[442]

Wragge continued promoting the Kosciuszko cause, despite See having found the other state premiers were disinclined to help reopen the observatory. In early June, Wragge's lantern slides of the Snowy Mountains spurred supporters in the farming town of Gunnedah to petition See to restore funding.[443] See took the observatory's likely demise to the premiers' conference in Sydney from May 13 to 15, but it was a minor matter compared with management of the waters of the Murray River, the case for a transcontinental railway, and selection of a federal capital site. Delegates dealt quickly with the Kosciuszko project, agreeing that unless a private backer emerged, nothing more could be done.

On June 17, Wragge returned to Brisbane uncertain of funding

---

442 *Brisbane Courier*, June 5, 1903, p10; *Goulburn Herald*, June 8, 1903, p2; *Brisbane Courier*, June 16, 1903, p3

443 *Evening News*, Sydney, May 15, 1903, p3; *The Daily Telegraph*, Sydney, June 8, 1903, p7

beyond June and planning to tour the Northern Rivers district of New South Wales in July. A week later, Premier Robert Philp announced the state's £700 weather-bureau subsidy would not be renewed, as See was no longer willing to finance the upkeep of Mr Wragge's agency. See told *The Sydney Morning Herald*: "We have got our own bureau and I do not see why we should be asked to contribute to another one." Later, Wragge complained of having been given only a week's notice, but must have guessed See was unlikely to have accepted the Kosciuszko debt without pruning the bureau's £300 hand-out. On June 27, Philp announced a remnant of the old bureau would be transferred to the Public Works Department under State Hydraulic Engineer John Baillie Henderson. The new weather office would simply record rainfall around the state and no longer issue forecasts. Edgar Fowles, who by then had worked with Wragge for 15 years, was appointed superintendent with an initial budget of £300. Fowles used his last newspaper column as acting bureau head to defend his skill as a meteorologist, stressing his forecasts during Wragge's most recent absence had been up to the Brisbane bureau's renowned standard of accuracy. Even allowing for absolute failures "a dozen times over", he had achieved a high success rate, considering he had issued roughly 160 forecasts in the previous seven months, including the unprecedented wet season of 1902–03.[444]

On July 1, writing in his final *Brisbane Courier* weather column, Wragge was adamant closing the bureau had been his

---

444   *The Telegraph*, Brisbane, June 18, 1907, p7; *Brisbane Courier*, June 27, 1903, p4; *Telegraph*, Brisbane, July 1, 1903, p5

own decision once he accepted the Queensland government was no longer able to assist the office on adequate lines:

> We have striven our very utmost in adverse circumstances, of which the people know nothing, to keep it and the forecasts afloat for their benefit, and again we thank the Governments concerned for all assistance rendered, but the limit has been reached.

He announced he would continue lecturing around Australia, trusting that his insights on the physical sciences would so elevate public opinion that no new weather service could ever be squandered like his. Ultimately, he intended to move back to England.[445] In Queensland, the bureau's closure was generally deplored. Brisbane's *Telegraph* condemned the vampire-like drain of Federation for diverting funds from meteorology, and the other states for having sponged on Queensland while enjoying the benefits of Mr Wragge's forecasting. But the *Evening Telegraph* in the gold-mining town of Charters Towers wished him good riddance, arguing neither Queensland nor Australia could afford his grandiose vision of a US-style weather service, nor his blatant "quackery". The best that could be said of him was that he had striven to establish a scientific system. The alleged greatness of the Queensland Meteorological Department was an advertising fraud, as Wragge's forecasting failures during the Federation Drought had frequently proved. In fact, a man with a rheumaticky leg would have done as well, at a lesser cost. The *Catholic Press*, in Sydney, observed Wragge had many enemies and many friends, "Those who believe in him believe in him very much, just as those

---

445 *Brisbane Courier*, July 1, 1903, p6

Chapter 19 Good Riddance, Australia, 1903–04

who snort at his prophecies take a very high key. Inevitably, his stakes were always highest among seafarers, untarnished by recalcitrant storms and contrary breezes."[446]

* * *

Wragge spent most of July travelling through the Northern Rivers district of New South Wales. His agent George Callender advertised him as "The Man of the Moment", laden with the latest astrophotography from Paris and the finest pictures yet of Mount Kosciuszko and the Grand Snowy Ranges. Admission two shillings, one shilling (children half-price). He visited seven towns before returning to Brisbane with an unspecified illness and started packing at Capemba.

On August 5, Nora wrote to Philp, asking him to find her estranged husband a job or at least employ one of their boys. Without discussing their separation, she said he had exhausted his independent means on scientific research and that beyond his itinerant lecturing he was unable to maintain their large family. She feared his health would buckle with the anxiety and strain of constant touring. For that matter, her own health had been chronically delicate since their Ben Nevis years. But if Mr Wragge or one of the boys could not be placed in a lucrative position, she would have to find herself a job and had already sought assistance from her husband's old friend James Drake, now Federal Postmaster-General. Nora signed off saying she had told no-one of her intended plea, "trusting your generosity

---

[446] *The Telegraph*, July 1, 1903, p4; *Evening Telegraph*, Charters Towers, p2; *Catholic Press*, Sydney, July 2, 1903, p12

to pardon my boldness". Philp replied a week later, regretting he was at present unable to help.[447] Eggie, then 22 years of age, moved to Sydney after the Brisbane bureau closure and left for London on August 21, possibly expecting to have to give evidence in the judicial review of the Kerry case. The six younger Brisbane Wragges were then aged between 12 and 20 years old.

* * *

In September, Wragge resumed touring after culling his Capemba treasures. The auctioneer's random inventory — terms cash, no reserves — is a core sample of his Queensland years. It included a large catamaran, 15 pairs of clam shells, four oriental busts, assorted native weapons, an acetylene gas generator, photographic materials, portable darkroom, three perambulators and a treadle sewing machine with fittings.

Callender retitled the lectures "Wragge's Farewell to Australia — a Grand Imperial Educational Entertainment", but attendances were poor in Darling Downs farming towns. On September 8, Wragge called off the second of his scheduled Dalby meetings saying he was not up to entertaining a would-be audience of five adults and a five-year-old child.[448]

Three weeks later, he re-advertised Capemba for sale (or for rental by a good tenant) and left for Sydney telling *The Telegraph* he was finished with Queensland. Between October and December, Wragge toured rural New South Wales and Victoria

---

447 Leonore Wragge to Queensland Premier, August 5, 1903, Queensland State Archives, PRE/A154, SR66402/1/147; COVR05900, 11/8/1903

448 *Darling Downs Gazette,* Toowoomba, September 15, 1903, p3

with an entourage, which, at times, included Edris, their toddler Kismet, one of the Kosciuszko dogs and George Callender.

An enterprising theatrical agent, Callender had made his name in Sydney and Melbourne by promoting artistes such as the illusionist Professor George Waldo Heller and his clairvoyant wife Maudeena, of Heller's Mahatma Co.[449]

In early November, Edris's presence on the western Victorian leg of the tour fascinated the *Melbourne Punch*. "Mrs Wragge ... is a genuine Indian Princess, but more of the Arab than the Hindoo type of beauty," a *Punch* columnist revealed. "She is a descendant of the famous Emperor Akbur, one of the great Mahommedan rulers of India. What a romance!" She was believed to have accompanied her husband all over Australia in the previous 18 months.[450]

In country Victoria, Callendar presented Wragge as an astronomical guru with a Grand Scientific Entertainment — "the most magnificent pictures and photographs of the heavens", received just a month earlier from the famous Paris Telescope and the Licks and Yerkes observatories in the United States. Wragge and entourage left a trail of bemusement as they rattled by steam train through a dozen or so Victorian towns in November and December 1903. A reporter in Beechworth, population 2500, described Wragge's treatment of scientific subjects as erratic,

---

449 See references to advertisements for Heller's shows managed by Callender, *Launceston Examiner*, May 19, 1900, p8; *Referee*, Sydney, August 29, 1906, p12; *Waikato Herald*, New Zealand, April 20, 1909, which calls Mme Heller an hypnotic medium

450 *Melbourne Punch*, November 12, 1903, p18 (reprinted in *Bendigo Independent*, November 19, 1903, p3)

his humour quaint, his comments on the rare astrophotographs unhelpful and his demeanour enigmatic:

> He kept us in a state of continual laughter by his quaint and forceful humour. And he looks so very sad and solemn, too. I do not think he ever laughs. Perched high on a stool beside his hydro-oxide light, he pops in the slides with one hand, whilst he bends low and harangues with the other.[451]

* * *

The Wragge caravan reached Tasmania on New Year's Eve for the final six weeks of their Australian pilgrimage. In mid-January, interviewed before catching a train from Launceston to Hobart, Wragge said he had spent the best years of his life in Australia, and was returning to England under circumstances he regretted. He blamed underhand tactics by interests hostile to his national vision, beginning with the other colonial meteorologists conspiring in 1888 to stop him forecasting for their arbitrary patches of the continent. He surprised his interrogator by producing, as proof of his weather-watching zeal, the Ben Nevis gold medal — "About the size of the big bronze one usually awarded to certain exhibitors at great exhibitions." The reporter noted Mr Wragge was touring Tasmania with his wife and their boy.[452] Beyond Hobart and Launceston, he visited more than 20 towns, generally hiring Mechanics' Institutes and Masonic halls. His Launceston shows were especially popular. The *Launceston Examiner* reported on

---

451 *Ovens and Murray Advertiser*, Beechworth, November 7, 1903, p11
452 *Mercury*, Hobart, January 27, 1904, p2

Chapter 19 Good Riddance, Australia, 1903–04

January 19, 1904, his gift of £17/2/6 — half of one night's takings — to an appeal to help the family of a skipper killed in a mishap on Bass Strait on Christmas Eve.

A few weeks later, during a return visit to Launceston, Wragge augmented forecasts issued by Tasmania's government meteorologist Henry Charles Kingsmill, to warn of two approaching Antarctic-generated low-pressure systems. He named them Propsting and Lyne, respectively, in honour of Tasmania's Quaker Premier William Bispham Propsting (1861–1937) and Railways Minister Carmichael Lyne (1861–1929), a brother of Federal Minister for Home Affairs, Sir William Lyne. Both forecasts proved accurate.

Wragge argued any future federal weather bureau must maintain the observatory he erected in 1895 on Hobart's sentinel, 4166-feet-tall Mount Wellington. Nearly a decade on, stories about his stamina on the summit seemed likely to outlast the flimsy wooden hut he and his team built. In 1908, after the station was accidentally burnt down by a picnicking sightseer, journalist Frank Morton recalled spending a chilly night in 1895 on high with Wragge singing sea shanties, self-accompanied on autoharp:

> The cowl blew off the chimney at midnight, and we were smoked-out until the abominable thing was fixed again. It was blowing a blizzard, and we were all colder than we had ever been in our lives. But not Mr Wragge. He sat there with the

flaps of his cap tied down over his ears, and he sang chants interminably to his own dolorous accompaniment.[453]

On February 19, Wragge and family sailed from Hobart, without much fanfare, for Bluff Harbour on the southern extremity of New Zealand's South Island.

* * *

Between late February and early August 1904, Wragge travelled New Zealand, giving at least 100 lectures at 60 country towns and the cities of Christchurch, Dunedin, Wellington and Auckland. His Grand Scientific Entertainment drew curious crowds everywhere, from the far-south hamlet of Riverton, population 850, to the North Island metropolis of Auckland, population 175,000. George Callender's PR was unchanged. He plugged the not-to-be-missed lantern slide pictures as "A Voyage Through the Universe, The Majesty of Creation, the Science of the Atmosphere and the Wonderland of Australia". And Wragge continued to punctuate his affirmations on Divine Nature with jabs at unholy injustices. In the South Island port of Oamaru, he argued Australia's niggardly federal government could have financed a people's weather bureau with an annual levy of just one shilling on income-tax payers. In Queenstown, he blamed ring-barkers for worsening Australia's recent massive drought by killing the native trees that retained moisture and courted the rain.[454]

---

453 *Manawatu Standard,* December 12, 1908, p6

454 *Oamaru Mail,* April 13, 1904, p1; *Lake Wakatip Mail,* March 25, 1904, p3

Chapter 19 Good Riddance, Australia, 1903–04

Everywhere, he stuck up for the controversial sacked Canterbury University Professor Alexander William Bickerton, whom he had visited while in Christchurch. Bickerton, dismissed in 1902 for neglect of duties, was best known for his partial-impact theory, proposing Earth's creation through a random collision of heavenly bodies, not divine inspiration. Wragge proclaimed to loud applause in the North Island railway town of Taihape that the world would yet recognise Bickerton as a second Sir Isaac Newton. Restating this view in the port of Napier, he warned the people of Christchurch would one day regret Bickerton's dismissal.[455]

Some journalists compared Wragge with the late British astronomer Richard Anthony Proctor who, in 1880, toured Australia and New Zealand with great acclaim. Proctor drew a crowd of 2000 in Dunedin, the second-largest city on New Zealand's South Island, on the rebound from New South Wales Premier Sir Henry Parkes' controversial banning of his farewell lecture in Sydney. Parkes was adamant Proctor's proposed Sunday talk at the Theatre Royal on The Birth of the Universe amounted to an illegal theatrical performance on the day of rest. An angry 50,000 supporters massed outside the astronomer's hotel in the biggest rally since anti-transportation days and only slowly dispersed after Proctor addressed them from a balcony. Summing up Proctor's popularity, an Auckland reporter praised his knack of expounding the theories and facts of astronomy in everyday language. He was a genial all-rounder, no one-track-mind

---

455 *Lyttelton Times*, July 12, 1904, p7, citing *Taihape Post*, July 8, 1904; *Otago Witness*, June 1, 1904, p32

science crank but dedicated to teaching the value of astronomy in everyday life. In 1888, after Proctor's death from malaria in the United States, aged 51, a friend wrote:

> Although a university man, and keenly sensitive to anything approaching rudeness, he disliked conventionality ... He was a genial man, and considered that everybody engaged in entertaining the public was doing useful work.[456]

In April 1904, reflecting on Wragge's popularity, contributors to the *Otago Witness* in Dunedin described his lecturing style as garrulous and less conventional than their memories of Proctor's brand of informality. "Mr Wragge isn't Mr Proctor: he is more and less," a *Witness* columnist wrote of the meteorologist's yen for lacing his lectures with God Talk, but noting Proctor, too, had acknowledged in his talks the immanence of The Almighty.[457]

Wragge's presentations in Dunedin attracted audiences not seen since Proctor's tour. Word of mouth from Invercargill, where he drew nearly 1000 people one night, portrayed him as a scientist with a quaint sense of humour and a fabulous set of celestial photographs. In mid-May, full houses in the capital, Wellington, allowed him to extend his visit there from three to seven shows, first in the city's dilapidated Exchange Hall, then the sturdy New Century Hall, recently opened by the Association of Spiritualists.

---

456 *The New Zealand Observer and Free Lance,* Auckland, December 11, 1880, p113; *New Zealand Herald,* Auckland, September 26, 1888, p5, quoting from tribute to Proctor by his friend and former New Zealand tour manager R S Smythe, first published in the *New Zealand Times,* Wellington

457 *Otago Witness,* April 13, 1904, p69

Callender returned to Australia after the Wellington shows to prepare for George and Maudeena Heller's latest tour. Wragge rewarded him with a golden, ruby-studded ring and hired another experienced manager, Victor Beck, in his place.

New Zealand audiences seem to have appreciated Wragge's innate *dagginess*, or "distinct originality" in the polite estimation of a journalist who heard him in early June in Palmerston North:

> His aptitude for raising a laugh at precisely the wrong moment is prodigious, but he is intensely in earnest, and even when knocking down Scriptural theories with one hand and expounding Scriptural dogma with the other, he is vividly interesting, principally because his hearers never know what is coming next.[458]

Beyond irritating fundamentalists, there's no evidence, at least in letters-to-the-editor columns, of disagreements over his critiques of White Australia,[459] wanton land-clearing and neglect of water conservation. In his lectures, Wragge illustrated his worldview by juxtaposing lantern slides of the Union Jack, Milan Cathedral and Fiery Firmament. In his mind, the flag symbolised loyalty and respect for the King of England, just as the famous cathedral symbolised reverence for the Great Architect of the

---

458 *Manawatu Times,* Palmerston North, June 10, 1904, p2
459 Wragge argued White Australia laws were impractical in the nation's tropics because white people were incapable of functioning properly in hot climates — the realm of black races. He asserted in early 1903, in a lecture in Bairnsdale, Victoria, the concept of a White Australia was "impossible, childish and absurd".

Universe. He felt astronomical observatories were ultimately like cathedrals — holy places of common ground for the stargazing brotherhood of man. But he conceded pride in Great Britain remained the bedrock of his identity, as he affirmed in an interview soon after landing in New Zealand:

> Of course, it will be understood that I have given you my views from the scientific standpoint and as an Imperialistic Englishman who has gained experience in Australia; proud of my native land and ever ready to welcome that which really tends to build up our great Empire.[460]

In early August, interviewed in Auckland and nearing the end of his tour, Wragge praised New Zealand as an interesting and romantic country and New Zealanders for being more intelligent and enthusiastic than Australians in scientific pursuits.[461] On August 10, he, Edris and Kismet sailed for Tahiti and the South Pacific to view an eclipse of the sun on September 9, with vague thoughts of touring Canada and the United States before settling in England for a more congenial intellectual climate than Australia's. He signed off with a lament widely reported across the Tasman: "I leave Australia a saddened and broken-hearted man. I shall never come back!"[462]

---

460 *Southern Cross,* Invercargill, p6, March 5, 1904
461 *New Zealand Herald,* August 3, 1904, p1
462 *Auckland Star,* August 10, 1904, p2

CHAPTER 20

THEOSOPHY, RADIUM AND THE ETERNAL DYNAMO, 1906–08

To observe Clement Wragge for a day on end is to be persuaded he has swallowed a large chunk of radium at some time or other. He doesn't stop. He is hustling about something or other all the time. — The Free Lance, Wellington, New Zealand, January 23, 1909.

In August 1906, Wragge was six months into a lecture tour of Western Australia when the Australian government began setting up a national weather service. He explained his unexpected return — so soon after wishing Australia good riddance — was chiefly to promote his new book *Romance in the South Seas*. But, he conceded, he had missed the climate, free and easy life, and breathing space of Australia, "reminiscent of camp-fires and redolent of the eucalypti". The book based on his experiences in New Caledonia and Tahiti received generally good reviews in Australia. In London, though, the *Pall Mall Gazette* quibbled at his forced humour, tedious use of common French phrases and excessive home-grown theology — offset by exceedingly good photographs.

In February 1906, Wragge arrived in Fremantle from London with Edris and five-year-old Kismet, telling reporters he had given

600 lectures during the recent Indian Colonial Exhibition at the Crystal Palace. This feat alarmed *The Sydney Stock and Station Journal*, which saw his interminable lectures on the Australian heavens as pathetic evidence of Queensland meteorology's fall from grace. The *British Australasian* concurred and pricked him by asserting the "erstwhile weather prophet of Australia and gold medallist of Ben Nevis Observatory" had become a mere showman.

Not true, Wragge replied. He was a meteorologist, no weather prophet, and a showman only in the sense of wanting to showcase the marvels of the universe: "To lift the masses to a more adequate conception of what noble astronomy really is, and their oneness with the infinite." Surely, there was more to life than bridge, horse-racing and pigeon-shooting. Anything capable of elevating and ennobling should be encouraged by an educated and enlightened press.[463] He brought with him from England tiny samples of radium, the sensational element isolated in Paris four years earlier by Pierre and Marie Curie. In March 1906, in the first of many radium lectures and exhibitions, he told a Perth audience he regarded the element's radioactivity as an emanation of God. He bewildered country town viewers by setting his glowing specks in the pattern of the Southern Cross.[464]

* * *

---

[463] *Daily Telegraph*, Sydney, November 16, 1910, p9; *The Sydney Stock and Station Journal*, June 27, 1905, p4; *The North Queensland Register*, July 7, 1905, p26, quoting from *British Australasian*, May 18, 1905

[464] *Swan Express*, Midland Junction, March 31, 1906, p2. Personal correspondence from Adrian Ingleby, quoting a story told to him by Peter Rogers of Merimbula, New South Wales, on his mother's first viewing of radium, at a Wragge lecture, circa 1906.

Chapter 20 Theosophy, Radium and the Eternal Dynamo, 1906–08

It's unclear if Wragge applied in August 1906 to head the proposed Commonwealth Bureau of Meteorology — the ambition that in 1900, on the brink of Federation, he confided to his friend James Drake, then-Queensland's Postmaster-General. Historian David Day believes Wragge was never a serious candidate for the position awarded in November 1906 to 40-year-old Henry Ambrose Hunt, then Acting Government Meteorologist of New South Wales. In Day's view, Wragge was probably sidelined by his footloose existence since 1902, when the Brisbane bureau folded. In June 1906, senators debating the second reading of the Meteorology Bill believed Wragge had moved to London and was hence ineligible. Day's research for *The Weather Watchers,* published in 2008, was inconclusive on the identity of applicants for the new position created when the Commonwealth Government's *Meteorology Act* became law in August 1906.

In later years, Wragge complained he had been unable to wait half a decade for the bureau's incarnation. But he knew some senators wanted him to lead the new show. During the second-reading debate, West Australian Labor Senator George Pearce described him as practically the first Commonwealth Meteorologist and complained weather forecasting had collapsed since the demise of his Brisbane bureau. Senator Robert Best, a Victorian Protectionist, argued Wragge — if available — would greatly benefit Australia with his experience and scientific knowledge. Home Affairs Minister Sir Littleton Groom, who steered the Bill through the Lower House, acknowledged Wragge had left a crucial blueprint in his detailed vision for a first-class federal meteorological service published in May 1901, in the *Pastoralists' Review*. In particular, the Government had followed his recommendation for a central weather bureau, recognising forecasts as invaluable for primary production and shipping. The government rejected the view of a Commonwealth meteorological conference, chaired by Sir Charles Todd in Adelaide in May 1905, that the central agency should be limited to theoretical and scientific meteorology, leaving forecasting to existing state bureaus.

Todd's biographer Denis Cryle contends Todd did his best in 1902 to ensure his upstart rival never headed any federal bureau. He found Todd had written to Pietro Baracchi in June 1902 stating his opposition to a central bureau under any circumstance, but more especially if placed under Wragge. He also lobbied his counterparts in New South Wales and Western Australia, H C Russell and William Ernest Cooke, to block any chance of Wragge's appointment. [465]

Wragge probably contacted friends in Melbourne after enactment of the Bill. In late August, the *West Australian* reported he was winding up his tour with farewell lectures in Perth before leaving for the eastern states and, on September 8, an unlikely source — the *Murchison Advocate*, 1000 kilometres north of Perth — speculated he would mingle with federal authorities while in the east, but doubted he could convince the powers-that-be of his claims on the plum job. By then, country papers in New South Wales had given the front-running to Hunt or Western Australia's Ernest Cooke. There's no evidence Wragge visited the east coast in September. Instead, on October 16, *The West Australian*'s shipping notices recorded Mr and Mrs Clement Wragge and child had left Fremantle for Colombo on the *RMS Orman*. Their seemingly

---

465 Cryle, Denis, *Behind the Legend, the Many Worlds of Charles Todd*, Melbourne, 2017, p218

Chapter 20 Theosophy, Radium and the Eternal Dynamo, 1906–08

sudden leaving suggests an impulsive decision on the rebound from Hunt's appointment. Yet Wragge had announced in Fremantle the previous February he had no wish to live permanently in Australia and intended to return to his "native place" — England — after arranging distribution of his book and going on a lecture tour. In the same interview, he described meteorology as his first love and true vocation for the previous 35 years. He continued, "If a man cannot do good to his fellows and benefit mankind as far as he is able, of what use is life except to himself?"[466]

On balance, I doubt he expected to lead the new bureau. It's likely he planned his Indian subcontinent trip months ahead. A precis of testimonials from his Indian expedition shows that within weeks of arrival in Ceylon in December 1906, he had official patronage from the colony's British Governor. Rather than being another good riddance to Australia, his stealthy departure was the unobtrusive launching of a new adventure. "You know there are two distinct classes of Britisher — the man who has travelled and the man who has not," he told a reporter in Perth at the start of the 1906 tour. He continued: "You cannot find a more tolerant, broader-minded man than the travelled Englishman."[467] Personally, he had first-hand knowledge of 32 countries and had just completed his seventh voyage to Australia. His memoir notes from these years are devoid of weather bureau politics. He summed up 1906–09 as: "Ceylon and India, Kis[met] in Calcutta, Northern India, Darjeeling, Shah Kot and The Yogi."

* * *

---

466 *Swan Express,* Midland Junction, March 24, 1906, p3
467 Ibid

In 2006, the national president of New Zealand's Ahmadiyya Muslim Association surprised one of Wragge's grandsons with a request from a visiting holy man to see Clement's grave in Birkenhead. The caller explained that Ahmadi followers believed Clement Wragge had converted to their universalist brand of Islam in 1908, in the then northern Indian city of Lahore. This was news to the grandson but he cooperated.[468] The visiting holy man, Hazrat Mirza Masroor Ahmad, was the international Ahmadiyya community's leader and great-grandson of the faith's founder, Hazrat Mirza Ghulam Ahmad. He offered as corroboration of Wragge's Ahmadiyyan faith an online version of his meeting with Hazrat Mirza Ghulam Ahmad, said to have been transcribed in Lahore in May 1908. Mirza Ghulam Ahmad (1835–1908), the bookish son of a wealthy Punjabi family, was a yogi in the general sense of being a holy man and, I think, was "The Yogi" in Wragge's jottings. However, he was anything but a mere yoga master. His followers revered him as the Promised Messiah, a claim denounced as blasphemy by other Muslims.

Wragge visited Lahore during a four-month stay in the state of Punjab in 1908. There are two online accounts of his meeting with Hazrat Mirza Ghulam Ahmad. In the Ahmadiyya Muslim Community's version, they met on May 12, 1908, a fortnight before the holy man's sudden death from dysentery.[469] The other transcript, on the Ahmadiyya Muslim Association's website,

---

[468] Personal correspondence from Stuart Wragge, Hamilton, New Zealand, April 4, 2006 and Dr Mohammed Shorab, former president of the Ahmadiyya Muslim Association in New Zealand, August 24, 2020.

[469] 'Professor Raig Meets The Promised Messiah, in Lahore, India, 12 May, 1908', https://www.alislam.org/library/malfoozat/professor-raig-meets-the-promised-messiah/, accessed 18 October 2022

Chapter 20   *Theosophy, Radium and the Eternal Dynamo, 1906–08*

records there were two audiences and that Professor Wragge's wife was present both times, one of them with their young son.

Edris was a native of northern India. Born in Karachi in 1864, she was at first raised in that city by her maternal grandmother and in 1870, aged six, joined her mother and stepfather in Multan, 200 miles south-west of Lahore. Her father, Edwin Horne, a civil servant of Eurasian ancestry, died six months before her birth. In 1868, her mother Mary Mansfield Horne married William Kelly, Chief Clerk at the British Commissioner's Office in Multan. Her life is a mystery between 1870 and 1890 when she and her older sister, Ida Blackwell, met Dr Emily Brainerd Ryder in Bombay. According to the Ahmadi Association's transcript, Edris sought confirmation from Mirza Ghulam Ahmad of her Spiritualist belief in communication with people who had passed away. He replied that a person who dug deep, spiritually, into clean and pure water might meet the dead through a vision. He said he had personally met Muhammed and Jesus.

The record of Wragge's questions on sin, Satan, atonement and eternity suggest he was still bothered by the spiritual anxieties evident in 1891 in his talks with his old friend William Henry Goss. The Stoke-on-Trent porcelain manufacturer and philosopher died in January 1906, aged 73, after a brief illness, leaving his family an estate of £59,603. In 1907, Eva Goss republished a selection of her father's letters, including one recalling his conversations with a friend from Brisbane — the astronomer and world traveller I'm sure was Wragge.

While meeting Mirza Ghulam Ahmad seems not to have entirely allayed Wragge's doubts, it's clear he at least concurred with the holy man that God was for everybody, never the possession of one community or religion. Mirza Ghulam Ahmad rejected the science of evolution, accepted by Wragge, saying he could not accept human descent from lower primates, but agreed humankind had fallen away from an originally close fellowship with God. This, he said, was why God sent reformers such as himself in every age and to every community. Wragge also raised his persistent worry over forgiveness of sins. How could a person do good in the world but still miss out on salvation through not having atoned and received forgiveness? Mirza Ghulam Ahmad's reported reply rejected atonement as a path to salvation. Knowing God as forgiving and merciful, he believed one might attain salvation through a life of righteous deeds. He defined sin as letting go of God in any way and said a virtuous life could really only be manifest in bright relief against the shadow of evil.

Touring India strengthened Wragge's view of death as the unveiling of a new life force, rather than the final curtain, a belief he later shared in his Australian and New Zealand lectures. He told a gathering in Invercargill in 1909 that the sight of corpses disintegrating in India's fiery funeral pyres had crystalised his belief in death as rebirth — the emergence of the eternal germ of life from one's bodily shell. His Invercargill speech had overtones of language used by his friend, the popular English novelist, Marie Corelli (1855–1924), whom he visited in England in 1905. Corelli wrote in the prologue to her novel *The Life Everlasting*, published in 1911, that she considered the "germ or embryo of the soul" a radioactive presence. She also explored what she described as the spiritual force of electricity in *The Mighty Atom* (1896) and *The Secret Power* (1921).

In 1905, Britain, the United States and, to some extent,

Australia were in the midst of a radium craze. In 1906, George Bernard Shaw complained the world had run raving mad on the subject of radium, which had excited the credulity of English folk, "Precisely as the apparition at Lourdes excited the credulity of Roman Catholics." Corelli, who welcomed Wragge as a house guest for a fortnight in 1905, shared with him the irritation of being mildly famous and misunderstood. He used her endorsement as an admirer of his work on the prospectus for his Indian tour.[470] The prospectus Wragge published in Mussoorie, introducing his lectures to "the public officials, princes and chiefs of chiefs of India, private residents, colleges, schools and messes", indicates he had intended to continue into 1909. But these plans changed when Kismet, then eight years old, became dangerously ill with smallpox and typhoid fever. Wragge and Edris took him back to their home base in Calcutta and abandoned the tour. They decided he would recover best in New Zealand's cool climate and reached Auckland on January 4, 1909. Wragge arrived with sensational predictions of another drought in Australia and likely earthquakes in New Zealand which inevitably helped promote the North Island tour he had conceived during the voyage.

* * *

The Old Guard of colonial meteorology in Australia ebbed away between 1906 and 1909. Todd retired in December 1906, five months after his 80th birthday, then still on the federal government's payroll as deputy head of the South Australian observatory. He died in January 1910, just after his meteorological

---

470 *Englishman's Overland Mail*, Calcutta, July 2, 1908, p14

successor in Adelaide, Edward Bromley, became divisional officer of South Australia's branch of the federal bureau.[471]

In 1908, George Bond (1874–1938), the only member of Wragge's Brisbane office still in government employment, took charge of the federal bureau's Queensland branch and stayed until his retirement in 1934. His son, Harold G Bond, joined the bureau in 1939 and eventually led Tasmania's divisional office. Edgar Lambert Fowles (1872–1952) resigned from the downsized Brisbane office in 1905 and made a career in business, including running Brisbane's Transcontinental Hotel in the 1920s. Archie Anderson died of tuberculosis in Brisbane in 1905, aged 42.[472]

In New Zealand, the retirements of head meteorologist Sir James Hector in 1903 and his successor Captain Robert Atherton Edwin in March 1909, left a leadership vacuum filled by Edwin's assistant, the Reverend Daniel Cross Bates (1868–1954). A retired Anglican minister with a flair for science, Bates was appointed director of New Zealand's government meteorological department in June 1909, and held that position throughout Wragge's Auckland years. His church career began in England with studies at St Augustine's Missionary College, Canterbury, followed by ordination in New South Wales in 1892, and country-curate travels culminating in 1898 with a posting to New Zealand. In 1902, he caught enteric fever while in South Africa serving in the Boer War as a New Zealand Forces' chaplain, was invalided

---

[471] State Records of South Australia, Post Office and Telegraphs, 1895, No 10023, Baracchi to Todd, 547, 548

[472] Steele, John, 'Jones, Inigo Owen (1872–1954)', *Australian Dictionary of Biography*, National Centre of Biography, Australian National University, http://adb.anu.edu.au/biography/jones-inigo-owen-539/text11915, accessed October 18, 2022.

home and retired from full-time ministry. He joined the Colonial Museum in 1903 as Edwin's climatological assistant and took charge of forecasting in 1906 after being made assistant director of the New Zealand weather bureau. A year later, Bates reluctantly supervised an unsuccessful rainmaking experiment in the South Island district of Oamaru. Drought-weary farmers chipped in for 50 pounds of gunpowder and 75 pounds of dynamite, which members of New Zealand's Torpedo Corps shot into unresponsive clouds for three days in August. Bates reported in 1908 the money would have been better spent on tree-planting.[473]

---

473 'Bates, Daniel Cross', *An Encyclopedia of New Zealand*, edited by A H McLintock, originally published in 1966, *Te Ara – the Encyclopedia of New Zealand*, https://teara.govt.nz/en/1966/bates-daniel-cross, accessed October 18, 2022; *Blain Biographical Directory of Anglican Clergy in the South Pacific ordained before 1932*, 2020 edition, entry for Daniel Cross Bates, http://anglicanhistory.org/nz/blain_directory/directory.pdf, accessed October 18, 2022; *Auckland Star*, April 28, 1909, p8; *The Free Lance*, Wellington, May 22, 1909, p4

CHAPTER 21

NEW ZEALAND, 1908-10

Mysteries ... or what to the ordinary mind are mysteries, form, if we may be excused this vulgarism, the stock-in-trade of the Theosophist. His great field of exploitation is the border between body and mind.
— Brisbane Courier, April 11, 1891, page 4.

News that an earthquake had razed the Sicilian city of Messina reached Auckland a few days before Wragge's arrival in early January 1909. The quake on December 28, 1908, and catastrophic tsunami in the Straits of Messina killed 80,000 people in southern Italy and left hundreds of thousands homeless. It flattened the historic heart of Reggio in the Italian province of Calabria. On arrival in Auckland on January 3, Wragge told journalists he had no doubt the quake had been caused by two massive storms on the face of the sun, which he detected six or seven days earlier. The present period of solar activity was in his opinion the most remarkable on record. He warned that when the sunspots inevitably waned to solar minimum conditions, earthquake activity would shift from the Northern to Southern Hemispheres:

> I am not saying this to frighten the people of New Zealand, but they must draw their own conclusions. I think we may

reasonably expect to hear of earthquakes in the Southern Hemisphere during the next few years. The time is coming when we will be able to forecast the earthquakes as well as probable weather two or three years ahead.[474]

The carnage inspired relief efforts around the world. An appeal by the Lord Mayor of London raised £95,000 for the Italian government, the Commonwealth of Australia sent £10,000, the governments of New Zealand and New South Wales £5000 each, and in England an appeal for clothing yielded 20,000 articles in three days.[475]

Wragge had put similar views after the disastrous San Francisco earthquake in April 1906. Interviewed in the Western Australian goldmining town of Laverton, he theorised that massive solar storms in the previous two years had also caused Pacific tidal waves, an eruption of Vesuvius, and earthquakes in India, Italy, Japan and Formosa. His study of sunspot cycles showed solar maxima periods such as that in early 1906 coincided with seismic disturbances and below-average rainfall in the Northern Hemisphere yet "comparative terrestrial quiescence, much rain

---

474 *Evening Post*, Auckland, January 6, 1909, p7; Hugh Clements, a London-based Irish amateur meteorologist and sunspot enthusiast, also predicted earthquakes and volcanoes around the world between 1900 and 1910, based on solar and lunar cycles. His views were promoted by British journalist William Digby in *Natural Law in Terrestrial Phenomena: A study in the causation of earthquakes, volcanic eruptions, windstorms, temperature, rainfall, with a record of evidence*, London, 1902. In 1905, while in London, Wragge ridiculed Clements' theory that sunspots were caused by "tidal action" of orbiting planets. *London Standard*, October 27, 1905, reprinted in *The Richmond River Express*, Casino, NSW, December 12, 1905, p3.

475 *Brisbane Courier*, January 8, 1909, p4; *Morning Post*, London, January 9, 1909, p8 and January 15, p6

and good seasons" in the Southern Hemisphere. By this pattern, Australia could expect a return to dry seasonal conditions by the end of 1912 and New Zealand faced greater likelihood of increased seismic activity. He concluded, "Why this should be is not at present clearly understood — it suffices that we have to do with fact and await further research."[476]

In January 1909, asked his opinion of Wragge's theories, Pietro Baracchi, was disdainful. "I do not care to discuss it," he said. "I will say, however, that there is no person in the world entitled to make such a statement as a sound and reasonable scientific deduction. No man can absolutely deny the possibility of connection between solar activity and seismic phenomena, but no man can say that such a connection has as yet been scientifically proven."[477]

The New Zealand press took notice. In June 1909, a Dunedin columnist quoted Baracchi when condemning Wragge's sunspot warnings as dangerously exaggerated. The *Otago Daily Times'* columnist, Civis, cited a derisory reference in the latest *Edinburgh Review* to theorists who linked every kind of variable to sunspot cycles, from Edinburgh's weather and the price of wheat to the results of Cambridge University exams.[478] Dunedin's *Evening Star* conceded Wragge had a fair track record as a "peril petrel", so why not make him honorary Minister for Earthquakes?[479] The South Island's Otago and Canterbury districts experienced 8 of the 22 quakes recorded in New Zealand between 1870 and

---

476 *The West Australian*, Perth, April 25, 1906, p7
477 *Bendigo Advertiser*, January, 1909, p7
478 *Otago Daily Times*, June 26, 1909, p6
479 *Evening Star*, Dunedin, January 9, 1909, p9, 'By The Way' column

1910.[480] Meanwhile, in Melbourne, Australia's Acting Federal Meteorologist Richard Griffiths (1857–1930) rejected Wragge's well-publicised prediction of a coming drought. He argued sunspot theories were unproven and that, in Australia, various attempts to correlate sunspots and rainfall patterns had invariably negative results. Typically, there were vast differences in rainfall around Australia under exactly the same sunspot conditions.

* * *

Wragge had intended revisiting New Zealand well before Kismet's illness hastened the family's exit from India. In May 1908, promoter George Victor Beck announced he had contacted him in Cawnpore, Uttar Pradesh, to discuss a second tour of the dominion. He had previously managed his hectic 1904 tour, a barnstorming 100 lectures in 94 towns squeezed into 16 weeks. Beck, whose previous clients included Annie Besant and Mark Twain, rated the 1904 tour a big success with unexpected profits.[481] In December 1908, he confirmed Wragge was on his way for an extended visit. He introduced him as a world-famous astronomer with a program of grand scientific entertainments illustrated by the finest and most recent astronomical photographs from the great observatories of the world.

Beck's ambitious 1909 itinerary entailed 146 lectures in 11 months

---

480 G A Eiby (1968) 'An annotated list of New Zealand earthquakes, 1460–1965', *New Zealand Journal of Geology and Geophysics*, 11:3, 630-647, www.tandfonline.com/doi/pdf/10.1080/00288306.1968.10420275, accessed October 18, 2022

481 *Wairarapa Daily Times*, Masterton, May 21, 1902, p4; *Otago Witness*, Dunedin, September 14, 1904, p61

on the road. The tour opened in late January, with full houses in the New Zealand capital, Wellington. As well as pictures of the solar system, Wragge gave audiences views of alpha waves pulsing from specks of radium, through his spinthariscope, a zinc sulphide-coated magnifying eyepiece.[482] Interviewed in Auckland before setting out, he rated Marie Curie's discovery of radium in 1898 the greatest find in the history of science. Radium's energy was evidence to him that matter of all kinds was electric and that the universe was an etheric electronic ocean. He believed the new element had the potential to create, prolong and also destroy life. Back in Auckland in late March, he offered to help any interested prospector test rocks for the presence of radioactive pitchblende. He told his audience at the city's Choral Hall there was big money in mining this source of radium, especially since Austria had prohibited its export. He estimated a ton of pitchblende might contain as much as £500 in radium, based on the going rate of £3,200,000 per pound.[483]

In 1909, Wragge's lectures on the majesty of creation, grandeur of the universe and marvels of radium conflated Spiritualist and scientific theories with those of the controversial British physicist Sir Oliver Lodge, whom he regarded as a kindred spirit.[484] In

---

482 The spinthariscope was invented in 1903 by British chemist William Crookes for viewing heat emissions from radium samples. Crookes derived the term from a Greek word for scintillation. Reference from Warner, Deborah Jean, 'The Spinthariscope and the Smithsonian', Smithsonian Institution Archives, 2018, https://siarchives.si.edu/blog/spinthariscope-and-smithsonian, accessed October 19, 2022.

483 *The Free Lance,* Wellington, January 30, 1909, p17; *Auckland Star,* January 7, 1909, p3 and March 26, 1909, p3

484 Sir Oliver Lodge (1851–1940) was a pioneer of electromagnetic physics and a Christian Spiritualist who was convinced, like Wragge and Arthur Conan Doyle, of having contacted the spirit of a dead son through a medium, https://en.wikipedia.org/wiki/Oliver_Lodge, accessed October 19, 2022

Man and the Universe, published in 1908, Lodge asserted the "manifest transcendence" of the soul and that telepathy and, to a lesser extent, clairvoyance were on the way to being scientifically proven. Dealing with immortality, Lodge wrote, "I want to make the distinct assertion that a really existing thing never perishes, but only changes its form." This became Wragge's mantra in following years. The *London Evening Standard*'s reviewer predicted Lodge's scepticism of salvation through atonement of sins would find bitter opposition from Evangelical Church of England theologians and orthodox Nonconformists.[485]

Wragge upset Evangelical Christians and rationalist atheists alike during his New Zealand tour. For example, while travelling through the South Island in August 1909, he engaged in a newspaper debate over his imagining of the soul as an electromagnetic spark. Richard William Pearse, a farmer and inventor from Temuka in the Ashburton district, ridiculed his faith in a sentient soul "wandering through space" as childish and outdated. How could anyone prove radium's alpha, beta and gamma rays were in any way connected with the existence of a soul? Pearse cited the disdain of German naturalist Professor Ernst Haeckel, "one of the greatest scientists of the present day",[486]

---

485 *Daily Telegraph and Courier*, London, November 6, 1908, p 15, book review and extracts from Lodge, Oliver, *Man and the Universe: A Study of the Influence of the Advance in Scientific Knowledge upon our Understanding of Christianity*, London, 1908; *London Evening Standard*, October 29, 1908, p5, 'Sir Oliver Lodge and Religious Reform'

486 In 1908, German zoologist Ernst Haeckel (1834–1919), won the Linnean Society of London's Darwin-Wallace Medal in recognition of his work as the leader on the Continent of the Old Guard of evolutionary biologists. Kutschera, U, Levit, G S & Hossfeld, U, 'Ernst Haeckel (1834–1919): The German Darwin and his impact on modern biology', *Theory in Biosciences*, 138, 1–7 (2019), https://doi.org/10.1007/s12064-019-00276-4, accessed October 19, 2022

for such endeavours. In reply, Wragge recommended Pearse buy Lodge's book, a bargain at seven shillings and sixpence. He continued:

> At present, we are but as kittens just beginning to open our eyes to the stupendous facts of Nature and the marvels around us; and our present science is as nothing to that which has yet to come in the course of mental evolution, and higher and nobler education. Not one electron can be destroyed.[487]

Wragge's array of lantern slides shown at Temuka included photographs by French physician and parapsychologist Hippolyte Baraduc (1850–1909), said to show "ethetic vibrations" from subjects at prayer and in mental torment.[488]

In another newspaper debate, in November and December 1909, Wragge challenged James Orr, a pious Queenstown district high school principal, over dogma that seemed to him wilfully unscientific. Wragge complained a local Presbyterian clergyman's sermon on Adam, Eve and the Serpent had been so ill-informed that he had fled from the church to Queenstown's beautiful gardens seeking harmony with the Divine Architect. "No wonder the churches are half empty," he signed off, with another endorsement of *Man and the Universe*. Orr replied by pitying "high-falutin scientists" for their agnosticism. Mr Wragge's

---

487 *Temuka Leader*, August 26, 1909, p2; Richard William Pearse (1877–1953) patented his first invention, a bamboo-framed bicycle, in 1902, and in 1906 patented, built and flew a monoplane aircraft, made from bamboo, tubular steel, wire and canvas. Source: 'Ogilvie, Gordon Richard William Pearse (1877–1953)', *The Encyclopaedia of New Zealand*, 1996, https://teara.govt.nz/en/biographies/3p19/pearse-richard-william, accessed October 19, 2022

488 *Temuka Leader*, August 14, 1909, p3

Architect of the Universe was, in his opinion, far too remote to care for fatherless children or to forgive a penitent and sin-burdened soul. In subsequent letters, they posed each other questions neither cared to answer. Wragge sought Orr's views on the origin of the stars and the likelihood of an everlasting hell. In reply, Orr asked if Wragge conceded mankind was mired in sinfulness, ruin, misery and death and, if so, his solution.[489]

Wragge's version of the Earth's ancient history and likely future fate was too grand for entanglement with quirks of human nature. In August 1909, he suggested in the *Auckland Star* that "the soul-wave of mankind" had reached Earth millions of years ago, via etheric vibrations from another planet or maybe the moon. He argued that, since then, a succession of dominant "root races" had been winnowed by massive climate-change events. In his view, the Aryans of Western Europe, kingpins of the 20th century, would meet this fate one day, when subsumed by a tidal wave from Antarctica. He believed Earth's southern ice cap was already being very gradually undermined by comparatively warm northerly winds, correlative to southerly blizzards encountered by British explorer Ernest Shackleton on his 1908 expedition. This theory was grounded in the occurrence of contrasting atmospheric pressure gradients over Earth's equatorial and polar zones. He argued the polar regions' severe low-pressure systems inevitably sucked in warm tropical air which, over millions of years, melted the polar ice caps, resulting in sudden and radical tilts in Earth's axis.

His geological observations, unaided by theories of continental drift and plate tectonics, suggested Earth's axis had tilted several

---

[489] *Lake Wakatip Mail,* Queenstown, November 30, 1909, p5, December 7, 1909, p4, December 14, 1909, p4, December 31, 1909, p5

times over millions of years from a vertical to horizontal plane to the sun. The idea of continental drift was still only a hunch in 1909 and the theory of plate tectonics half a century away.[490] He believed the ice caps' next collapse, which he hoped was still some thousands of years off, would flood Western Europe and tip Earth's axis horizontal to the sun. In March 1909, asked his opinion of Shackleton's reported discovery of Antarctic coal deposits, Wragge declared it evidence of an ancient axial shift. He believed Earth's present polar regions had once had equator-like exposure to the sun.[491] His views on Lemuria combined Theosophist dogma from Helena Blavatsky's *The Secret Doctrine* with his own ideas on realignments of Earth's axis, possibly influenced by Jules Verne. In his 1889 novel, *The Purchase of the North Pole,* Verne imagined a scam by a tycoon who, having bought sovereign rights to the Arctic, used a giant cannon to jolt Earth's axis into horizontal alignment with the sun, causing the ice cap to melt and uncovering vast coal reserves.[492]

---

490 Alfred Wegener (1880–1930) a meteorology lecturer at the University of Marburg, Germany, proposed the theory of Continental Drift in 1912, based on his observation that Earth's large landmasses looked like jigsaw puzzle pieces. He found strong geological similarities on the western and eastern Atlantic coastlines, https://en.wikipedia.org/wiki/Alfred_Wegener, accessed October 19, 2022. The theory of plate tectonics — dealing with the dynamics of Earth's outer shell — emerged in the 1960s from geological and geophysical research on mountain-building processes, volcanoes and earthquakes. From Andel, Tjeerd H van and Murphy, J Brendan, 'Plate tectonics', *Encyclopaedia Britannica*, https://www.britannica.com/science/plate-tectonics, accessed May 21, 2021.

491 *Otago Daily News*, Dunedin, March 27, 1909, p5; *Auckland Star*, August 14, 1909, p11, 'The Romance of the Earth'

492 Verne, Jules, *Topsy-Turvy* (also known as *Purchase of the North Pole*), New York, 1890, precis from Wikipedia, https://en.wikipedia.org/wiki/The_Purchase_of_the_North_Pole, accessed October 19, 2022

In June 1909, the Reverend Paul Wynyard Fairclough (1852–1917), long-time astronomy correspondent for Christchurch's weekly *The Press*, called Wragge's fancy of wild polar shifts as far-fetched as his efforts to correlate local storms with sunspot activity. In November 1909, Shackleton's paper on his expedition, published in *The Geographical Journal*, made only passing reference to coal.[493] Forty years later, in 1961–62, New Zealand scientists reassessed what, by then, were known as the Buckley Coal Measures and found them unlikely to have been formed by a warm temperate or subtropical climate. Instead, based on fossil evidence and glacial varves — annually deposited layers of clay and silt — the coal was more likely the product of a relatively cool to cold temperate, pre-glacial climate, early in the Permian age (299 to 251 million years ago) before Antarctica's separation by continental drift from the supercontinent of Pangea.[494] By 1962, geologists agreed the continents of Australia and Antarctica as well as the New Zealand islands had once been part of the south-eastern Gondwanaland segment of Pangea.

* * *

Wragge felt at home in "God's Own Country". On the lecture trail in 1909, he speculated on New Zealand's prehistory with

---

[493] *The Press*, Christchurch, Astronomical Notes June 30, 1909, p6 and January 31, 1912, p8; *Monthly Notices of the Royal Astronomical Society*, Vol 78, Issue 4, February, 1918, obituary of Rev P W Fairclough, p245, https://academic.oup.com/mnras/article/78/4/245/1053247, accessed October 19, 2022

[494] Grindley, G W (1963), 'The Geology of the Queen Alexandra Range, Beardmore Glacier, Ross Dependency, Antarctica; with notes on the correlation of Gondwana sequences', *New Zealand Journal of Geology and Geophysics*, 6:3, 307-347, DOI: 10.1080/00288306.1963.10422067

audiences who were open to his flurries of science and sci-fi. He declared the dominion's mountains, plains and rivers carried messages from the ancient continent of Lemuria. Every ravine whispered of aeons-past ice-melts when brief tropic summers marinated the glacial Southern Hemisphere. His imaginings of Earth's primitive Lemurian and advanced Atlantean races were direct from *The Secret Doctrine*, but attributing their genesis and fate to massive, cyclical climate change was his own work. He saw climate change as the Great Adjuster of human history. The reporters who chronicled his 1909 tour rated the lectures disconcertingly hilarious yet earnest. Wellington's popular weekly, *The Free Lance*, commented:

> To observe Clement Wragge for a day on end is to be persuaded that he has swallowed a large chunk of radium at some time or other. He doesn't stop. He is hustling about something or other all the time.

His audience in the North Island gold-mining town of Waihi applauded his questioning of New Zealand's just-announced pledge to meet the £1.7 million cost of Britain's latest battleship. Why not send money to feed the poor and starving multitudes? They also applauded his declaration that God might be found everywhere, from a flea to the Milky Way. And they laughed with him at his aside that all scientific men were a little bit off their heads.[495] He joked that his meteorological peers Baracchi, Hunt and Bates would love to silence his heretical views, especially

---

495 *Waikato Independent*, Cambridge, May 18, 1909, p6, quoting from an undated issue of the *Waihi Daily Telegraph*

Baracchi, his "Lord High Executioner".[496] In the South Island town of Oamaru, North Otago, Wragge had some fun with a reporter who quizzed him on mysterious lights seen over the town in late July 1909. Wragge replied that a spaceship of refugees from drought-ravaged Mars had landed, then fled in alarm at the complacency of Earth people. For example, their indifference to the glories of the universe, unconcern for tortures endured in the name of religion, and acquiescence to "bloody wars, political and social intrigues and the awful gap between rich and poor". But, given the Martians' environmental predicament, he predicted they would fly on to London to ask Prime Minister Joseph Ward to waive writing and dictation tests and let them settle in New Zealand to improve the human race and teach the people wisdom.[497]

Wherever he travelled in 1909 and 1910, Wragge argued a contentious link between the spiral-patterned tattoos he saw on Maori faces everywhere and the ubiquitous spiked and spiralled finials he saw on scores of the dominion's rooftops. He believed the swirling lines were a sun-worshippers' symbol, handed down through Mongoloid and Aryan races, originating aeons ago when Earth's axis was horizontal to its solar orbit. In this ancient era, he argued, the sun would have spiralled to and from the vertical in polar regions. During July 1909, he parried with a Dunedin newspaper columnist over these views. He challenged the *Otago Daily Times*' columnist Civis to prove spikes on houses were not ancient sun-worship symbols and that the first people of New

---

[496] *Auckland Star,* August 14, 1909, p11, Wragge, Clement L, 'The Romance of the Earth'

[497] *Oamaru Mail,* August 4, 1909, p6

Zealand — Maoris and Morioris — were not lineal descendants of the original pre-Aryan root race. In reply, Civis sought proof that spikes on Prussian helmets had not originated the same way, and that the moon was not made of green cheese.[498]

In March 1910, after touring the Bay of Islands district, Wragge rushed to the press with photographs he declared were evidence of a lost civilisation. His home-developed and printed pictures showed massive stone columns that he argued had been inscribed by ancient sun-worshippers, possibly kin of the creators of Easter Island's enigmatic statues. He said he also had found "a very ancient stone instrument like an axe" in the same area, but declined to say exactly where.[499] Doubters guessed he meant a peculiar formation of columnar basalt beside the Kerikeri Inlet Channel, about 150 miles north of Auckland. Professor Algernon Thomas, Professor of Geology at Auckland University College and a keen bushwalker, rejected his interpretation of pits and pock marks in the stone as ancient symbols. "The onus lies on Mr Wragge to prove they are other than natural markings," he told the *New Zealand Herald*. "We have no evidence whatever in New Zealand of occupation by prehistoric people."[500]

* * *

Within a month of Wragge's arrival from India, the Brisbane gossip paper, *Queensland Figaro,* reported he planned to live in

---

498 *Otago Daily Times,* July 19, 1909, p6 and July 24, 1909, p6
499 *Auckland Star,* March 7, 1910, p6
500 *New Zealand Herald,* March 9, 1910, p8

Auckland. In June 1909, the same journal reported he had made his base in that city and was pursuing weather forecasting and "model farming". His luggage was said to have included several cases of photographic negatives, a similar number of cases of lantern slides and personal papers, testimonials, presents from Indian Maharajahs and 11 cases of remarkable sea shells.[501] Many of these treasures remained in storage until 1916, when he displayed them in his private museum at Birkenhead. The cities of Nelson, Hamilton and Birkenhead Borough all declined his offers of the collection on permanent loan.

In May 1909, an *Auckland Star* columnist likened Wragge's finding snug harbour in the city to the much-travelled Brisbane journalist William Lane's arrival in 1900. New Zealand offered a kindly home for stray dogs and lost causes, Alfred George Stephens wrote. Lane had found refuge, "ambition thwarted, health and fortune shattered", after seven years struggling to sustain an egalitarian paradise in Paraguay. In short, he had packed up his dreams for a regular job on the *New Zealand Herald*. Stephens, an Australian expatriate too, felt Wragge was wasting his time on the lecture circuit — "His lean and pathetic visage displayed in every grocer's shop." Why hadn't he been recruited after Captain Edwin's retirement as weather chief?

In August 1909, Wragge replied without much tact in a critique of New Zealand meteorology and Edwin's successor, the Reverend Daniel Cross Bates. Describing the government's funding of meteorology as absurdly meagre, he had a dig at Bates for abandoning his clerical calling for "the opera of the winds" and advised him to try helping London's starving poor instead

---

501   *The Free Lance,* Auckland, January 23, 1909, p4

of grappling with isobars. In October 1909, the *Star* reported circulation of a petition asking Prime Minister Sir Joseph Ward to appoint Wragge as Government Meteorologist. But I've found no evidence he was interested. By then, he was supplying daily forecasts to the press in Auckland and Wellington.[502]

Wragge spent half of 1910 touring Australia and rattling rural audiences with controversial sunspot science. He visited 65 Tasmanian and Victorian towns between May and November, assisted by an energetic new manager, Norman Carryer.[503] By careful design, they opened in Launceston on May 18, a day before the Earth's brief scheduled transit through the tail of Halley's Comet. A curious crowd packed the city's Mechanics' Institute hall to hear Wragge's message in a lecture titled "A Voyage Across the Universe". The *Launceston Examiner* reported Wragge's reassurance that there was nothing to fear and praised his "bold, rugged eloquence":

> Mr. Wragge has added to his scientific attainments the platform skill of the professional lecturer. He speaks at a terrific rate, crowding into two hours an enormous mass of information about his subject, which is illustrated by beautiful photographs and punctuated by a very happy wit ... He had the house in his grip all through, and had the lecture been twice the length he could easily have sustained the interest to the end.

---

502 *Auckland Star*, August 18, 1909, p6; *New Zealand Herald*, October 6, 1909, p6

503 Wragge's friend Norman Carryer — an architect and accomplished clarinet player — travelled with him to Tasmania in April, 1910, and later settled permanently in Australia. A native of Newcastle, Staffordshire, he served in the Boer War with the North Staffordshire Regiment and later became a delegate for the Universal Esperanto Association for Hobart, see *Daily Telegraph*, Launceston, July 1, 1915, p7; *Daily Post*, Hobart, March 21, 1917, p2; *The Mercury*, Hobart, September 15, 1950, p3

Word spread from his next Launceston lecture, a night later, that he had predicted an imminent severe drought. Farmers and graziers still craved his long-range forecasts, regardless of the new federal weather bureau's disdain for sunspot-guided seasonal forecasting. He obliged with a very carefully worded warning that Australia was in for several years of below-average rainfall up to 1914, "due to solar maxima and minima", and suggested — to laughter and applause — the wise should begin buying tanks to be in harmony with cosmic law.[504] But "What does Wragge say?" still jangled on the bush telegraph. Before long, squatters and cocky farmers alike swore Wragge had divined another drought — even after he wrote to Melbourne's press insisting he had avoided the word "drought" and that the coming long dry spell would be nothing like the Federation Drought. When fearful squatters began selling off stock, Victoria's Agriculture Minister George Graham condemned any talk of drought as mischievous and Hunt issued a statement debunking sunspot-based seasonal forecasting. He avoided mentioning Wragge but echoed his advocacy for water and fodder conservation in readiness for future arid seasons.

Before returning to New Zealand in late November, Wragge argued the millions likely to be wasted on building a national capital should be invested instead in new dams and irrigation schemes. He described the Yass-Canberra district as an absurd site and declared if Australia could afford a federal metropolis, which he doubted, it would best be built in the Snowy Ranges, "Where

---

504 *Launceston Examiner,* May 19, 1910, p6 and May 20, 1910, p3.

the bracing atmosphere and pure water supply would breed a race of gritty statesmen."[505]

In December 1910, the Auckland weekly *New Zealand Observer* reported he had bought land in the Northcote district and intended to become a permanent resident. Other sources suggest this was the tract of native bushland in the adjoining borough of Birkenhead that Wragge later praised as the sweetest nook in New Zealand. He and Edris seem to have moved onto their quarter-acre block by May 1911, when, in a letter to a friend, he referred to his home in Birkenhead. Their hillside land at number 8 Awanui Street was within walking distance of the ferries to and from the city. Named by a whimsical land agent, Birkenhead was roughly as close to Auckland over Waitemata Harbour as the Lancashire port of Birkenhead's two miles across the Mersey River to Liverpool. In 1910, the *New Zealand Herald* described it as a rapidly growing settlement of 1600 people, blessed because of its elevation with some of the best views around Auckland.

In a tribute published in 1944, an old, unnamed resident wrote Wragge had been enchanted by the wealth of native bush that had survived European settlement on Birkenhead Point. He had loved the restless presence of New Zealand's ubiquitous, metallic-sheened tui birds: "Many times I have seen him stand in silent reverence, listening to the tui's glorious notes." The writer described Wragge as a nature-lover well-known for his

---

505 *Thames Star*, Thames, May 31, 1911, p2; *The Age*, Melbourne, August 13, 1910, p10 and August 30, 1910, p7; *The Argus*, Melbourne, September 14, 1910, p13; *The Age*, September 15, 1910, p8, quoting from Wragge's lecture in Moe on September 14, where his views on the national capital were applauded. He gave 65 lectures during his six months in Australia in 1910.

kindness and courtesy.[506] Interviewed while digging his garden overlooking Little Shoal Bay, Wragge praised North Auckland's mild climate and potential for luxurious growth of tropical plants. In 1913, he gave 1500 seeds of the Brazilian palm, *Cocos flexuosa*, and a number of other varieties, to Auckland's city landscape gardener for planting in the Domain and other public parks.[507] The Waiata Botanic and Acclimatisation Gardens, which he and Edris opened in 1916, included a plantation of 40 species of palm tree from around the world, propagated from seed and planted out in the preceding five years.

In 2011, a survey and report commissioned by Auckland City Council rated 17 surviving trees as having national and perhaps international significance. Cultural Landscapes and Historic Gardens consultant John Adam recommended permanent recording and scheduling of the trees by GPS technology. These included species from North and South America, Africa, Australia, China and the Pacific. The Phoenix and Date Palms standing in Myers Park in central Auckland are believed to have been gifts from Wragge.[508]

* * *

In March 1911, Wragge took Edris and 10-year-old Kismet with him to the Tongan islands, 1250 miles north-east of Auckland, for a total eclipse of the sun. He avoided British and Australian scientists

---

506 *The Press*, Canterbury, December 6, 1944, p7. The letter is signed DES, of Birkenhead.

507 *New Zealand Herald*, February 14, 1913, p8

508 Adam, John P, *The Waiata Tropical Gardens at Birkenhead*, Auckland City, May 2011

who had set up in the Vava'u islands group and landed instead on Lifuka Island, 80 miles south. "It seemed a mistake for all to assemble at one point ... within the eclipse zone," he wrote later. His old friend Pietro Baracchi was in charge of a small group organised by the Australasian Association for the Advancement of Science. Astronomer William Lockyer led a British team comprising 14 Royal Naval officers and 107 men from the *HMS Encounter*.

Leaving Edris and Kismet on Lifuka, Wragge ventured 15 miles north-east to Mo'Unga'One Island with friends Dr Henry Bolton and the Reverend E S Harkness, both Lifuka residents. Clear skies on the morning of April 28 yielded ten photographs in the two minutes of total eclipse. In contrast, Lockyer and Baracchi were frustrated by heavy cloud. A disappointed Lockyer reported his expedition had achieved little. Wragge boasted his shots were unique and magnificent. He was happy, too, with his delvings into rock carvings on Mo'Unga'One Island and the nearby uninhabited island of Telekivavau that, in his opinion, showed the islands had been inhabited by an ancient race, "Long before the Maoris and Tongans came into existence."

Back in Auckland, he showed off his pictures of the eclipse and mysterious petroglyphs for the next couple of months, then announced plans for another extensive Australian tour, this time through New South Wales and South Australia. In late August, *The Bulletin* in Sydney welcomed the arrival of Australia's versatile old friend, "Inclement Wragge".[509] A review of his opening lecture at Sydney YMCA Hall complimented his talent for "warming up cold science on the stove of enthusiasm":

Wragge ... is a cataract of tremendous facts and theories; and

---

509   *The Bulletin,* Sydney, August 31, 1911, p8

most of his time he is standing on tiptoe, pointing up-up-up into Space with a long, thin arm, and talking reverentially in a sharp, high key of that snowstorm of which we form one flake.

The Bulletin recommended precious politicians, mayors and aldermen should, for the good of their souls, sample his stimulating perspective on immense things.[510]

---

510   *The Bulletin*, Sydney, September 7, 1911, p8

CHAPTER 22

AUSTRALIA, 1911–14

My forecast for Australia? There is a probability that we will have some statesmen in the country some day. — Clement Wragge, interviewed in September 1911, by *The Sun*, Sydney, after an eight-year absence from New South Wales.[511]

In late December 1908, as Wragge steamed for New Zealand, Henry Ambrose Hunt was sloshing through cyclonic southern India. Hunt's tour of British Raj weather stations had, by then, taken six or so weeks, through Bombay, Poona, Calcutta and Simla in northern India, as well as Madras in the south. On December 29, Hunt experienced his first tropical cyclone while inspecting an observatory near the Tamil Nadu hill station of Kotai, 7700 feet above sea level. London-born Hunt had spent his entire meteorological career in temperate Sydney. He was described a few years later as a pleasant, meek and well-fed man with a mane of black hair and mammoth moustache — in demeanour more like a musician or poet than a weather prophet of Wragge's ilk.[512]

On January 3, Hunt wrote from Ceylon to Home Affairs

---
511   *The Sun*, Sydney, September 2, 1911, p1
512   *Punch*, Melbourne, August 3, 1911, p6

Minister Hugh Mahon that, after sheltering from the deluge for a day, he had walked 14 miles through rain, wind and mud to catch a steamer to Colombo from the port city of Tuticorin. On his return to Australia in February 1909, he pledged to improve reliability of daily and long-range forecasting, but added this hinged on improving telegraph links with Pacific and South Asian weather stations. "I have visited many countries, but in none of them did I find anything approaching the general interest in weather matters that is displayed in Australia," he told Sydney's *Evening News*.[513] The Indian Meteorological Service's problems seemed to him similar to Australia's, in particular, the vagaries of forecasting the fickle south-west monsoon. He mentioned Edinburgh specifically when listing 43 weather stations visited in Britain, Canada, the United States, Germany, France, Austria, Italy, India, Ceylon, Singapore, Hong Kong, the Dutch East Indies and Pacific islands. His detour to Scotland was to learn for himself the history of the Ben Nevis observatory, abandoned in 1904.

* * *

Hunt treated Wragge with caution and condescension, rarely mentioning him by name in official documents or press interviews. In 1913, Hunt overlooked Wragge when listing Australia's revered meteorological pioneers as Russell, Ellery and Todd. Their strenuous and untiring labours had yielded significant advances in antipodean climatology, Hunt asserted in his book, *The Climate*

---

[513] *Ballarat Star*, January 5, 1909, p2; *Daily Telegraph*, Sydney, January 28, 1909, p8; *Evening News*, Sydney, February 24, 1909, p7; *The Sydney Morning Herald*, February 24, 1909, p6

and Weather of Australia, co-authored with South Australian geographer Griffith Taylor and Assistant Commonwealth Meteorologist E T Quayle. Hunt's self-styled first textbook of Australian meteorology included a detailed list of Western Australian hurricanes between 1872 and 1912, but none of the Coral Sea cyclones named by the Brisbane bureau between 1894 and 1903. I doubt there was any spite in this omission. Queensland's meteorology had languished between 1903 and 1908 when under supervision of the Department of Public Works. In 1907, Hunt reported after inspecting each of the state weather offices that Queensland's was shambolic. He found meteorological records in disarray and the only two permanent staff members closeted in Brisbane's Treasury building, some distance from the telegraph office and observatory. Hunt feared the records had been moved so many times that some had been lost completely.

Wragge travelled Australia far and wide from 1910 to 1914, giving 350 lectures on four separate tours. During this time, he and Hunt bickered over maritime storm warnings, prompted by two shipping disasters on Australia's tropical eastern and western coasts. On March 24, 1911, the Adelaide Steamship Company's *SS Yongala* vanished in a storm between Mackay and Townsville with 122 passengers and crew aboard. Wragge, then in Tonga for the eclipse, wrote immediately to *The New Zealand Herald* in amazement the *Yongala* had not been detained at Mackay. He argued the federal bureau should have managed at least a 48-hour warning, considering the daily flow of data from New Caledonia by cable. In June 1911, the Marine Board of Queensland found the weather bureau's Brisbane office had telegraphed a signal station near Mackay on the afternoon of March 23, warning of cyclonic conditions further north. This was signalled to the passenger ship *SS Cooma*, but the *Yongala* had left some hours previously. During the inquiry, the Brisbane officer-in-charge George Bond gave evidence he had sent out special warnings on the afternoon of March 23 as soon as the storm's intensity was confirmed. Hunt wrote to Australia's metropolitan papers in April, pledging to press shipping companies to install wireless telegraph sets on all passenger steamers.[514]

In late March 1912, Wragge was touring southern New South Wales when he heard another passenger ship had strayed disastrously into an apparently unheralded cyclone, this time off the Western Australian coast. The Adelaide Steamship Company's *SS Koombana* was reported to have vanished with 150 passengers and crew aboard on March 30, in a howling hurricane — known locally as a *willy-willy* — between Port Hedland and Broome. He wrote at once to the Sydney press that, in pre-Federation days, his chief bureau in Brisbane gave three-day warnings of Western Australian cyclones. Not so now that the West's forecasts were issued in Melbourne. Both Hunt and the New South Wales divisional officer Stewart Wilson — another H C Russell protégé — replied immediately, blaming a telegraph breakdown. Wilson chided Wragge for "talking at random". He would have been in the same fix if cut off from contact with the West. The weather stations along Australia's north-west coast were well equipped with

---

[514] *New Zealand Herald*, April 26, 1911, p5; *Brisbane Courier*, April 4, 1911, p5, 'Report of an Inquiry held by the Marine Board of Queensland into the circumstances attending the loss of the *SS Yongala* of Adelaide, on or about the 23rd day of March, 1911, between Mackay and Townsville', issued June 20, 1911, http://yongalarevisited.blogspot.com/2016/09/the-marine-inquiry.html, accessed October 19, 2022

mercurial and aneroid barometers and had known the storm was coming, which would have enabled a warning to be given locally.[515] Later in April, Hunt conceded data had been missing from crucial centres for some time before the storm burst.[516] He made lengthy field trips along the West Australian and Queensland coasts later in 1912 and 1913, double-checking weather stations.

That April, Wragge's forward agent Harry Ockenden died suddenly in Wagga, aged 49. Ockenden, a professional photographer from Auckland, had assisted him since the previous September. His sister, Nettie was a member of the Auckland Theosophical Society. Wragge pressed on but in May collapsed during a lecture in Goulburn, reportedly feeling the strain of the tour. He admitted himself to a private hospital in Sydney and returned to New Zealand for several months to recuperate. He resumed touring New South Wales in November.

Meanwhile, in May 1912, Hunt wrote to all major Australian newspapers to explain his scepticism of seasonal forecasting based on sunspot and lunar cycles. He said modern meteorologists had concluded after many years of study that astronomical cycles were of no practical assistance in forecasting. Without naming Wragge, Hunt questioned certain claims made for accurate long-range predictions based on "astronomical considerations". In fact, such predictions were complicated by the existence of four or five climate zones over the massive continent. Any pronouncement on the likelihood of universal drought or universal rainfall

---

515 *Evening News*, Sydney, April 1, 1912, p8; *Telegraph*, Brisbane, April 12, 1912, p7; *The Daily Telegraph*, Sydney, December 5, 1917, p6, obituary of Stewart Wilson

516 *Barrier Miner*, April 15, 1912, p5

was bound to fail. Furthermore, there was no evidence of a set periodicity either in the return of good and bad seasons or volcanic and earthquake phenomena. Neither was there evidence of quakes alternating between the Earth's Northern and Southern Hemispheres. Hunt concluded:

> Every meteorologist realises that variation in solar activity is a cause and probably the prime cause of weather changes, but the local results depend on a complex of other factors which are only guessed at in present-day meteorology, and for which no law is yet found to hold.[517]

Wragge felt slighted and replied at once complaining of Hunt's injustice in implying he only recognised one Australian climate when, in truth, he had identified a dozen major zones and innumerable minor ones while head of Queensland's weather bureau. In making seasonal forecasts for any region, he considered solar, lunar and planetary factors as well as modifying agencies such as latitude, altitude and physiographical features. He signed off boasting again that his old bureau would have given ample warning of the *Yongala* and *Koombana* hurricanes. In contrast, the federal bureau devoted too much time to purely statistical work and too little to the eminently practical.[518]

* * *

Beginning in August 1913, Hunt spent five weeks inspecting coastal weather stations in tropical north Queensland, Darwin,

---

517 *Express and Telegraph*, Adelaide, May 20, 1912, p4
518 *The Sun*, Sydney, May 22, 1912, p4

and Port Moresby in Papua. He returned promising to improve cyclone forecasting by insisting on daily data from Port Moresby and asking shipping companies to telegraph him their weather observations. As for storm warnings, he had two proposals for the Postmaster-General's Department — placement of post office barographs in public view and extending to regional ports the big city practice of hoisting weather warning flags.

Interviewed in Rockhampton on September 8 while waiting for a train south, Hunt said the trip had been successful in terms of improving weather data from the north.[519] But chambers of commerce in Cairns and Townsville were unhappy over his blunt rejection of their case for a north Queensland weather bureau. Hunt told them he believed he had fine-tuned the existing service by installing barographs at all principal ports and insisting on thrice-daily telegraphing of weather data through the storm season. In a letter to the Cairns group in January 1914, he blamed difficulties in tracing the north's cyclones on "a want of historical facts" about past disturbances, stemming from Queensland's neglect of meteorology. He asserted detailed data had only been collected since the federal office opened. Besides, there was no money for a bureau destined to be ever-vulnerable to telegraphic disruptions in the Wet.[520]

Meanwhile, one of Wragge's north Queensland coastal shipping friends, Captain George Irvine, skipper of the *SS Innamincka*, had written to him in Auckland canvassing his interest in setting up an independent bureau. In January 1913, Irvine and his officers had spent 38 hours on the bridge of the passenger steamer battling through a cyclone north of Port Douglas. They had just scraped

---

519 *Morning Bulletin*, Rockhampton, September 9, 1913, p5
520 *Brisbane Courier*, February 10, 1914, p6

over the Alexandra Reef and grounded safely on sand near the mouth of the Mowbray River, all 39 passengers unscathed.

In early 1914, Wragge replied with a proposal for a special wet-season forecasting service and began negotiations with Townsville City Council for the summer of 1914–15, on a weekly salary of £20. He agreed to finalise the deal in June 1914, during the north Queensland leg of a proposed massive Australian lecture tour. But this scheme came unstuck in Brisbane when Nora's solicitor tracked Wragge down and served a writ for £1500 in unpaid maintenance.

Recalling this shock when canvassing divorce from Nora in 1921, he wrote that she had "hidden" during his visit to Brisbane and all but two of the children had refused to see him. He pressed on north as far as Rockhampton, where he collapsed with a nervous breakdown and admitted himself to a private hospital. He abandoned his trip to Townsville, still another 400 miles away, quietly returned to Brisbane, then Auckland, and this time quit Australia for good.[521] Seven years later, in November 1921, the federal government opened a cyclone-warning station on Willis Island, 280 miles east of Cairns, in radio contact with Cooktown. The meteorologist and two radio-telegraph operators based on the coral atoll detected their first cyclonic system and raised the alert in March 1923.[522]

---

[521] *Townsville Daily Bulletin*, January 13, 1914, p3; *Week*, Brisbane, March 6, 1914, p4; *The Queenslander*, Brisbane, February 8, 1913, p38, 'Another Tropical Disturbance'; *The Northern Miner*, June 24, 1919, p3, 'Thrilling Fight with a Submarine' – biographical details on George Irvine; *Morning Bulletin*, Rockhampton, June 9, 1914; draft of letter from Wragge to J F W Dickson Esq, solicitor, Auckland, September 12, 1921, headed 'The Tangle of My Life', original in possession of Wragge family, New Zealand

[522] Day, David, *The Weather Watchers: 100 Years of the Bureau of Meteorology*, Melbourne, 2007, pp 136–38

Chapter 22 Australia, 1911–14

* * *

In August 1914, the cavernous old homestead in Morningside that Nora had leased since parting from Wragge in 1899 was still home to their eight adult children. They were registered, with Nora, as residents of "Cannon Hill near Morningside", eligible to vote in the federal election on September 5, 1914, a month after Australia entered World War I. By that election, Bert had left home for the army, followed by Eggie a few weeks later. On enlistment, they gave their occupations as public servant and student, respectively.

Bert, 32, had worked since 1909 as a stream gauger in the Queensland Public Works Department's water supply branch. In late September, he left Brisbane for Egypt as an ammunition column bombardier in the first Australian Imperial Force. Eggie, 34, had spent 1905 to 1911 in Britain. His enlistment papers gave his military experience as membership of the Queen's Volunteer Rifles in Edinburgh, where he is said to have studied medicine. He embarked from Sydney to Egypt in December 1914, as a trooper in the 2nd Light Horse Regiment of the AIF.

That left George, 30, Reginald, 28, Lindley, 23, and their sisters May, 36, Violet, 27, and Margaret, 25, at home with Nora. George, Lindley and Nora were listed as clerks, Reginald as a teacher, May and Margaret as music teachers, and Violet as a telephonist. Nora first gave her occupation as a clerk in 1912, aged 60, some years after unsuccessfully seeking help from Robert Philp. In 1903, she signed off her plea to the Premier with a peculiar aside, given her separation from Wragge was common knowledge: "I have been in very delicate health, having taken a chill while helping my husband on Ben Nevis many years ago. Otherwise, I should help him now ..." She told Philp she doubted Clement would make enough from lecturing to cover the family's needs and feared he would not survive incessant touring. If the government could not re-employ him or find one of their sons a lucrative position, she vowed to find paying work for herself.

On July 19, 1900, a fortnight after Wragge's departure for England with Edris, *The Queenslander* disclosed Nora's move down the river from the Taringa ridge to rural Morningside. Mrs Wragge would, during her husband's absence in Europe, reside at Cannon Hill, Morningside, with her family, the weekly reported.[523]

The steep-gabled weatherboard mansion Nora called Sunny Brae had a patchy history. Richard Warren Weedon, a wealthy English widower, built it in 1867, a year after migrating to Brisbane with his ageing mother, four teenage sons and two unmarried sisters. He chose a rise beside the Old Cleveland Road that other settlers had nicknamed Cannon Hill in honour of the silhouette of a gun barrel-like burnt log tilted against a stump on the hilltop.[524] Suburban Brisbane crept south-eastward in the 1880s, following the railway towards Cleveland, 20 miles from the inner city, on Moreton Bay. In 1888, anticipating the opening of two new railway stations, big landholders in the Morningside district carved their properties into 600 housing lots. The Weedon family sold the

---

523 *The Queenslander*, July 7, 1900, p46

524 'Cannon Hill, What's in a Name?', State Library of Queensland Blog, February 26, 2015, https://www.slq.qld.gov.au/blog/cannon-hill-whats-name, accessed October 20, 2022; *Ancestry*, Richard Warren Weedon (1819–1894), https://www.geni.com/people/Richard-Warren-Weedon/6000000000670055605, accessed October 20, 2022; *Brisbane Telegraph*, June 10, 1930, p13

six-bedroom house and most of their 120-acre farm to a Sydney syndicate which subdivided 400 blocks around the proposed Cannon Hill station. Morningside's station opened in 1888 as planned, serving a separate 200-lot development, but work on the Cannon Hill whistle-stop faltered when the developer folded. The house on the hill stood alone for nearly two decades under changing ownership and nicknames.

Nora taught music with daughters May and Margaret. In 1903, *Queensland Figaro* mentioned her among guests at a vice-regal welcome to British organist and composer Dr William Creser, an examiner for Trinity College of Music in London. In 1904, Nora and the girls advertised plans to conduct a school in the recently opened Hill's Hall, a popular venue in Cleveland Road, Morningside. They pledged special attention to delicate and backward children, "terms very moderate", and also tuition for Trinity College of Music exams. Their advertisements in Brisbane's *Telegraph* newspaper vanished in 1906 after notice of a change of venue from Mornington to Cannon Hill. Nora gave her occupation as "home duties" in the 1905 and 1908 electoral rolls. Lindley, the youngest of the children, left school in 1906, aged 15.

During 1911, land agents and lawyers corralled Sunny Brae into suburbia, on news the US meat-processing company Swift and Co planned to build a massive export-beef abattoir and holding yards on 200 acres between the railway and the river. Between 1911 and mid-1914, Cannon Hill's population grew rapidly. The suburb's newly formed progress association reported fifty new houses, three shops and a bakery had materialised around the finally completed railway station. Opened in April 1914, the new meatworks was the largest of Brisbane's six abattoirs. Swifts'

Australian Meat Exporting Company boasted a daily slaughtering capacity of 1000 head of cattle and anticipated employing 800 people — from butchers and boilermakers to clerks and caterers. It's possible Nora's electoral-roll listing as a clerk from 1912 to 1919 was connected with the company, given her apparently straitened finances and evidence in newspaper job advertisements of occasional clerical vacancies.

* * *

As if wanting to conceal his parents' separation, Eggie when enlisting in 1914 named his father as next-of-kin, care of Nora's address, Cannon Hill, Morningside. By then, Wragge and Edris were well established in Birkenhead. Eggie willed Nora his property and personal effects — a song book, wallet and book of poems — and she took charge of his remembrance. His death was commemorated in 1916 at Nora's church in Brisbane and in 1917 at the parish church in Oakamoor. In December 1916, Morningside Anglican Church erected honour boards to 28 members who had, by then, volunteered to fight overseas. Eggie and Bert were both on this roll.

In October 1917, Oakamoor's Anglican parish church was full for the unveiling of Nora's memorial to Eggie, recording his birth in nearby Farley in August 1880, and death at Lone Pine in May 1915. The rector, the Reverend Arthur Wellesley Greeves, recounted the circumstances of his fatal wounding. Hit by shrapnel while helping wounded men after the capture of a Turkish trench, Trooper Wragge had died while being evacuated to Alexandria and was buried at sea. Mr Greeves, who lost two sons on the Western Front, said he detested the 10-month Gallipoli campaign, which had cost the lives of 44,000 Allied Troops and 86,000 Ottoman Empire soldiers.[525] Archie Anderson's eldest son, Sergeant Archibald Colledge Anderson, was killed in action in France in 1917.[526]

In 1922, Wragge gave the army his Auckland address during correspondence on Eggie's record for the Lone Pine Memorial then being built at Gallipoli. At the time, hundreds of Greek stonemasons were engraving the memorial's Turkish limestone slabs with the names of 4900 Australian and New Zealand soldiers without known graves. Wragge died before completion of the monument in 1923, not knowing the authorities had erred while trying to correct his eldest son's next-of-kin details. In 2020, the London-based Commonwealth War Graves Commission's online database gave Birkenhead, Auckland, New Zealand, as Nora's home address as well as Clement's.[527]

---

525 *The Beaudesert Times,* March 1, 1918, p8, said to have been reprinted from a newspaper found in a dug-out on the Western Front. Egerton Wragge once worked as a surveyor's assistant in the Beaudesert district, south-west of Brisbane. An edited version of the Reverend Wellesley Greeves' comments, omitting his criticism of incompetent statesmen, was published in *The Staffordshire Advertiser,* October 27, 1917, p7.

526 Wildman, Owen, *Queenslanders Who Fought in the Great War,* Brisbane, 1919, p3

527 Commonwealth War Graves Commission website, 'Lone Pine Memorial', https://www.cwgc.org/find-war-dead/casualty/719793/wragge,-clement-lionel-egerton/ accessed October 20, 2022

CHAPTER 23

NEW ZEALAND, 1916–22

Mr. Clement L. Wragge infers that praying for rain [is] grovelling in the dust. The expression calls for protest. His weather predictions may be remarkably accurate ... but when he implies that praying for rain is dust-grovelling, he is out of his latitude — all at sea. — Letter to the editor of the *New Zealand Herald,* December 2, 1919.

Between 1916 and 1918, Wragge toured New Zealand with a rationale for the carnage of Gallipoli and the Western Front described by a Dunedin newspaper as gloriously optimistic. Addressing Dunedin's crowded Burns Hall on March 1, 1916, Wragge predicted the war would demolish class distinctions, racial prejudices, religious conflicts and the power of money. Men would be brought nearer to God and the spirit of Christ would prevail through the Almighty's purging lesson. He called his lecture "The Eternal Universe and The War", and augmented his homilies with startling lantern slides of the moon, sun and Milky Way. Dunedin's *Evening News* advertised his visit as a grand scientific entertainment, admission from one shilling, viewings of radium for a small additional fee. That week, other diversions in the South Island city included newsreels of "War in the Air" over the Dardanelles and a travelling vaudeville show led by Daisy Jerome — "The Greatest and Most Magnetic Artist of the Present Day".[528]

Wragge mentioned in passing Eggie's death, Bert's service in France and his own contribution to the war effort, a manual titled *Practical Hints to New Zealand Troopers*, published by the government in May 1915. Based on the Royal Geographical Society's *Hints for Travellers* and Francis Galton's *The Art of Travel*, this pamphlet had advice on subjects such as compass-reading to water-proofing one's clothing. During his tour of both islands, Wragge complained of having seen young men idling in the streets instead of doing their bit at the Front, and said he had tried to enlist as a meteorologist but had been rejected because of his age — 62, in September 1914.[529]

* * *

On October 19, 1916, Wragge and Nora's second-youngest son, Reginald, told a Brisbane Police Magistrate his father had been away from home for 12 years and did nothing to support their family. Reginald was among 30 applicants before the Commonwealth Government Exemption Court at Wynnum South, set up under a new law imposing compulsory military training on all unmarried, able-bodied men aged between 21 and 35. Then aged 30, he argued he was needed at home because of Eggie's death, Bert's active service and their father's absence. The

---

528   *Evening Star*, Dunedin, February 29, 1916, p5 and March 2, 1916, p8
529   *Patea Mail*, Wellington, May 12, 1916, p2; Clement Wragge, *Practical Hints to New Zealand Troopers*, Wellington, 1915

Brisbane Courier reported the magistrate's sympathy for bereaved families, his regret that Reginald lacked any recognised grounds for exemption and certainty he had a strong case for review.

Twice in the next 18 months, voters rejected government proposals to extend conscription to overseas service — narrowly in a plebiscite on October 28, 1916, and decisively in another on December 20, 1917. More than 46,000 Australian soldiers died and 132,000 were wounded in three years of war in France and Belgium. One day's fighting in the Battle of Fromelles, July 1916, cost 5533 Australian casualties.

Reginald enlisted for training on October 20, 1916, the day after his exemption court hearing, appealed immediately and was discharged in July 1917, when serving at Brisbane's Enoggera army base. Clement and Nora's youngest son, Lindley, aged 25, was automatically exempted, having married in June 1916. His brother George, 32, was possibly exempt too, as an essential services worker. Electoral rolls in 1917 and 1919 listed him as a teacher and tutor in the Emerald district of central Queensland. George and Lindley both distanced themselves from their father by inventing hyphenated surnames — George Paulson Ingleby-Wragge and Lindley Herbert Musgrave Egerton-Wragge — culled from their array of given names.

* * *

In 1917, Wragge's analysis of the war was increasingly in line with the socialist ideals of the Theosophist leader Annie Besant. In 1915, in a widely reported condemnation of German barbarism, Besant argued the Allies must prevail to establish a new civilisation, "cooperative, peaceful, progressive, artistic, just and free — a brotherhood of nations". All Theosophists should stand for right against might, law against force, freedom against slavery, and brotherhood against tyranny.[530] Wragge's wished-for reforms flowed from his conviction the war had been brought about by Higher Powers to regenerate 20th century civilisation. He argued at Invercargill, on November 1917, that all nations were paying for indiscretions:

> The British Empire and America for grasping capitalism and commercial greed, France for excessive frivolity, Russia for the horrors of Siberia, Belgium for the Congo scandals, Serbia for the murder of their King and Queen, Italy for invading Tripoli, Turkey for the Armenian massacres, and Germany and Austria for the debasing of Science.

To his list of hoped-for benefits from the war, Wragge added the end of infant mortality, guaranteed simple justice free to all, just payment of labour, conscription of wealth and breaking of combines, eradication of disease and state control of liquor — in his mind, sanctioning its use while limiting abuse.[531]

During 1915, Daniel Cross Bates, then chaplain of New Zealand's territorial forces as well as dominion meteorologist, gave two apolitical talks in Wellington on the vagaries of weather in famous historical battles. In September 1919, he visited Paris for an international meteorological meeting and returned home less sanguine than Wragge about the Great War's likely legacy. He doubted the League of Nations could extinguish the hatreds and

---

530 *The Suffragette,* June 18, 1915, p5
531 *Southland Times,* Invercargill, November 9, 1917, p5

jealousies he observed in Europe and warned history would repeat itself, unless human nature radically changed.[532]

William Lane, editor of the *New Zealand Herald* from 1913 until his death in 1917, also believed in the war's divine purpose. "War is the natural, that is, the Divine process by which the inhuman is rooted out and the human given room to expand," he wrote in 1900, soon after joining the paper. In 1915, his eldest son was killed in action on the first day of the Gallipoli campaign, along with 15 Auckland Infantry Battalion comrades.

In September 1916, the *Herald* argued the war had exposed social, industrial and political shortcomings, especially neglect of secondary education. Soldiers of English stock would ultimately need more than innate idealism, initiative and physical health to prevail over enemies who were their superior in formalised knowledge:

> In view of the higher intelligence and technical knowledge required of all workers by reason of the increasing complexity of modern industrialism, all children must receive some form of secondary education.

Wragge seemed to concur. In a letter published two days later, he advocated introducing high school courses in agriculture, horticulture and climatology. "The value of scientific instruction in Agriculture and, I may add horticulture, cannot be over-estimated," he wrote. He was in earnest, as his youngest son intended adopting scientific farming as a profession in "God's Own Country".[533]

---

532 *Maoriland Worker*, May 28, 1915, p4; *Evening Post*, Wellington, May 5, 1915, p2; *Northern Advocate*, Whangarei, January 22, 1920, p3

533 *New Zealand Herald*, September 27, 1916, p6 and September 29, 1916, p 10; Corporal Donald Bennett Lane, of the Auckland Infantry Battalion, was killed in action at Gallipoli on April 25, 1915.

In 1918, Wragge curtailed his touring and war talks and began regular limelight lectures on astronomy at his Birkenhead institute. All his productions hinged in varying ways on the mystery of God, the Infinite Dynamo. In April 1918, he reaffirmed in a Good Friday lecture at Auckland's Empress Theatre his belief the war had been brought about for the betterment of humanity. I don't know who, if anyone, disagreed with this, but later in 1918, a new Australian drought reopened an old debate over God's imprint on human affairs. Wragge's hope in a divine plan behind the horrors of the Great War did not extend to belief in supernatural control of the weather, or that God's arms could be twisted to make it rain.

In November 1919, hearing more than 1000 people had prayed for rain at a special Sydney church service, he sent Australian and New Zealand papers his own prayer for wisdom in adversity. Anything beyond that was, in his opinion, grovelling before the Infinite. Three-quarters of New South Wales was drought-affected by the end of 1919, the driest year since 1902 in many regions. An estimated 7 million of the state's 39 million sheep starved or were culled, and the wheat harvest was the lowest since 1902.

Wragge's prayer was published widely and critiqued suspiciously by some readers. Writing in December 1919, Adelaide businessman and philanthropist William Herbert Phillipps (1847–1935) agreed water conservation was desirable but not much use without rain. Building dams would never save large-scale grazing and agriculture from drought. These endeavours relied on rainfall beyond man's creation or regulation, subject solely to God's control. Mr Phillipps, founder of the South Australian Employers' Federation and a Congregational Church lay preacher, believed in the effectiveness of praying for rain with the assurance of response, as set down in the Bible:

> ... it is impossible to conceive that the Author of this vast universe of ours — the magnitude of which is each year being increasingly brought to our knowledge, and of which this earth of ours is so infinitesimal a part — should be arbitrarily limited by the laws of His own creation.[534]

In November, an unnamed civil engineer wrote to the *New Zealand Herald*, pointing out Australia possessed only two perennial streams and that every possible attempt at water conservation and irrigation was already being tried as funds and circumstances allowed. Farmers in New South Wales and Western Australia knew much better than any casual visitor from Brisbane that it was hard to reduce theory to practice. In the writer's experience, every time the nation had "with one voice" asked God for relief in time of disaster, these prayers had been answered. A reader pen-named "Farmer" replied days later, "The world could be Eden if we had faith, the weather paradisiacal." And another correspondent, from the Waikato farming settlement of Putarura, questioned Wragge's insistence on praying only for wisdom: "I say it should take the form of supplication for anything one wants, subject to God's will."[535]

* * *

---

534 *The Register*, Adelaide, December 9, 1919, p9
535 *New Zealand Herald*, November 25, 1919, p9, and December 4, 1919, p9

In July 1918, Bert was discharged from the AIF with chronically defective eyesight and came home from England where he had been serving as a Company Sergeant-Major. He had married that February but arrived without his young wife, Winifred, who was stuck in London until early 1919 awaiting approval of a free passage. They stayed through 1919 with Nora and sundry siblings while building a house at Coorparoo, four miles from Brisbane's heart.

Bert had, by then, returned to his job with the Public Works Department. In 1920, perhaps prompted by Winifred's pregnancy, Nora moved from Cannon Hill to Coorparoo. She rented a big old house called St Helen's, on Bennett's Hill, within cooee of Bert's new place. An eyrie with superior views, like Capemba and Sunny Brae, St Helen's had been built in 1887 by licensed surveyor John Smith Bennett and named after the town of his birth in Lancashire. In 1918, after retiring as chief clerk of Queensland's Lands Department, Bennett subdivided and developed the large paddock surrounding his family home. Bert's block in the newly formed Fryar Street was the first sold, in January 1919.[536] Bennett seems to have vacated St Helen's and let it to Nora before the death of his wife, Emily (nee Fryar) in August 1920.

Bert and Winifred's first child, Raymond Lindley, was born on December 28, 1920. Incidentally, their son wasn't Nora and Clement's first grandchild. In Edinburgh in July 1906, Eggie was identified in a paternity suit as the father of a baby boy put up for adoption four months after birth. Named Clement Egerton Wragge Anderson on his birth certificate, the child was baptised

---

536 Brisbane City Council, Local Heritage Places, 'St Helens', https://heritage.brisbane.qld.gov.au/heritage-places/464, accessed October 20, 2022

Andrew Horsburgh and known only by this name through his long life as a coalminer at Loanhead, south of Edinburgh. He died in 1994, aged 88.[537]

In January 1923, a month after farewelling Wragge as a courageous trier, the *Brisbane Courier* reminded readers his widow, presently a resident of Bennett's Hill, Coorparoo, had lived in Brisbane for the whole 20 years since his dismissal. Two of their sons had reportedly left for New Zealand in late December to attend to business matters connected with their late father's estate. Rupert and Lindley travelled to Auckland hoping to extract from his estate the £1528 arrears in maintenance payments he was ordered to pay Nora in 1914. They also claimed a further £1292 unpaid since 1914.

On his return to New Zealand in 1914, Clement transferred all of his interest in the Birkenhead property to Edris and appointed her and lawyer Thomas Dawson trustees of his will. Later, he gave Kismet all of his personal belongings, valued at £800. In May 1923, probate found his estate was bankrupt — £2557 in the red, with total liabilities of £3680 against assets of £1122. That October, New Zealand's Public Trustee called a creditors' meeting to finalise disbursements. It's unclear how Nora fared as an unsecured creditor, but on their return to Brisbane, Bert and Lindley bought St Helen's for her.

* * *

---

[537] Personal correspondence from Kevin Stevens, Andrew Horsburgh's grandson, email July 1, 2020, and from Terry Mortlock, Rupert Lindley Wragge's grandson, email August 16, 2019.

Nora died at St Helen's on September 21, 1942, aged 90, remembered by Brisbane's *Telegraph* as her late, famous husband's helper in their Ben Nevis adventure, long ago. The *Telegraph's* short obituary identified her as the fifth daughter of Adelaide barrister Edward Thornton and sketched her role in the intrepid meteorological project that brought Wragge his renown in Britain. Her observations at the low-level meteorological station at Fort William had been of great assistance while her husband was engaged in his high-level observations on the Ben. The *Telegraph* noted her husband's death in 1922 and reported Nora was survived by six of their eight children — two daughters and four sons. A brief tribute in *Queensland Country Life* incorrectly described Inigo Jones, Wragge's protégé and fellow seasonal forecaster, as her son-in-law.

None of the three Wragge girls married. May died in Dalby in 1929, aged 51, and younger sisters Violet and Margaret lived for many years with their mother at St Helen's. Her eldest surviving son, Bert, still lived nearby. Divorced from his wife, Winifred, in 1928, Bert remarried a year later, to Emily Grace Henderson, a granddaughter of the late Queensland Government Engineer John Baillie Henderson, his former boss in the Department of Public Works' stream gauging branch.[538] In May 1942, he passed himself off as 53 years old when enlisting in the Volunteer Defence Corps. At the time, his eldest son, Raymond, an Australian Infantry Forces gunner, was missing in action after the Japanese invasion of Singapore. Bert, who turned 60 in August 1942, served for three years as a corporal

---

538   *The Telegraph*, Brisbane, September 23, 1942, p4; *Queensland Country Life*, September 24, 1942, p4; *Brisbane Courier*, December 5, 1929, p22

in one of the light anti-aircraft batteries set up around Brisbane during World War II. Raymond survived three years as a Japanese prisoner of war and returned to Brisbane in October 1945.

EPILOGUE, 1922–2022

If humankind knew ... what profound inner pleasure await those who gaze at the heavens, then France, nay, the whole of Europe, would be covered with telescopes instead of bayonets, thereby promoting universal happiness and peace. — Nicholas Camille Flammarion, *Popular Astronomy*, 1880.

Edris died suddenly in an Auckland hospital on Sunday 2 November 1924 of complications from a hysterectomy. Her funeral was held the next day, followed by burial in Wragge's plot in the Pompallier Cemetery at Birkenhead.[539] In a lazy tribute on November 8, a Taranaki daily, the *Hawera and Normanby Star*, remembered her as a meteorological expert like her late husband. Kismet advertised in Auckland's papers the same day that the Waiata gardens were still open, old and new visitors welcome. Edris had been a mainstay of the family's Waiata enterprises. Her guided tours of the museum sparkled with insights from her own world travels. Her love of saris, shawls and tiaras seemed evidence of her rumoured royal blood. In 1918, a reporter from the *Gisborne Times* nicknamed her the "delightful Princess Wragge" who was "conversant with books, plays, cults and religions". In 1922, the

---

539 'Edris Wragge (1924)', Certified copy of New Zealand Death Certificate, Edris Wragge, 2 November 1924, Registration No 1924008658, Auckland

New Zealand Observer asserted she was a "Hindoo high-caste lady".[540] A next-door-neighbour, Francis Hayman, remembered her as a skilled teacup reader:

> [The] tearooms were always open and to those anxious about their future, the prophet's wife, a lady of considerable grace and charm, reputed to be a princess, officiated over the reading of cups. On these occasions, she looked almost regal, dressed always in bright Indian apparel.[541]

Another observer writing to Australian friends in 1922 grumbled most visitors found "old Wragge" hard work, while his wife was delightful company.[542] By then, Wragge had curtailed his lecturing, was supplying the *New Zealand Herald* with daily weather forecasts from his home bureau, and running a popular long-range forecasting service for clients in Australia and New Zealand. Two days before his death, the *Herald* received his final dispatch, warning of an approaching storm he'd named Xamion.

From 1916, he and Edris promoted the Wragge Institute and Waiata Botanic Gardens as a holiday hideaway within reach of the city by ferry and bus — "English, French, Italian and Hindustani spoken, one shilling admission". Entry to the museum, art gallery and observatory cost another shilling. Later they offered evening astronomy sessions, offering glimpses of the famed radium specks if clouds obscured the planets. Their enterprise was among Birkenhead's first tourist attractions. The harbour-side suburb

---

540 *Gisborne Times*, Gisborne, December 18, 1918, p13; *The New Zealand Observer* and *The Free Lance*, Auckland, December 16 and 23, 1922

541 Francis Hayman, quoted by Hadden, Kathy, *Birkenhead: The Way We Were*, Birkenhead, 1993, p108

542 *Yackandandah Times*, January 20, 1922, p1

was then also famed for strawberry farms.[543] The maze of palm and banana trees between their place and the sea was, like Wragge's backyard in Brisbane, a curated wilderness. Today, remnants of his plantings in the former grounds of Waiata and Capemba tell of his restlessness as much as any yen for rest and peace in nature. His itch for affirmation was inescapable, even in the potentially relaxing hobby of horticulture. In July 1914, showing off Waiata to a reporter, he boasted he had confounded naysayers by growing bananas from Abyssinia and Fiji and 40 varieties of tropical palms. He scorned the timid replication of staple English trees and flowers then popular in Auckland, and advocated experiments with tropical varieties in the city where he asserted Antarctica and the tropics virtually shook hands. Truant palms sprouted in neighbours' yards to prove his point.[544] Helped by local children, he built tracks down the escarpment for would-be bathers and rowers, using shells and stones scavenged from the shoreline. His zigzagging paths were whimsically named, like the trails he shovelled at Capemba 20 years earlier, with schoolboy Latin titles like *Via Lata, Via Media* and *Via Amatoria*.

* * *

Edris surely loved their house on the hillside above Little Shoal Bay. It was her own beautiful home, a sanctuary after many footloose years. At least that was the way her elder sister, Ida Blackwell, saw it in a speech to Theosophist friends in England during World War I. Ida recalled that, in 1890, she and Louisa

---

543 *North Shore Heritage*, Volume 2, North Shore Area Studies and Scheduled Items List, Auckland, 2011, p329
544 *Auckland Star*, July 17, 1914, p6

—as she then still knew her — had shared a house in Bombay with an American journalist, Anna Ballard, who introduced them to Dr Emily Ryder. They met through the Sorosis Club of Bombay, founded by Dr Ryder in 1889 for young women of all nationalities, inspired by similar groups in Europe and the United States. The club's motto was "And Earth Was Made for Women Too."[545]

Anna Ballard was a close friend of Theosophist Society founder Helena Blavatsky, who lived in Bombay in the 1880s. Ida's reminiscences found among family papers include a story that Madame Blavatsky once produced a set of jewels for Louisa so real she had believed them part of a long-lost inheritance, never suspecting they were an illusion.[546] Louisa shared with Wragge a strong interest in spiritual delving. He discovered Theosophy in 1891, when the Theosophical Society's president Henry Steel Olcott introduced the movement to Brisbane as an embryonic universal brotherhood devoted to studying Eastern thought, the unexplained laws of nature and "the psychical powers latent in man".[547] Wragge embraced Theosophy with such enthusiasm that, by 1897, *The Bulletin* described him as an ornament to the fraternity of the Sydney Theosophical Society.[548]

---

545 *The Harvest Field, A Missionary Magazine*, edited by the Reverend Henry Haigh, Methodist Episcopal Publishing House, Madras, 'A Talk With Dr Emma Brainerd-Ryder, November 1890', 1891, pp 171–74, https://findit.library.yale.edu/bookreader/BookReaderDemo/index.html?oid=11051435&page=4#page/14/mode/2up, accessed 29 October 2022

546 'Lotus Day, a talk in honour of Madame Helen P Blavatsky', by Mary Elizabeth Ida Clara Cecilia Blackwell (nee Horne), elder sister of Louisa Horne, from a collection of Louisa's Theosophical writings compiled by her grandson, Christopher Wragge.

547 *The Telegraph*, Brisbane, April 6, 1891, p3

548 *The Bulletin*, Sydney, August 7, 1897, p13

In nicknaming Louisa "Edris", a feminine rendering of Idris, a prophet and patriarch in the Muslim tradition, Wragge seems to have affirmed her as his wise woman.[549] She called him either "Azaul" — for his blue eyes — or "Lin", from his second name, Lindley.[550]

In photos taken after Kismet's birth in 1900, Wragge looks, at 48, stringy and bemused, while Edris, then aged 36, is serene. In 1903, Wragge wrote in his magazine a homily on what he termed connubial bliss, arguing most men to be truly contented needed the company of a twin soul, "the wife of the bosom".[551] In 1921, in his letter canvassing divorce from Nora, he said Edris had made a great sacrifice in becoming his de facto wife. Considering his wretched state when they met, their relationship owed everything to her compassion. He had found her Nora's opposite in her capacity for kindness.

Accounts of talks he and Edris gave in 1919 suggest their life experiences left them with a disciplined spirituality. In a lecture titled "Your Occult Nature", given to the Auckland Spiritual Scientists in May 1919, and repeated to the Auckland Theosophical Society in July 1923, Edris said her study of major faiths while in India from 1906 to 1908 had convinced her religion was "most necessary for

---

549 Wikipedia, Idris (prophet), https://en.wikipedia.org/wiki/Idris_(prophet), accessed October 20, 2022

550 Fragment of a letter from Edris Wragge, April 1, 1908, opening with "Lindley Darling," found among family papers of Stuart Wragge, Hamilton, New Zealand.

551 *Wragge*, January 29, 1903, 'Meteorgraphia No VII', 'The Blackall Rages Again', January 29, 1903

anyone craving greater closeness to their maker".⁵⁵² In those years, she had visited ancient Buddhist ruins at Benares, Uttar Pradesh, the Sikhs' Golden Temple of Amritsar, and unidentified Muslim mosques, Hindu temples and a Zoroastrian fire temple. In Goa, she and Wragge observed sick and invalid pilgrims seeking healing from the remains of the Catholic missionary, Saint Francis Xavier.⁵⁵³ Edris saw that each religion had developed practices to suit the special needs of diverse peoples. In her view, cultural differences invalidated comparative value judgements on standards of morality. There were surely many pathways to ultimate atonement with God. She recalled that while in Ceylon she had enjoyed an extraordinary privilege in meeting the leader of the celebrated Buddhist monastery at Anuradhapura, home of the ancient Jaya Sri Maha Bodhi Tree — said to have been propagated from the Bodhi tree in Bihar, India, under which The Buddha attained Enlightenment. The leader of the monastery had granted her and Clement half an hour of learning and blessed them as they left.

Dealing with her own spiritual practices, Edris said she had developed a mind of conscious, awakened power in her quest for occult or hidden knowledge. Attaining such a mind was of more use than any analytical, philosophical or religious talent. She was not a crank to be humoured or pitied, nor was she capable of unfathomable depths of learning in a quest for mystical

---

552 *Your Occult Nature*, transcript of lecture given at Orange Hall, Sunday, May 4, 1919, from Louisa's Lectures, access granted by Christopher Wragge, Auckland; Auckland Star, May 3, 1919, p10, advertised Mrs Wragge's talk to the Spiritual Scientists' Church, Orange Hall, on 'Your Occult Nature'; *New Zealand Herald*, July 28, 1923, p6, advertised Mrs Edris Wragge's public lecture on 'Your Occult Nature', Auckland Theosophical Society.

553 'Who Carry the Signs', transcript of paper given by Clement Wragge to the Spiritual Scientists' Church, Orange Hall, Auckland, March 30, 1919.

unapproachable beings — as some believed of Theosophists. She concluded, with underlining in her transcript:

> This is the age of <u>mind power,</u> so if you would not be left behind, rouse up and use yours. There is a wonderful <u>exhilaration</u> in mind exercise, it lightens the body and makes one young with joy, so that all work becomes pleasure and increasing years are robbed of their age. Will you try it?

In a notebook inherited by her grandson, Christopher Wragge, Edris listed rules to help concentration, beginning with regularly setting aside 30 minutes daily, on an empty stomach. Her routine comprised five minutes of rhythmic breathing, five minutes of cleansing breath, five minutes dwelling on the life of Christ, and five minutes sending out peace and love to North, East, South and West, and dwelling on the word Peace. She finished with 10 minutes of visualisation and listening: "For the final 10 minutes, picture yourself looking into a deep blue lake (your mind) and get it so still there is not a ripple on it. Waiting beside it, watch for ripples from its depth. Know then your Soul is casting its reflection on your mind — heed this well and follow where it leads. Try to grasp the teachings that follow. Have NO FEAR, knowing your soul is pure and God's gift to last through eternity. Your Soul is Spirit and God the creator of it."

* * *

In January 1931, Kismet wrote to the Department of Science in Wellington asking to be considered if, as rumoured, a vacancy in the meteorological service was imminent. It was a bold letter prompted by talk that Dr Edward Kidson, the service's highly credentialled director, was in line to succeed Henry Ambrose Hunt as head of Australia's Bureau of Meteorology. Kismet called the existing agency almost useless and argued certain changes, undisclosed in his letter, were needed to bring practical benefits to all New Zealand. "The name of Wragge is synonymous with weather and at this stage I do not think it's necessary for me to enlarge on that aspect of the matter," he wrote, adding, "Believe me, it can be done." The Prime Minister's office promised to pass on his offer at an appropriate time.[554] However, Hunt, who retired in February 1931, aged 65, was succeeded by his New Zealand-born assistant William Shand Watt (1876–1958), and Kidson reigned until his untimely death in 1939.

Kismet signed himself as a Fellow of the Royal Meteorological Society, a recognition of professional standing obtained in May 1922 after several years assisting in the Waiata observatory. His father coached him in daily and seasonal forecasting — the former a somewhat intuitive exercise using limited local observations, the latter entailing correlations of solar, lunar and geographic data. These were informal, dovetailed lessons, since Kismet often travelled away as an itinerant stockman. In June 1919, aged 18, he advertised himself in a situations-wanted column as a strong and able youth, with three years' experience

---

554 Draft of letter from Kismet Kent Wragge to Department of Science and Industrial Research, Wellington, January 23, 1931; New Zealand Prime Minister's office, to Kismet Kent Wragge, January 27, 1931

as a shepherd, good dogs and the highest credentials. In 1928, reflecting on the connection between his disparate occupations, he acknowledged every farmer's thirst for weather news, be it a day, a week or 12 months away.[555] Kismet began long-range forecasting some months before his father's death in December 1922, and for a while took over his two daily Auckland and district newspaper weather columns. He continued forecasting for the *New Zealand Herald* until April 1923, and provided four-day outlooks for the *Northern Advocate*, in Whangarei, until August 1925, when replaced by Howard Milo Vincent (1896–1973), who was appointed Auckland's government meteorological observer in 1928.

* * *

Vincent discontinued Kismet's quirky identifications of the eastward-bound Antarctic disturbances that shaped New Zealand's weather. In 1944, US armed services' meteorologists revived Wragge's cyclone-naming practices, but it was not officially readopted in Australia until 1963. One of the American weathermen is said to have been inspired by George Stewart's 1941 novel *Storm*, in which a maverick meteorologist names tropical storms after ex-girlfriends. Stewart acknowledged Clement Wragge as his inspiration. In 1975, Australia's Science Minister Bill Morrison declared the fickle-and-wilful aura of "feminine"

---

555 *Auckland Star*, June 28, 1919, p1 and September 29, 1919, p4; *New Zealand Herald*, May 3, 1928, p12

cyclones were old hat, and the Bureau of Meteorology began alternating male and female names, standard practice since then.[556]

* * *

In June 1928, a fire destroyed the Wragge Institute and all of its uninsured contents — six rooms of curios, meteorological instruments, personal papers and research records. Neither Kismet nor his maternal aunt, Edris's younger sister Kathleen Burton, were home.[557] Birkenhead Burrough councillors suspended their meeting to help the Northcote Fire Brigade save the nearby family residence at 10 Arawa Street. Coincidentally, in Brisbane two years earlier, March 5, 1926, a shower of steam train cinders ignited and razed the mansion Nora called Sunny Brae. The cedar-walled villa was still widely known as Clement Wragge's old house, despite his never having lived there and Nora's departure for St Helen's in 1919. The glow could be seen all over Brisbane and drew an enormous crowd to Cannon Hill. Neighbours rescued a grand piano from the inferno for the absent owners. The next day

---

556 *Hamersley News*, Perth, November 20, 1975, p1; Lawrie Zion, *The Weather Obsession*, Melbourne, 2017, p18, citing Chris Landsea and Neal Dorst, 'How and Why are Tropical Cyclones Named?' National Oceanic and Atmospheric Administration, 2016, https://www.aoml.noaa.gov/hrd/tcfaq/TCFAQ_B.txt, accessed October 20, 2022; Personal Correspondence, email from Andrew Burton, Manager Tropical Cyclone and Extreme Weather, Bureau of Meteorology, May 11, 2020

557 *Auckland Star*, June 21, 1928, p8; *New Zealand Herald*, June 21, 1928, p10, "Besides the collection of curios from all parts of the world, the personal records of the late Mr Clement L Wragge, together with his memoranda in connection with his meteorological work and research were destroyed."

only four brick chimney stacks remained.⁵⁵⁸ On Christmas Day 1913, Wragge's Hut on Mount Kosciuszko went up in flames after a lightning strike. Visitors in 1914 found nothing but ashes on the summit. The wooden hut had been maintained by the New South Wale Tourist Bureau since the observatory closed in 1902.⁵⁵⁹

In contrast, Wragge's prophetic aura, especially among Australian and New Zealand farmers, was for decades quite imperishable. In September 1942, when Nora died, aged 90, at St Helen's, "What did Wragge say?" was still a standard applied in country newspaper debates over the likely length of the latest drought. His legacy and legend were sustained by Inigo Jones and Kismet.

Kismet continued long-range forecasting for years after discontinuing daily newspaper reports. He was encouraged by his father's clients such as Gisborne MP and farmer William Douglas Lynsar (1867–1942), who wrote to him in 1928, complaining of the New Zealand bureau's antipathy to seasonal predictions. Lynsar trusted Kismet to sustain Wragge's record of valuable assistance to farmers. During the 1930s, after marrying and moving to Hamilton, Kismet became an insurance agent, with what was left of the Wragge bureau as a sideline. His aunt, Kathleen Burton, died in 1929 and the sale of Waiata — operated under leasehold in 1931 as Wragge's Gardens — was finalised in 1935.⁵⁶⁰ In 1938,

---

558   *The Telegraph*, Brisbane, March 6, 1926, p9; *Brisbane Courier*, March 6, 1926, p7

559   *The Queenslander*, April 11, 1914, p29

560   *New Zealand Herald*, January 2, 1931, p12, Mr L Castleton reported to be conducting Wragge's Gardens at Birkenhead; *New Zealand Herald*, November 9, 1935, p1; *Ancestry*, Kathleen Mabel Burton died October 17, 1929, https://www.ancestry.com.au/discoveryui-content/view/2854353:60528, accessed October 20, 2022

Kismet championed his father's long-range forecasting methods in speeches to the Rotorua Club and the district's young farmers. In view of the Queensland government's financial support for his father's one-time student and assistant Inigo Jones, Kismet hoped New Zealanders would demand government recognition of astronomically based forecasts. He was certain he could attain 100 per cent accuracy by using father's methodology.[561]

* * *

Inigo Jones reinvigorated Wragge's legend in Australia during the 1930s. In 1934, writing as director of Queensland's Bureau of Seasonal Forecasting, he remembered him as a man of character and originality, a bold thinker and a great organiser. He owed his career in meteorology to Wragge's having seen in him "something that might be of use" and hiring him as an assistant when 15 years old. He stressed his mentor's success in reorganising the colonial weather service, evident in much-improved observation records. Jones traced his own seasonal forecasting studies to Postmaster-General Theodore Unmack's request to Wragge in 1892 to investigate the Bruckner cycle. Writing in 1950, Jones inferred Wragge's guiding spirit had helped him develop astronomical correlations for seasonal forecasting. His first long-range efforts in the 1920s were for his farming neighbours in the Upper Stanley district north of Brisbane.[562] He continued:

---

561   *Rotorua Chronicle,* August 2 and September 14, 1938; *Northern Advocate,* October 12, 1946, p8, Kismet Wragge, FRMS, advertises seasonal forecasting, yearly, for local areas and districts, £5/5/– in advance.

562   *The Telegraph,* December 10, 1934, p15

It seems in this connection a very strange yet curious fact that it was so soon after his [Wragge's] death, that I, his favourite pupil, began to actively pursue the studies that were begun during my first years with him.[563]

He asserted Wragge had believed in benign powers who, through the ages, lovingly sustained certain human insights and inspirations. I've not found this notion in Wragge's writings, but Marie Corelli held a similar belief. In 1905, Corelli told the US academic magazine *The Journal of Education* that any person having decided to accomplish a certain goal simply had to place themselves "in accord with the universe" and everything in the world would tend toward the accomplishment of that object. It seemed to her humans were aided by "things far above and beyond us".[564]

Jones, like Wragge, was often at odds with the Bureau of Meteorology. In 1924, director Hunt cold-shouldered his unsolicited application to join the bureau as a seasonal forecasting researcher. Recalling this rebuff, Jones wrote he felt he had proved his credentials by successfully forecasting drought-breaking rain for the closing months of 1923 — based on the Bruckner Cycle and winter temperatures in 1922. Hunt replied the rain had arrived 20 days later than forecast.[565] But Jones was backed by the Brisbane Chamber of Commerce and Chamber of Manufacturers and the

---

563 Jones, Inigo, 'The Life and Work of Clement Lindley Wragge', *Queensland Geographical Journal*, No 40, Vol LIV, 1951, p48
564 *The Journal of Education*, Vol 61, No 19 (1529) May 11, 1905, pp 511–12, 'Authors Who are a Present Delight (XV)'
565 *Nambour Chronicle and North Coast Advertiser*, 'Why I Built the Crohamhurst Observatory', by Inigo Jones, October 16, 1938, p16; *The Daily Mail*, Brisbane, 'The Long Drought Decisively Broken', January 1, 1924, p8

state's Council of Agriculture, representing various commodity boards and producers' associations.

In October 1926, the Council of Agriculture appointed Jones its honorary director of seasonal forecasting and joined efforts to form a trust to sustain his work. Established in 1929, the Inigo Jones Seasonal Forecasting Trust won state government support, culminating in 1935 with the opening of a seasonal forecasting observatory on his property, Crohamhurst, near Peachester in the Glass House Mountains. The Queensland government acknowledged his standing by publishing his scientific papers. Public confidence in his complicated analysis of sunspot cycles and planetary orbits rested on well-publicised successes; for example, his three years' warning of a drought that parched the Riverina district of New South Wales in 1936, and later his accurate forecast in 1944 of torrential rain that flooded many areas of New South Wales and Queensland in 1950.[566]

Between 1938 and 1950, primary producer groups and other supporters tried in vain to win federal government recognition for Jones's forecasts. In 1942, his trustees reconstituted themselves as the Australian Long Range Weather Forecasting Trust with oversight of the observatory. In April 1954, Federal Interior Minister Wilfred Kent-Hughes told the then 81-year-old Jones he was sorry the government could not accept his theories but granted him £1000 in recognition of his toils.[567] Jones died six months later, leaving his assistant, Robert Lennox Walker, to carry on as Crohamhurst's research director. The long-range trust deemed

---

566 Day, David, *The Weather Watchers*, Melbourne, 2007, p147

567 *The Land*, Sydney, July 21, 1950, p41; *The Courier-Mail*, January 22, 1953, p3; *Central Queensland Herald*, Rockhampton, May 6, 1954, p3; Day, David, *The Weather Watchers*, Melbourne, 2007, pp 289–90

Walker, a 29-year-old ex-serviceman and forestry roads' surveyor, had been adequately trained in his 15 months as Jones's assistant.

In 1956, the trust, which was then broke and winding up, sold the observatory to Walker, who in the next four decades built a commercial long-range forecasting service grounded in monitoring solar flares, analysing historical data and observing planetary relationships and orbital patterns.[568] Since Walker's retirement in 2000, his son, Hayden Walker, has used the same long-range methodology in his own business. On his *Hayden Walker's Weather* website, he describes himself as a fourth-generation forecaster in the tradition of Wragge, Jones and his father, and asserts the primal influence of the sun on the world's weather patterns.[569] In 2017, the value of Australia's private weather-forecasting businesses was put at $50 million annually. That year, Australia's largest private weather enterprise, The Weather Company, reported six million Australians had either downloaded its Weatherzone app or used its website.[570]

* * *

In 1985, Australia's Bureau of Meteorology established the National Climate Centre to study cyclical changes in Pacific Ocean currents, sea-surface temperatures and trade winds believed linked to Australia's widespread drought in 1982–83. Combined, these linked periodic changes are known as the El

---

[568] *The Australian Women's Weekly*, December 12, 1954, p31; *The Argus*, Melbourne, July 20, 1956, p1

[569] Hayden Walker's Weather website, 'Methodology', https://www.haydenwalkersweather.com.au/methodology, accessed October 19, 2022

[570] Zion, Lawrie, *The Weather Obsession*, Melbourne, 2017, p67

Niño Southern Oscillation (ENSO). Typically, droughty El Niño years are counter-balanced by a cycle of cold, wet La Niña seasons when trade winds push warm water towards Asia.[571]

By 1985, meteorologists were convinced nine of Australia's ten post-war droughts had been associated with the El Niño effect — a cyclical warming of the typically cold ocean currents off South America's Pacific coast. The bureau issued its first seasonal forecasts in June 1989, based on the climate centre's research and historical records.[572]

Today, temperature and rainfall variability caused by global warming is also factored in. Meteorologists now use massive super computers for climate modelling based on the physics of the atmosphere, oceans, ice and land surface combined with millions of observations from satellites as well as from land and sea. In 2020, the bureau reported Australia's climate had warmed by about 1.4 degrees centigrade since 1910, and in southern Australia there had been in recent decades a reduction of between 10 and 20 per cent in cool-season rainfall.[573] Discussing the challenges

---

571   Miller, Julia, 'The Fall of an Angel: Gendering and Demonizing El Niño', *World History Connected,* June 2007, https://worldhistoryconnected.press.uillinois.edu/cgi-bin/cite.cgi, accessed October 2022, outlines the gendered origins of the names El Niño and La Niña — the young boy and the young girl. El Niño, the 'Christ Child', was traditionally used by Peruvian fishers for a periodically occurring, beneficent warm north-south coastal current. In the 1970s, climate scientists linked both with the cyclical Southern Oscillation weather system. El Niño — now the bad boy of global weather — was blamed for Australia's recurring droughts. In the late 1970s, scientists coined the name La Niña — El Niño's calming feminine consort — for the cool phase of the Southern Oscillation that recurrently brought flooding rains to Australia and drought to the west coast of North America.

572   Day, David, The Weather Watchers, Melbourne, 2007, pp 416-18

573   Bureau of Meteorology climate outlooks, http://www.bom.gov.au/climate/outlooks/#/rainfall/median/seasonal/0, accessed September 3, 2020

of global warming, a bureau spokesman told me forecasting imminent climatic changes was far more difficult than long-term projections. He was prepared to predict with great confidence that in future Australia would be hotter and have harsher fire seasons, but struggled to make precise predictions of the climate six months to a year ahead.[574]

The United States' National Oceanic and Atmospheric Administration (NOAA) reported its estimate of global land and ocean surface temperature for January 2020 was the highest on record, in 141 years of observations — on average 1.14 degrees centigrade above the 20th century average.[575] In 2019, greenhouse gas pollution continued to increase heat trapped in the atmosphere, according to a report by the NOAA in May 2020. The US organisation's annual greenhouse gas index, tracking fluctuations in concentrations of chiefly human-caused greenhouse gas emissions, showed a 60 per cent increase in the impact of carbon dioxide on the atmosphere's heat-trapping capacity.[576]

* * *

Wragge condemned indiscriminate land-clearing as an assault on the balance of nature. He argued through the Federation Drought that settlers who ringbarked forests for farmland were to blame

---

574 Personal correspondence, email from James Ashley, acting manager, Tropical Cyclone and Extreme Weather Bureau of Meteorology Community Services Group, August 7, 2020

575 Global Climate Report, January 2020, National Oceanic and Atmospheric Administration, September 9, 2020, https://www.ncdc.noaa.gov/sotc/global/202001, accessed October 21, 2022

576 *NOAA Research News,* May 28, 2020, https://www.esrl.noaa.gov/gmd/ccgg/covid2.html, accessed October 21, 2022

for widespread micro-climate change, evident in reduced rainfall. In December 1902, lecturing in Horsham, western Victoria, he warned any future clearing must be minimal and supplemented by compensatory tree-planting, "If our rainfall, already sufficiently scant and unreliable, is to be preserved." And in Melbourne, the same month, he inferred landholders had contributed to the "howling wilderness" of Queensland's parched Warrego district by failing to preserve sheltering timber belts. In May 1906, he told an audience in Perth, Western Australia, the state was being denuded of native forests in an appalling manner and asserted all meteorologists knew forests tended to equalise climates by increasing rainfall and preventing excessive evaporation.[577] In Wragge's "Gospel", man's despoiling of the land was unforgivable, a desecration of Eden. His imagined paradise irredeemably lost:

> Before man appeared on the scene, Nature, with its unerring law, clothed the surface of Australia with luxuriant vegetation in the shape of trees of all kinds. These have been ruthlessly destroyed, and their moisture-conserving properties lost to the country.[578]

Wragge recoiled from the doctrine of original sin. His Creator would not stoop to engineering a breach with humanity to secure grovelling obedience. As Wragge saw it, the only real hope of atonement for man, the prodigal waster, was using his God-given intelligence to make amends.

In 1908, New Zealand's home-grown meteorologist, the Reverend D C Bates, also advocated tree-planting as a long-term remedy for drought in the South Island's Oamaru district. In a

---

[577] *The Horsham Times*, December 19, 1902, p3; *The Argus*, Melbourne, December 2, 1902, p9; *The West Australian*, Perth, May 10, 1906, p5

[578] *Swan Express*, Midlands Junction, Western Australia, March 24, 1906, p3

report on unsuccessful rainmaking experiments, he wrote that the dug-up remains of ancient tree roots suggested the Oamaru's bare hills had been once luxuriously forested. He believed replanting deep-rooted trees would, by increasing transpiration, create a "beneficial humidity".[579]

* * *

In 1908, the popular press barely noticed speculation by the Nobel Prize-winning Swedish scientist Svante Arrhenius (1859–1927) on the global impact of industrial growth on climate. In the English translation of his book *Worlds in the Making — The Evolution of the Universe*, published that year, Arrhenius theorised that industry's enormous combustion of coal could contribute to prolonging Earth's present warm interglacial period, in addition to the effect of volcanic activity such as the eruptions of Krakatoa in 1883 and Martinique in 1902. Scottish geologist James Geike (1839–1915) introduced the concept of interglacial and pre-glacial periods in 1874, based on his study of glacial sedimentation in Britain and the Continent. He argued the Great Ice Age of Earth's latest geological era — the Pleistocene — had featured an alternating sequence of warm and cold periods, and that the world was in the midst of the latest warm interglacial period.[580]

Arrhenius's hypothesis owed much to the insights of French mathematician Joseph Fourier (1768–1830) and Irish physicist John Tyndall (1820–1893). In the 1820s, Fourier likened the earth's atmosphere to glass panes in a hothouse, simultaneously

---

579 *Wairarapa Daily Times*, Masterton, NZ, April 15, 1908, p7
580 *The Scotsman*, March 3, 1915, p8, obituary of James Geikie

transparent and insulating — admitting sunlight and corralling reflected infra-red radiation. In the 1860s, Tyndall calculated the heat-absorbent and radiative properties of various atmospheric gases — water vapour, carbon dioxide, nitrogen and methane. Arrhenius conceded concern that there might be no coal left for future generations. All the same, if a warming trend favoured cropping, then coal-powered industrial growth might yield a serendipitous climate change "for the benefit of rapidly propagating mankind". He theorised a rapid increase in atmospheric carbon dioxide would outstrip absorption of carbon dioxide by sea water and contribute to further, potentially beneficial, warming. The shorthand "Greenhouse Effect", anticipated by Fourier's glasshouse analogy, was coined in 1909 by English physicist John Henry Poynting (1852–1914), influenced by Swedish meteorologist Nils Gustaf Ekholm (1848–1923).[581]

The *London Evening Standard* heeded Arrhenius's view of carboniferous-era coal deposits as the bounty of an ancient atmosphere very rich in carbon dioxide and concluded: "It is probable that the present annual increase of the gas will, in a few hundred years, result in our having an appreciably warmer climate."[582] But Arrhenius stirred much greater interest with his theory of life on Earth having originated by chance from living

---

[581] Arrhenius, Svante, *World in the Making —The Evolution of the Universe*, translated by Dr H Borns, London, 1908, p61, https://archive.org/details/worldsinmakingeo1arrhgoog/page/n85/mode/2up, accessed October 21, 2022. Professor Steve Easterbrook, of Toronto University's Department of Computer Science, gives a concise summary of the genesis of "Greenhouse Effect" in his Serendipity blog, 'Who first coined the term "Greenhouse Effect"?', August 18, 2015, https://www.easterbrook.ca/steve/2015/08/who-first-coined-the-term-greenhouse-effect/ accessed October 21, 2022

[582] *London Evening Standard*, April 30, 1908, p12

fragments flung billions of miles through space from an exploded, formerly populated, world. He conceded that while life could in this hypothesis be considered eternal, like matter and energy, it was also evident that life might be suddenly annihilated without necessarily giving rise to other life. In June 1908, *The Age* in Melbourne, credited Lord Kelvin with first imagining the germ of life arriving on earth from a meteor broken from some other world. But the newspaper acknowledged Arrhenius had boldly advanced this idea by reimagining the universe as self-renovating. Recent research showing germs capable of surviving for six months in liquid air had given his hypothesis some plausibility.

* * *

In 1908, Ernest Shackleton became the first explorer to use a motor vehicle in Antarctica. Scottish shipbuilder Sir William Beardmore (1856–1936) gave him the custom-built car to speed him to the South Pole and — he hoped — bring fame to his struggling subsidiary, the Arrol-Johnston Car Company. In 1908, the Nimrod Expedition coincided with Henry Ford's launching in the United States of his Model T automobile — the catalyst for a transport revolution that relied on aeons of stored carbon and iron ore, similar to the steam-train revolution 70 years earlier. Ford sold 15 million mass-produced Model T vehicles around the world between 1908 and 1927. In the US, motor vehicle registrations surged from 14,800 in 1901 to 667,000 in 1911, then on to 8.5 million in 1920. Existing US oil fields were supplemented during these two decades by finds in Canada, Sumatra, Persia, Peru, Venezuela and Mexico. In

1914, Ford dealers in New South Wales reported selling 100 new vehicles per month. In New Zealand, car registrations more than quadrupled between 1915 and 1925, from 17,000 to 71,403, while imports of petroleum spirits, including kerosene, nearly doubled between 1919 and 1923 — from 1.2 million to 2.3 million cases. Shackleton was generally happy with the four-cylinder, 15-horsepower Arrol-Johnston car, but found it unsuitable for his trek to the geographic South Pole. While having run well on sea ice when carrying stores between depots, the vehicle's heavy back wheels reportedly floundered in snow and its air-cooled engine was prone to over-heating.

* * *

In 1955, Scottish mountaineering enthusiast Benjamin Hutchison Humble (1903–1977) read Wragge's Ben Nevis notebooks and rated his trekking record in 1881 and '82 as Britain's finest climbing feat. Humble — a deaf, retired dental surgeon and mountain-rescue pioneer — wrote in the popular monthly magazine *Scottish Field* that Clement Wragge's nearly 180 hikes to and from the summit were unbeatable for courage and enthusiasm. Based on his own 40 years of hiking and climbing experience, Humble said Wragge had set the bar extremely high. No other mountaineering or fell-walking achievement surpassed this obscure Englishman's conscientious endeavours, not even a famed ascent of the highest mountains in England, Scotland and Wales in 24 hours. After all, The Ben's wide summit plateau was Britain's most exposed eyrie, often swept by furious gales, usually covered with dense clammy mist, and likely to be snowbound at any time of the year. In 1881, any ascent to the Ben's rocky crown had still been reckoned an achievement, yet Wragge had volunteered, sight unseen, simply in return for lodgings at Fort William, the services of an assistant and use of a pony. Humble was astonished at his perseverance in writing up 12 pages of data daily, every reading to the third decimal point. Although the statistics made dry reading, he had enjoyed the personal glimpses found in his asides. For example, his devotion to duty was evident in his notes on October 14, 1881, when caught in a snowstorm at the Red Burn, halfway to the top. Despite a 70-miles-per-hour gale and blinding snowdrifts, Wragge and his guide Colin Cameron had hung on for 10 minutes to take the 9 am observations as required at the burn crossing:

> So we lay in the pitiless snow — I with thermometer in my numbed hand — waiting for 9 o'clock and wondering if we would ever get back. Our clothes were hard frozen and coated with ice and ice lumps like large eggs had formed on our beards.[583]

On July 20, 1882, Wragge scrawled "the great value of a permanent observatory" after explaining the impossibility of properly drying instruments with benumbed hands, and hence being unable to take his required dry-bulb readings. His widely reported zeal for the Scots' great cause inspired the public appeal that financed construction of the Ben Nevis observatory that opened in 1883. He took such pride in his rite of passage on the mountain, as well as the Scottish Meteorological Society's gold medal pinned permanently on his jacket, that obituary writers

---

583   Humble, B H, *Scottish Field*, republished by *The Beaudesert Times*, July 8, 1955, p11

understandably included founding the observatory among his foremost achievements. "Eighteen carat," he reportedly told a Dunedin audience when showing off the medal in 1916. "It's not every Englishman who carries gold out of Scotland."[584]

Nearly 140 years on, data from the observatory is being used as a benchmark in British global warming studies. In 2017, Professor Ed Hawkins, a climate scientist at the University of Reading, appealed on the Zooniverse web portal for volunteers to digitise weather data recorded at the Ben observatory from 1884 until its closure in 1904. Hearing that the figures would be used as a benchmark for climate-change modelling, nearly 4000 people volunteered to transfer the figures into an electronic database, each of them adopting sections to type in, column by column. Completed in three months, the project gave Reading University's National Centre for Atmospheric Science a detailed snapshot of late 19th century and early 20th century rainfall, air-pressure, temperature and wind speed on the Ben. In 2017, discussing the project in *Scientific American*, Professor Hawkins said the climate of the Scottish Highlands was warming at one of the fastest rates on Earth. The historical data from the Ben could now be used to provide context for present-day weather extremes.[585]

* * *

---

584 *Otago Witness*, Dunedin, March 8, 1916, p9
585 *Scientific American*, November 7, 2017, '"Slightly Crazy" 19th-Century Weathermen Who Braved Formidable Conditions Could Aid Climate Predictions', by Andrea Thompson, https://www.scientificamerican.com/article/ldquo-slightly-crazy-rdquo-19th-century-weathermen-who-braved-formidable-conditions-could-aid-climate-predictions/ accessed October 21, 2022

Records of Wragge's two well-publicised, high-altitude, weather-observing exploits, on Ben Nevis and Mount Kosciuszko, have never been properly studied. His legacy in 21st century Australian meteorology rests solidly on conscientious spadework more than inspired innovation. For example, his planting and nurturing of hundreds of major and minor weather stations in Queensland between 1887 and 1902. His sometimes rivals Charles Todd, Henry Russell, Robert Ellery, Pietro Baracchi and William Cooke did likewise in their colonies. In 1908, the Federal Bureau of Meteorology picked up the pieces of these colonial endeavours and, by 1910, had established a uniform network with a body of data now the benchmark for global-warming studies, both by the bureau and CSIRO.

Wragge's 77 notebooks from Ben Nevis and Achintore in 1881 and 1882 are now held in the archives of the National Records of Scotland, along with data recorded in the summer of 1883 by William Whyte and Angus Rankin. Scottish Meteorological Society secretary Alexander Buchan was sure the data would help refine storm forecasting, but made little progress in the next decade. In 1892, Scottish Supreme Court Judge and amateur astronomer Lord John McLaren alerted a meeting of the British Association in Edinburgh to Mr Wragge's "remarkable series of observations". Lord McLaren, then-vice-president of the Royal Society of Edinburgh, said the society's youngest fellow, 20-year-old mathematician Robert Mossman was well advanced in discussing these figures. Buchan, who suggested the project to Mossman, later guided him into a career in meteorology that began on the Ben and ended as a staff member of the Argentine Meteorological Office in Buenos Aires, run by his once-assistant Angus Rankin.

Retired Edinburgh meteorologist Marjory Roy has written in an appreciation of Buchan's career that two factors hindered his applying the Ben data to forecasting. His approach to the problem was ultimately too climatological. Furthermore, the depth of atmosphere sampled — roughly 4000 feet — was a relatively small section of the troposphere. Dr Roy wrote that Buchan had still been working hard on this study at the time of his death in 1907, aged 78.[586] She strongly advocates digitisation of all the archived notebooks, having found, as a historian of Scottish meteorology, that, as well as the numerical observations, they contain a wealth of comments on weather conditions encountered during the daily climbs. Her advocacy for wider use of the data dates back to 1983, when she helped organise celebrations in Fort William to commemorate Wragge's endeavours and those of the meteorologists who staffed the observatory from 1883 to 1904.

* * *

At daybreak on June 25, 1983, a party of 75 professional and amateur meteorologists, climbers, walkers and skiers left Fort William for the summit of the Ben to retrace the steps of "the extraordinary Clement Wragge". Their outing was the highpoint of a two-day centenary celebration organised by the Scottish Centre of the

---

586 *Glasgow Herald,* August 5, 1892, p4; *The Scotsman,* August 7, 1929, p9, obituary of Angus Rankin; Wikipedia, https://en.wikipedia.org/wiki/Robert_Mossman, accessed October 21, 2022; Roy, Marjory, 'The Contribution of Alexander Buchan to the development of Climatology and Synoptic Meteorology', Scottish Centre, Royal Meteorological Society, paper presented to the Commission on History of Meteorology, Polling, Germany, 2004, pp 3–4 http://www.meteohistory.org/2004polling_preprints/docs/abstracts/roy1_abstract.pdf, accessed June 30, 2017

Royal Meteorological Society. Marjory Roy, then Superintendent of the Edinburgh Meteorological Office, in a preview story in *New Scientist*, expressed disappointment that the Ben data had been so lukewarmly received in London. She argued then for the data to be stored on computer file, for analysis and comparison with the automatic high-altitude weather station at Cairn Gorm, which was set up in 1977, fifty miles from the Ben.

Wragge's spirit is present in Fort William on the first Saturday of September every year when fell runners from across Britain dash to the top of the Ben and back. I was there in 2012, when the race was nearly called off because of inclement weather — a 70-miles-per-hour gale — and scores of the 600 entrants pulled out. On the ascent, I chased Cosmic Hillbashers and Highland Hill Runners up heathery slopes, plodded through rain and sleet over the scree beyond the snowline, reached the foggy top and stopped. The Lochaber Mountain Rescue team gave me a plastic poncho and Mars Bar, which I chewed while watched by a friendly Scotch Terrier. This helped, but my stumbling return downhill alarmed the team, who then deemed me exhausted and flew me to hospital in Fort William with other casualties. I wondered what Wragge would have thought of this humiliating rescue. A day later, an Aberdeen paper reported the airlift of exhausted and injured Ben Nevis racers had included an Australian suffering from mild hypothermia. That was 10 years ago.

Chasing Wragge since then, through the electromagnetic ocean where his virtual traces still roam, I concede the purpose of my pursuit changed along the way. I first heard of him more than 20 years ago, while researching British and Chinese agriculture in tropical north Queensland. He was a bit player, the colony's conscientious, much-travelled 19th century meteorologist. Then a friend of a friend, who had thought better of attempting Wragge's biography, gave me a slew of leads, including Keith Dunstan's portrait of a harmless ratbag.

What set me going was imagining his childhood, the grief and the uncertainties of not knowing one's parents and the potential burden of a fabulously large inheritance as compensation. But now, in parting, I can see I also soon placed him in a 21st century climate-change frame, as if the petri dish of his words and deeds might explain why many of us resent having weather joys and sorrows complicated by science and politics.

Eight years on, I accept what Wragge's friends and foes said about his contradictory character: he was demanding, boastful and blind to the selfishness of his drive for fame, but he was also kind, sometimes wise, and absolutely true to himself in his quest to know God. In this sense, he was — in his own way — everyman.

I feel his attitude to my endeavour, both in running after him up the Ben and attempting to pin him down in a biography, might — if he's listening on an interstellar line — be akin to his patience with curious tourists he met on the mountain. His logbook entry for July 10, 1881, includes this slightly peeved note: "Delayed by a French visitor who could not keep pace with me on the ascent, and, owing to thick fog, I could not leave him [alone] till near the summit."

Soon after, news of his derring-do prompted an adventurous Scottish traveller to shadow Wragge up and down the mountain. In August 1881, recalling their several encounters that day, the anonymous tourist wrote he had been grateful for Mr Wragge's "hallooing" guidance through low cloud and willingness to share his smoky fireplace on the summit. Not only had he been a guide, philosopher and friend in between two-and-a-half hours of observations, he had also later offered a guided tour of his sea-level station at Fort William. Mr Wragge, in the opinion of this new-found friend, was a great enthusiast and, indeed, a genial and sociable gentleman.[587]

---

587 *Hawick Express,* Hawick, Roxburghshire, Scotland, 'Notes on a Trip to the Highlands', August 20, 1881, p4

ACKNOWLEDGEMENTS

I was acting editor for a string of small newspapers in north Queensland when I first heard the genteel name of Clement Wragge. Inevitably, in fretful Wet Season stories, some reporter would laud Wragge's inventiveness in nicknaming Coral Sea cyclones. It was a matter of pride for Queenslanders that in the 19th century their first government meteorologist's quirky practice had spread around the world. It seemed to me the nub of his fame had been confirming an age-old fear of cyclones and hurricanes as capricious, almost human, entities. He seemed to have had a sense of humour in calling some of his storms after politicians.

It has taken me 25 years to write Clement Wragge's biography. Many people have helped in its making. First, my wife Diane, who scoured archives with me around Australia and NewZealand, and now rules our mass of digital records. She is so happy that publication is finally at hand.

Queensland historians, Dr Russell McGregor and Dr Jonathan Richards, guided my early research. Russell introduced me to Wragge as a prototype conservationist, and Jonathan alerted me to his friendship with Archibald Meston, a one-time Protector of Aborigines in southern Queensland. About 2003, Jonathan gave me a reading list discarded by a student who had begun, then thought better of attempting, Wragge's biography.

In 2011, I decided to take up the challenge and met Christopher Wragge — one of Clement's New Zealand grandsons — who put me in touch with John Williams, a local historian from Cheadle, North Staffordshire, in northern England. During 2012, I took long-service leave, flew to the UK and tramped with John around Clement's haunts in Oakamoor and the Churnet Valley. John has been of constant assistance with the Wragge and Ingleby family histories, as well as background on the Oakamoor copper-wire works.

Next stop was Fort William, Scotland, for the 2012 Ben Nevis Race — thanks to the organisers tolerating my minimal mountain-running nous. The sleet that pulled me up on the summit with mild hypothermia gave a painful insight into Clement's spartan trekking. Donald Cameron from the Ben Race Association has helped greatly with the Fort William chapters, as has retired Edinburgh meteorologist Marjory Roy. John Williams introduced me by email to Adrian Ingleby, grandson of Wragge's nephew and early Kosciuszko team leader Bernard Ingleby. The chapters on the ambitious, under-funded and ill-fated observatories on Mount Kosciuszko and at Merimbula owe much to Adrian's research. I'm grateful for many discussions with him on the project's significance in Snowy Mountains history and share his indignation at disgraceful government neglect of years of carefully measured data.

Wragge's New Zealand grandsons Christopher and Stuart Wragge have both died since our several meetings, Christopher in 2016 and Stuart in 2020. This biography would have stalled long ago without their friendship and hospitality, reminiscences and generous sharing of family records. Likewise, his great-granddaughter Lynette Wragge of Brisbane, and great-great grandson Terry Mortlock of Tura Beach, New South Wales, have provided insights into his family life in Australia.

In 2017, I retired from full-time journalism, visited North America and chased some other loose ends. Janet Olson, volunteer archivist of the Frances K Willard Memorial Library and Archives in Chicago, directed me to background on the Little Wives of India Mission and its founder, Dr Emily Brainerd Ryder — employer of Wragge's de facto wife, Louisa Horne, in the 1890s. Reference librarian Gina Bardi, of the San Francisco Maritime National Historical Park Research Centre, went the second mile in finding details of the coal ship *Melpomene* that took Wragge from New South Wales to San Francisco in 1874.

Thanks to retired Bureau of Meteorology severe weather specialist Jeff Callaghan for constructive criticism of my precis of Wragge's early years in Brisbane. The string of cyclones and floods between 1887 and 1894 provided a crash course in tropical meteorology. Adelaide historian Peter Adamson helped with his research on Wragge's innovative naming of cyclones — published in the Royal Meteorological Society Journal Weather, in 2003. Peter also alerted me to Wragge's account of his voyage to South Australia in 1883 and subsequent two years in Adelaide before moving to Brisbane.

David Day's history of the Bureau of Meteorology, *The Weather Watchers*, and Denis Cryle's biography of Sir Charles Todd, *Behind the Legend: the Many Worlds of Charles Todd*, have been invaluable sources on weather-forecasting politics immediately before and after Australia's Federation in 1901.

The staff of state archives and public record offices in Brisbane,

Sydney, Adelaide and Canberra, as well as the Auckland War Memorial Museum Library, have been invariably cheerful and helpful. And thank goodness for the National Library of Australia and National Library of New Zealand and their sponsors for the boon of free online access to their vast digital collections of newspapers, magazines and journals, respectively, through Trove and Papers Past websites. By the way, not one of the country newspapers I edited 25 years ago has survived as chattels of big business, a real loss for future grass-roots historians and biographers.

Finally, many thanks to my copyeditor, Dr Euan Mitchell, for his careful, excellent work on the manuscript and help and advice on my first independent-publishing enterprise.

Clement Wragge made his mark in Australia, New Zealand and Britain in diverse endeavours, way beyond his cyclone-naming renown or controversial rainmaking experiments. I hope my biography spells out his true legacy. His zeal on Ben Nevis hastened establishment of an official weather station on Britain's highest peak, yielding observations now used as a 19th century benchmark in global-warming studies. He and other mountain-top meteorologists refined use of high-altitude data for advanced storm warnings. In the 1890s, as Queensland's pioneering Government Meteorologist, he nurtured volunteer observers around the colony, leaving an invaluable resource for the national Bureau of Meteorology in 1908.

Australian and New Zealand farmers and graziers trusted his seasonal predictions despite the national bureau's scorn of sunspot cycles. Today, weather services around the world satisfy our age-old hunger to know what's written in the clouds, using data from satellites, supercomputers and ocean current cycles. You can take your pick of good and bad tidings from any number of oracles in traditional media or online. In the 1890s, Australian newspaper readers had a choice of often conflicting forecasts from Queensland, New South Wales and South Australia. In this era, Wragge set himself apart by issuing his mostly astute predictions in fantastically purple prose. His followers loved his brashness. A generation on from his death in 1922, his true believers were said to invariably read the official forecasts, then ask, "But what does Wragge say?"

Ian James Frazer, Townsville, November, 2022

BIBLIOGRAPHY

Books

Ackroyd, Peter, *London: The Biography*, London, 2000

Adam, John P, *The Waiata Tropical Gardens at Birkenhead*, Auckland City, May 2011

An Encyclopaedia of New Zealand, 'Wreck of the Ariadne, 1901', Wellington, 1966

Anderson, Katharine, *The Weather Prophets, Science and Reputation in Victorian Meteorology*, York University, Ontario

Arrhenius, Svante, *World in the Making —The Evolution of the Universe*, translated by Dr H Borns, London, 1908

Bettany, George Thomas, 'Robert Christison', *Dictionary of National Biography, 1885–1900*, Volume 10, https://en.wikisource.org/wiki/Christison,_Robert_(DNB00)

Blavatsky, Helena Petrovna, *The Secret Doctrine*, Volume Two, Anthropogenesis, London, 1888

Bruckner, Eduard, *The Sources and Consequences of Climate and Variability in Historical Times*, (KlimaSchwankungen seit 1700: nebst Bemerkungen über die Klimaschwankungen der Diluvialzeit), Vienna, 1890

Browne, Janet, *Darwin's Origin of Species, A Biography*, New York, 2006

Clarke, Kim, *Professor Porta's Predictions*, M Heritage Project, University of Michigan, 2021, https://heritage.umich.edu/stories/professor-portas-predictions/

Coulter, John, 'Norwood: From 19th century Common Land to City Commuters', from *Ideal Homes: A History of South-East London Suburbs*, http://www.ideal-homes.org.uk/case-studies/norwood

Crook, Malcolm and Leach, Stephen (translators), *Rousseau in Wotton*, a translation of Courtois, L J, *Le Sejour de Jean-Jacques Rousseau en Angleterre (1910)*, https://philosophyk.files.wordpress.com/2008/10/rousseauinwootton_courtois.pdf

Croucher, John S and Croucher, Rosalind F, *Mistress of Science*, Stroud, UK, 2016

Cryle, Denis, *Behind the Legend – the Many Worlds of Charles Todd*, Melbourne, 2017

Day, David, *The Weather Watchers: 100 Years of the Bureau of Meteorology*, Melbourne, 2007, pp 416–18

Dell, R K, 'Hector, James', *Dictionary of New Zealand Biography*, first published in 1990, Te Ara — the Encyclopedia of New Zealand, https://teara.govt.nz/en/biographies/1h15/hector-james

Dick, Thomas, *The Solar System with Moral and Religious Reflections in Reference to The Wonders therein Displayed*, Vol X, Philadelphia, 1854, preface, p12 (accessed online from State Library of Queensland)

Digby, William, *Natural Law in Terrestrial Phenomena: A study in the causation of earthquakes, volcanic eruptions, windstorms, temperature, rainfall, with a record of evidence*, London, 1902.

Doyle, Arthur Conan, *The Wanderings of a Spiritualist: On the Warpath in Australia, 1920–1921*, Chapter VIII; originally published in London, 1921; A Project Gutenberg of Australia eBook, http://gutenberg.net.au/ebooks13/1307001h.html

Douglas, Kirsty, 'Under Southern Skies: understanding weather in colonial Australia, 1860–1901', Australian Government Bureau of Meteorology, Metarch papers, No 17, May 2007

Dunstan, Keith, *Ratbags*, Melbourne, 1979

Egeson, C, *Egeson's Weather System of Sunspot Causality: Being Original Researches in Solar and Terrestrial Meteorology*, Sydney, 1889, National Library of Australia digital copy, https://nla.gov.au/nla.obj-2558311020/view?partId=nla.obj-2558321668#page/n8/mode/1up

Evans, Raymond, *A History of Queensland*, Melbourne, 2007

Garden D, 2010. 'The Federation Drought of 1895–1903, El Niño and Society in Australia', in *Common Ground: Integrating the Social and Environmental in History*, Massard-Guilbaud G, Mosley S (eds), Cambridge Scholars Publishing, United Kingdom, pp 270–92

Goss, Eva Adeline, *Fragments from the Life and Writings of William Henry Goss*, Stoke on Trent, 1907

Hadden, Kathy, *Birkenhead: The Way We Were*, Birkenhead, 1993

Kilgour, William T, *Twenty Years on Ben Nevis*, facsimile of second (1906) edition, published Holyhead, Gwynedd, Wales, 1985

Jensen, Harald Ingemann, *Reminiscences of a Geologist* [unpublished manuscript], circa 1957, State Library of Queensland, Harald Ingemann Jensen Records, 1905–65, Call No OM69-29

Landsea, Chris and Dorst, Neal, 'How and Why are Tropical Cyclones Named?' National Oceanic and Atmospheric Administration, 2016, https://www.aoml.noaa.gov/hrd/tcfaq/TCFAQ_B.txt

Lodge, Oliver, *Man and the Universe: A Study of the Influence of the Advance in Scientific Knowledge upon our Understanding of Christianity*, London, 1908

Loyau, George, *Notable South Australians, 1885*, Adelaide, 1885

Mackenzie, Dana, *The Universe in Zero Words – The Story of Mathematics*, Princeton, 2012

Macleod, John, *Highlanders, A History the Gaels*, Edinburgh, 1996

Niven, W D, (ed), *The Scientific Papers of James Clerk Maxwell*, Vol II, 1890, online edition, Cambridge, 2013: https://doi.org/10.1017/CBO9780511710377

Paterson, Lorraine, 'New Caledonia The Penal Colony', in *Convict Voyages: A global history of convicts and penal colonies*, School of History, University of Leicester, UK, 2018, http://convictvoyages.org/expert-essays/new-caledonia

Roy, Marjory, 'The Contribution of Alexander Buchan to the development of Climatology and Synoptic Meteorology', Scottish Centre, Royal Meteorological Society, paper presented to the Commission on History of Meteorology, Polling, Germany, 2004, http://www.meteohistory.org/2004polling_preprints/docs/abstracts/roy1_abstract.pdf

Pugin, Augustus, *The Collected Letters of AWN Pugin, Vol 5, 1851–1852*, edited with notes and an introduction by Margaret Belcher, Oxford, 2015, (letter to 'Lord Shrewsbury, Alton Towers, June 30, 1841', p672)

Shoesmith, Dennis, 'Holt, Joseph Thomas (Bland)', *Australian Dictionary of Biography*, National Centre of Biography, Australian National University, published first in hardcopy, 1972

Steele, John, 'Jones, Inigo Owen (1872–1954)', *Australian Dictionary of Biography*, National Centre of Biography, Australian National University, http://adb.anu.edu.au/biography/jones-inigo-owen-539/text11915

Steiner, Rudolf, *The Submerged Continents of Atlantis and Lemuria: Their History and Civilization, being chapters from the Akashic Records (authorised translation from the German)*, London, 1911

Turner, Gerard l'Estrange, *Nineteenth Century Scientific Instruments*, Berkeley, 1983

Twain, Mark, *The Innocents Abroad*, Connecticut, 1870

Verne, Jules, *Topsy-Turvy* (also known as *Purchase of the North Pole*), New York, 1890

Wheeler, Malcolm, *History of the Meteorological Office*, Cambridge, 2012

Whitmore, R L (ed), *Eminent Queensland Engineers*, Brisbane, 1984

Wildman, Owen, *Queenslanders Who Fought in the Great War*, Brisbane, 1919

Woods, Gregory D, *A History of Criminal Law in New South Wales: The Colonial Period, 1788–1900*, Sydney, 2002

Wolff, Joshua D, *Western Union and the Creation of the American Corporate Order, 1845–1893*, Cambridge, 2013

Wragge, Clement, *Experiences of a Meteorologist in South Australia*, pp 21–24, No 1 in the Pioneers Books Reprints Series, Warrandale, South Australia, 1980, from a three-part series published in *Good Words*, London, 1887, pp 621–26, 685–90, 754–61

Wragge, Clement L, *The Romance of the South Seas*, London, 1906, Internet Archive: https://archive.org/details/romanceofsouthseo0owragiala/page/n7/mode/2up

Zion, Lawrie, *The Weather Obsession*, Melbourne, 2017

Journals, periodicals

Adamson, Peter (2003) 'Clement Lindley Wragge and the naming of weather disturbances', Weather, *Royal Meteorological Society Journal*, Volume 58, Issue 9, pp. 359-363, https://rmets.onlinelibrary.wiley.com/doi/abs/10.1256/wea.13.03

Bean, Michael (1975), 'Heinrich Samuel Schwabe, 1789 – 1875', *Journal of the British Astronomical Association*, Historical Section, Vol 85, 1975, pp 532–33

Campos, Luis (2007) 'The Birth of Living Radium', *Representations*, Vol 97, No 1, University of California Press, p15, https://online.ucpress.edu/representations/article-abstract/97/1/1/95737/The-Birth-of-Living-Radium

Eiby, G A (1968) 'An annotated list of New Zealand earthquakes, 1460–1965', *New Zealand Journal of Geology and Geophysics*, 11:3, pp 630–47

Haigh, Rev Henry, (ed), 'A Talk With Dr Emma Brainerd-Ryder', *The Harvest Field, A Missionary Magazine*, Methodist Episcopal Publishing House, Madras, November issue, 1890, pp 171–74, https://findit.library.yale.edu/bookreader/BookReaderDemo/index.html?oid=11051435&page=4#page/14/mode/2up

Halberg, F, Cornelissen, G, Bernhardt, K H, Sampson, M Schwartzkopff, O and Sontag, D, 'Egeson's (George's) transtridecadal weather cycling and sunspots', *History of Geo and Space Sciences*, 1, 2010, pp 49–61, www.hist-geo-space-sci.net/1/49/2010/doi:10.5194/hgss-1-49-2010

Jones, Inigo (1951) 'The Life and Work of Clement Lindley Wragge', *Queensland Geographical Journal*, No 40, Vol LIV, p48

Miller, Julia (2007) 'The Fall of an Angel: Gendering and Demonizing El Niño', *World History Connected*, https://worldhistoryconnected.press.uillinois.edu/4.3/miller2.html

Miller, Julia (2014), 'What's Happening to the Weather? Australian Climate, H C Russell and the Theory of a Nineteen-Year Cycle', *Historical Records of Australian Science*, 25, p24

Theses

Astore, William Joseph, *Observing God: Thomas Dick (1774–1857), Evangelicalism and Popular Science in Victorian Britain and Antebellum America*, MA thesis submitted to the Faculty of Modern History, University of Oxford, for Degree of Doctor of Philosophy, 1995

Jensen, Judith A, *Unpacking the Travel Writers' Baggage: Imperial rhetoric in travel literature of Australia 1813-1914*, PhD thesis, James Cook University, https://researchonline.jcu.edu.au/10427/4/04part3.pdf

INDEX

Aboriginal artefacts 75, 89, 157, 226 -27, 253
Abercromby, Ralph 200
Ahmad, Hazrat Mirza Ghulam 360 - 61
Alleyne's Grammar School 41
Anderson, Archibald 168, 171, 191, 201, 218, 229, 254, 257- 62, 364
Anderson, William 113 -16
Allen, Murray Leith 335
Alton Towers 35-36
Archibald, Edmund Douglas 189, 213 - 14
Arrhenius, Svante 3, 431 - 33
Auckland Higher Thought Centre 21
Auckland Spiritual Scientists' Church 13, 22, 417 - 18

Ballard, Anna 416
Baracchi, Pietro 245, 281, 308, 358, 369, 385, 437
Barton, Edmund 268
Bates, Daniel Cross 18, 364 -65, 380
Baraduc, Hippolyte 373
Beck, George Victor 353, 370
Ben Nevis 105-132
Ben Nevis race 439
Bickerton, Alexander William 351
Billington, William 295 -298
Birch, Reverend Wickham Montgomery 49
Birkenhead 383, 409, 414, 422
Blackwell, Ida 246, 361, 415
Blavatsky, Helena 10, 209 – 10, 375, 416
Bond, George Grant 229, 284, 289, 364, 390
Borchgrevink, Carsten 162 - 163
Brough, William Spooner 81 - 82
Brough and Boucicault Comedy Company 211, 228

Bruckner, Edward 186, 213, 302
Buchan, Alexander 96, 109, 120, 126, 131
Buninyong Hurricane 217 – 19
Bunsen, Robert 184
Burcher, Robert Leslie 357
Burton, Kathleen 409, 423
Burton, Sarah 199

Callaghan, Jeff 220
Callender, George 345 - 346
Cameron, Trooper Allan 241 - 43
Cameron, Colin 116 – 17, 122
Cannon Hill 396 - 98
Capemba 170, 176 177, 227, 251, 278, 285, 346
Carr, Noel Alfred 337
Carryer, Norman 381
Causer, John Junior 31
Chataway, James 248 - 49
Cheadle Copper and Brass Co 28, 33
Chermside, Sir Herbert 283 - 84
Christison, Sir Robert 47, 206
Churnet Valley 28, 34 - 35
Cracknell, William John 265 - 66
Crystal Palace 44 – 45, 356
Conroy, Alfred Hugh Beresford 289 - 91
Cook, Joseph 291
Cook, Thomas 51 – 52, 55
Cooke, William Ernest 142, 281, 358
Cooktown 150 – 151, 155, 157
Coorparoo 408 -09
Corelli, Marie 262 – 263, 425
Cyclone Leonta 310 -311
Cyclone Mahina 230
Cyclone Sigma 229

De Bort, Teisserenc 244
Deakin, Alfred 209, 286 – 87,
Dickson, James Robert 263
Downing, Isaac 31 - 32
Downing, Mary [nee Causer] 30
Downing, Elizabeth 30
Doyle, Sir Arthur Conan 9 - 13
Drake, James 246 – 247, 254, 260, 263, 266 - 67

Egeson, Charles, 94 – 98, 163 – 164, 237 – 38
Ellery, Robert 174, 81, 36,
El Nino 71, 308, 310, 428
Espy, James Pollard 274
Eurydice, sinking of 85
Ewens, John 75
Eyemouth Hurricane 115

Fairclough, Paul Wynyard 376
Faraday, Michael 184
Federation Drought 232, 310
FitzRoy, Captain Robert 37 – 38, 45 – 46
Flammarion, Camille 18, 413
Forrest, Sir John 15, 169
Fort William 105 - 106
Fourier, Joseph 431 - 32
Fowles, Edgar Lambert 168, 229, 257, 262, 310, 312, 341, 343
Foxton, Justin 278
Fraser, Reverend Donald 254
Fraser, Sir Simon 287
Frere, Sir Henry, Bartle Edward 66 - 67

Gaw, William 18
Geike, James 431
Good Words 107
Goss, William Henry 123, 200, 206 - 210
Goyder, George Woodroffe 75
Greenhouse effect 432
Gregory, Augustus Charles 147, 279
Gregory Downs Station 155
Griffith, Sir Samuel 165, 196 – 97, 355
Griffiths, Richard 370
Groom, Littleton Ernest 290, 357

Halberg, Franz 304 - 305
Hart, Mary 120
Hawkins, Ed 436

Hector, Sir James 137
Henderson, John Baillie 343
Herschel, William 183
Higgs, William Guy 286
Hogben, George 14
Home, David Milne 96- 98
Horne, Edwin 252
Horne, Mary Mansfield 252
Huggins, William 207
Humble, Benjamin Hutchison 116, 434 -35
Humboldt, Alexander von 63
Hunt, Henry Ambrose 225, 357, 387 – 88, 390 - 92
Hurford, Captain W 271
Hutcheson, David 94

Iliff, Captain Charles 326
Ingleby, Bernard 158, 322, 328 – 29, 334 - 335
Ingleby, Clement 27
Ingleby, Margaret (nee Dutton) 78 - 79
Ingleby, Margaret (nee Thornton) 72 – 73, 78 – 79, 111
Ingleby, Rupert 71 – 73, 79, 111
Ingleby, John 72
International Arbitration and Peace Association 21
International Meteorological Committee 197, 199
Irvine, Captain George 393 - 94

Jack, Robert 17
Jensen, Harald Ingermann 262, 322 - 24
Jones, Inigo Owen 168 – 69, 175, 178, 213, 219,
Jones, Owen 169, 410, 423, 424 - 27

Kerry, Chas 326
Kerry, Thomas Caradoc 338 - 39
Kelly, William 252
Kingsmill, Henry Charles 349
Kirchhoff, Gustav 184
S S Koombana disaster 390

Lahore 360 - 361
Lane, William 380, 405
La Nina 220, 308, 428
Little Wives of India Mission 252, 445
Livingston, Colin 98, 107, 118, 243
Lemuria 10, 375, 377

Index

Lockyer, Joseph Norman 184 – 85, 188, 300 - 301
Lockyer, James 300, 302 - 03
Lodge, Sir Oliver 18, 371
Lynsar, Douglas 423

MacDonnell, Edmund 138 – 39, 153 – 54, 167, 265
Marconi, Guglielmo 15 - 17
McIlwraith, Sir Thomas 196, 223
McLure, Ernest 338
Maranoa 131, 149 – 51
Markham, Clements Robert 66 - 67
Maxwell, James Clerk 37, 64, 184
Messina tsunami 367
Melba, Dame Nellie 17
Mein, Charles Stuart 266
Meldrum, Charles 187 - 88
Melpomene 57 - 60
Merimbula 231, 325
Meston, Archibald 173 – 74, 209
Milne-Home, David 96 – 100,
Mossman, Robert 437
Mount Kosciuszko 230 – 31, 279 – 81, 289 – 94, 314 - 39
Mount Wellington, Tasmania 136, 229, 349
Muirhead, Charles Mortimer 171 -72
Munn, Armstrong Lockhart 322
Musgrave, Sir Anthony 148 - 49

Negretti, Henry 88
New Caledonia 191 – 94, 196 – 98, 207, 222, 224, 264 – 65, 281
Newth, Basil de Burgh 325 – 26, 330 – 32, 336 38
North Staffordshire Naturalists' Field Club 82, 84, 86, 92, 123

Oakamoor 25 – 26, 28 – 29, 34 – 35, 38
Ockenden, Harry 391
O'Connor, Richard Edward 287
Ogg Forbes, Henry 148
Olcott, Henry Steel 209, 416
O'Mahoney, Percival George 229
Orr, James, 373 -74
Oxenham, Humphrey 228

Paris Exposition Universal 244
Paul, William Sheffield 209
Pearce, George 357
Pearse, Richard William 370
Pearsall, Ann 30
Philp, Sir Robert 129, 174, 248, 264, 268, 286, 342 - 43
Phillipps, William Herbert 406
Porta, Professor Alberto Francisco 14
Powers, Charles 174 - 75
Powers, General Edward 274
Proctor, Richard Anthony 351 - 52

Radium 355 – 56, [ch21]
Rangatira Health Institute 21
Rankin, Angus 110, 244, 437
Reid, Sir George Houston 21, 280, 314 – 15, 327
Richardson, Captain Thomas 218 - 19
Russell, Rev Alexander 79, 159
Russell, Henry Chamberlain 136, 138, 179, 182, 190, 197 – 98, 221, 225, 230 – 31, 241 -42, 281, 301 – 02, 308, 358, 437
Russell, William 142, 147, 223
Royal Charter storm 38, 226
Royal Meteorological Society 18, 85 – 86, 88, 91, 96, 102, 131, 151, 156, 159, 168, 336 [ch23]
Roy, Marjory 130, 439, 444
Rutledge, Arthur 267
Ryder, Emily Brainerd 251 – 52, 261, 416, 445

San Francisco earthquake 368
Saville-Kent, William 153
Schwabe, Heinrich Samuel 187
Scott, Robert Henry 46, 197, 199,
Scott, Robert Townley 267
Scratchley, Sir Peter 147 – 48, 152
See, Sir John 312, 314 – 15, 319 – 20, 342 – 43
Shackleton, Ernest 374 – 76, 441 – 42
Shelton, Edward Mason 153, 161, 174
Smith, Robert Barr 324
Sorosis Club of Bombay 424
Southern Oscillation Index 220
Sowden, William 201 - 02
Steiger, [aka Stiger] Albert 273, 295
Steiger Vortex Gun 273 – 74, 295

Stevenson, Thomas 95, 98 – 99,
Stewart, George 412
Stokes, Captain Oliver Haldane 274

Tait, Peter Guthrie 119
Tardent, Henry Alexis 274
Tay Bridge Disaster Hurricane 100, 119
Taylor, Janet 68
Theosophy 10 – 11, 209 – 10, 416
Thomas, Algernon 379
Thomson, Sir William [Lord Kelvin] 120 -
   21, 185, 207, 433
Thornton, Anna 77, 140, 158, 211
Thornton, Blanche 78 - 79
Thornton, Edward 77 – 78,
Thornton, Edward Jun 111, 129, 140, 158, 172
Thornton, Frances 78
Thynne, Andrew Joseph 224, 228
Todd, Sir Charles 73, 133, 137, 197, 290, 358
Tonga 19, 384, 389
Townsville 172, 217, 229, 262, 311, 389,
   393 - 94
Tozer, Sir Horace 247, 249
Tyndall, John 65

Unmack, Theodore 197, 212, 213 – 14, 223 – 24

Vereker Lloyd, John 74, 158
Verne, Jules 375

Waiata Gardens 22, 384, 413 – 415, 420, 423
Walker, Matthew 20, 22
Walker, Robert Lennox 426 -27
Walker, Hayden 427
Warburton, Colonel Peter Egerton 67, 73 - 74
Warburton, Philip Egerton 121 - 22
Watt, William Shand 420
Whelan, Philip Sydney 279, 335
Whyte, William Mackenzie 109
Wilson, Walter Horatio 168, 224
Wolf, Rudolf 187

Wragge, Clement Lindley
Character 112 – 13, 207 – 209, 323 – 24, 440
Lincoln's Inn 28, 50, 55, 66, 68
Meteorological training 26, 58 -59, 84, 86 –
   88, 92 - 95

Midshipman training 69, 142
Séances 11, 206
St Bartholomew's Hospital 68
Some Reminiscences of an Eventful Life 19
South Australian Surveyor-General's
   Department 74 – 77
Ringbarking 3, 300, 353, 429, 430
Voyage on the Wimmera 69
Wragge, Anna Maria Marguerite 30
Wragge, Charles John 29, 47
Wragge, Clement Ingleby 25, 28 - 30,
Wragge, Clement Lionel Egerton ['Eggie']
   74, 101, 112, 211 – 12, 289, 291 – 93, 319
   -22, 336 – 39, 398 - 99
Wragge, Edmund 47, 63
Wragge, Edris [Louisa Horne] 9, 13, 43,
   229, 246, 250 - 53, 260, 278 – 79, 285, 347,
   361, 385
Wragge, Emma [Clement Wragge's
   grandmother] 26, 33 – 37, 41 - 43
Wragge, Leonora May Emerline Ingleby
   [known as May] 397
Wragge, Frances Anne 29, 43,
Wragge, George [Clement Wragge's
   grandfather] 26, 28, 33 – 34, 51
Wragge, George Paulson [Clement
   Wragge's uncle] 42, 51, 66
Wragge, Lindley Herbert Musgrave
   Egerton 211, 397
Wragge, Nora [nee Thornton] aka Leonore
   Eulaliecia Edith Florence d'Eresby 71 –
   82, 90, 101 – 03, 109 – 12, 120, 129, 132,
   135, 140 – 41, 158, 170, 199, 251, 285, 345
   – 46, 394 - 97
Wragge, Rupert ['Bert'] Lindley 112, 140 –
   41, 211, 285, 325, 337 – 38, 395
Wragge, 'Violet' baptised Leonore, Eulalie,
   Clementine, Adelaide 170,
Wragge, William Henry 70, 73, 206
Wragge, Winifred 408, 410

Vincent, Howard Milo 408

Yongala 389 - 90

Zambra, Joseph 88

www.ingramcontent.com/pod-product-compliance
Lightning Source LLC
Chambersburg PA
CBHW031227290426
44109CB00012B/195